Principles of
Animal Taxonomy

Principles of
Animal Taxonomy

Ashok Verma

Alpha Science International Ltd.
Oxford, U.K.

Principles of Animal Taxonomy
404 pgs. | 67 figs. | 14 tbls.

Ashok Verma
Associate Professor of Zoology
SPM Government Degree College (University of Allahabad)
Phaphamau, Allahabad

Copyright © 2015

ALPHA SCIENCE INTERNATIONAL LTD.
7200 The Quorum, Oxford Business Park North
Garsington Road, Oxford OX4 2JZ, U.K.

www.alphasci.com

ISBN 978-1-84265-944-1

Printed from the camera-ready copy provided by the Author.

Dedication

The book is dedicated to my parents who
taught me that sincerity, honesty and
hard devoted work always pays,
and to my teachers who
led me to the fascinating realm of Zoology
and to my students who
made the subject often challenging.

"..........endless forms most beautiful and most wonderful have been and are being evolved".

"..........The affinities of all living beings of the same class have been represented by great tree....As buds give rise by growth to fresh buds, and these if vigorous, branch out and overtop on all sides many a feebler branch, so by generation I believe it has been with great Tree of Life, which fits with its dead and broken branches the crusts of the earth, and covers the surface with its ever branching and beautiful ramification."

Charles Darwin, 1859

PREFACE

The present book is an outcome of my interest in Zoology, which arose over four decades, to augment the long felt need of the students pursuing their studies in animal taxonomy. The study of classification and classifying animals is both fascinating and challenging. The fascination is generated by a natural inquisitiveness towards animals and the challenge is to assess the vast array of animals and give them a proper place in animal kingdom depending upon their similarities and differences.

I have observed during the course of my study as well as teaching that majority of students feel that study of taxonomy is tedious, therefore, this book is being presented with a hope that it will enthuse and invigorate interest among students who are novice to the subject. Since the students hardly get comprehensive resource material; this book addresses the problems of taxonomy in easy and lucid manner. The book has been written in simple language as far as possible, by conveying important aspects of taxonomy without neglecting the current status of development.

In compiling this book, I have made an attempt to present different topics associated with animal taxonomy in 23 chapters sequentially. I have tried to be careful by describing certain topics within limits, which otherwise would require lengthy explanation. The figures being simple have been drawn by the author himself. Throughout the book, the key terms have been presented in bold face. The subject index is comprehensive and especially designed to locate the important terms, topics and concepts. In addition to subject index, genus and author index has also been added. The references cited in the chapters are listed under bibliography and would provide sufficient details for those interested in learning more about a particular topic. The contents of the book are summarized as under:

Chapter 1 begins with a brief resume about the history of animal taxonomy. It also provides an idea about the way the foundation for classifying animal life was laid by Aristotle, Linnaeus and Darwin. It also includes a brief note about population systematics and phylogenetic systematics.

Chapter 2 explains the basic concept of animal taxonomy, systematics, hierarchical classification, taxonomic characters and different attributes for classifying animal life along with associated characters. It also deals with the task of systematist and myths and conceptions of taxonomy and its significance.

Chapter 3 introduces the principles of zoological classification and types of classification along with their merits and demerits.

Chapter 4 is in sense the core of the text. It provides an account of binomial nomenclature, different types of nomenclature, necessity for rules of nomenclature, nomenclature Code and general principles of nomenclature. It also deals with the International Code of Zoological Nomenclature, the principles of priority, homonymy, synonymy, typification and its preamble and features. An account of International Commission of Zoological Nomenclature its powers, nomenclature versus taxonomy, taxonomic inflation and modern developments in classification have also been given.

Chapter 5 gives a brief account of micro and macro taxonomy, phenon, taxon and category.

Chapter 6 explains recapitulation theory, phylogenetic systematics, features and methodology of cladistic analysis, interpretation of cladogram and groupings in cladistic analysis. It also attempts to differentiate between evolutionary trees, phylogenetic tree and cladogram. Biosynthetic phylogeny and the significance of phylogenetic systematics have also been discussed.

Chapter 7 extends the discussion on features and different types of phylogenetic tree, the way to construct phylogeny, limitations in the constructions of phylogeny, common genes and new family tree and gene tree versus species tree.

Chapter 8 highlights the concept about molecular phylogeny, essentials of molecular phylogenetic analysis, database for phylogenetic analysis, tree building, measurement of immunological distances, protein sequencing, significance of hemoglobin and cytochrome. The chapter reviews the importance of molecular phylogeny, nucleic acid sequencing, DNA hybridization technique, comparison of DNA sequence and chromosome painting including Zoo FISH and BLAST and molecular clock. It also highlights the importance of molecular phylogeny.

Chapter 9 outlines the features of evolutionary taxonomy, taxon, concept of evolutionary taxonomy and its relationship with cladistics and phylogenetics. Limitations of evolutionary taxonomy along with its significance have also been discussed.

Chapter 10 signifies the features, basis and procedures of numerical taxonomy including character assessment. It also deals with essentials of taxonomic system, nested hierarchy, and frequency distribution of similarity coefficient, differential shading of the similarity matrix, data, factor and cluster analyses along with limitations and importance of numerical taxonomy.

Chapter 11 gives an account of population taxonomy, characteristics and dynamics of population, different types of animal interaction, sources and types of variability in populations. Differentiation of population, genetic drift and gene pool has also been described. A brief note about mutation, natural selection and adaptations in populations has been added at the end of the chapter.

Chapter 12 deals with taxonomic characters, its role in systematics, measurement of characters, homology and homoplasy, weighting and analysis of different types of characters. The chapter discusses the significance of characters in biodiversity studies.

Chapter 13 adequately explains the species concept and the different types of species and sub-species. Detailed account describing the way new species occur in nature, the process and pattern of speciation, types of speciation, isolation and its role in speciation has also been given. Further, a note has been added on natural selection, environment and speciation in the establishment of new species along with speciation and biodiversity.

Chapter 14 describes the types and principles of typification in taxonomic studies and significance of type specimens.

Chapter 15 reviews the Linnaean system of hierarchical classification in brief, different types of taxonomic category and its features, origin of categories and demerits of higher categories.

Chapter 16 discusses about the data analysis, morphological variability in natural populations, meristic and non-meristic characters in fishes and insects and other characters of importance during identification of specimens.

Chapter 17 introduces the concept of biodiversity its distribution and assessment, different types of biodiversity, species richness and different indices used to assess the vast array of populations in an ecosystem. A brief note has also been added on phylogenetic measurements of biodiversity, taxonomic surrogacy, biodiversity hotspots, genetic pollution and role of biodiversity in ecosystem, climate change, evolution, food security, agriculture and sustainable development, human health, trade and commerce, in refreshing moods and developing aesthetic sense, informatics and threats to biodiversity. Different legal provisions for conservation of nature and natural resources also find mention.

Chapter 18 explains about the construction of taxonomic keys, its features, presentation and applications for the identification of animals and the use of computer generated keys, description of genera along with merits and demerits of taxonomic keys.

Chapter 19 is devoted to different types of taxonomic publications and its various features. The form and style of taxonomic publications has been given to enable the worker to know the method a taxonomic paper is submitted in different leading journals on systematics and the way a new name of any specimen is registered in the records, the zoological records, index to organisms names, electronic publications and significance of taxonomic publications have also been given.

Chapter 20 presents a detailed account of collection and preservation of natural history specimens including different collection procedures and equipments, chemicals used as narcotizing agents, preparation of killing bottle and how to display preserved specimens. A note has been added regarding the manner in which a specimen is sent to the authority/agency for identification. Significance of natural history specimens has been described in detail.

Chapter 21 gives an account of information retrieval and its working along with the concept of biological names and taxonomies, meta-ontology of biological names. It also describes the ways to manage the vast array of animal diversity digitally, including object oriented taxonomic databases. Description about different internet resources providing wealth of information for information retrieval including universal register, zoological record, animal diversity web, the electronic zoo, DELTA, global names architecture, index to organism names, the global biodiversity information system, taxon tree, SysTax Database system, BAMBE, Mantis, TreeBASE, Pandora, BioNet International, Diana, Platypus, Taxis, UBio Portal, OCEAN, *Systeam Naturae* 2000, Taxonomicon, Tree of life project, Species 2000, ITIS , ZooBank and the Catalogue of life.

Chapter 22 attempts to enrich reader's understanding of taxonomy by describing the current trends in taxonomy. The chapter includes role of morphology, embryology, ecology and behaviour on the basis of current tools. Importance of cytotaxonomy, chemotaxonomy, molecular phylogeny, chromatography, electrophoresis, infrared spectroscopy, histochemistry, and comparative serology, numerical and molecular taxonomy has also been discussed in detail. A basic of DNA barcoding along with its applications has been explained. Description about Molecular operational taxonomic units, integrated operational taxonomic units, cataloguing ancient life, constraints of DNA taxonomy, convention on biodiversity, the global taxonomic initiative and integrative taxonomy earn special mention. The chapter ends with a brief note regarding difficulties encountered while totally relying on barcode taxonomy.

Chapter 23 summarizes the discussion about the significance and applications of biosystematics and its role in conservation biology. The role of taxonomy *vis-a vis* biodiversity has been described adequately along with the present scenario of taxonomy.

I have also tried to make this presentation exact but definitely it is not, and if I had to wait for this, it would never have come to this shape as it is. I have very much enjoyed the process of putting the scattered information together and hope that you would enjoy using it to the same degree. I am confident that the present book will enable the students pursuing taxonomic studies to understand the fundamentals of animal taxonomy with greater ease.

I would appreciate it very much, if this book creates continuing curiosity and interest in the students and researchers alike in the field of animal taxonomy. Though, every care has been taken while preparing this book, but shortcomings are inevitable; therefore I would welcome receiving valuable comments, constructive criticism and suggestions.

Ashok Verma

ACKNOWLEDGEMENTS

I wish to put on record my sincere gratitude to my teachers for having shaped my thinking and creating my interest in Zoology particularly in insect taxonomy, that lead me to opt for Entomology as a special paper during my post graduate studies at the University of Gorakhpur (1973-75) on Govt. of India National Merit Scholarship.

When I thought to write a book on this subject, I was sure I would be able to complete it in a few months, but the task was not easy at all. During the course of preparing the manuscript, I realized that writing a book on a subject like this is far more difficult than teaching. The mission became even more difficult due to scanty information available in the subject. It is quite difficult for me to express my sincere gratitude to every one who should be acknowledged in presenting this book as such, I wish to put it on record that I am extremely grateful to all the individuals for their sustained assistance which has made the task of undertaking this project more a pleasure than a burden. I am also obliged to all the authors whose excellent work made it possible for me to present this book and to many publishers who have allowed me to use their publications as and where needed. Certain websites have also been consulted and I am grateful to the contributors for the information provided.

Special thanks are due to different experts of taxonomy for having allowed me to quote certain text from their research papers, viz., Dr Quertin Wheeler, Arizona State University, USA, (Chapter-2 and 23); Prof. Gerardo Canfora, University of Italy, (Chapter-21);Prof. Jouni Tuominen, University of Helsinki, (Chapter-21); Prof. AG Valdecasas, Madrid, Spain, (Chapter-21); Prof. Zhi-Qiang Zhang, New Zealand (Chapter -4 and 23); Prof. Alesandra Minelli, University of Padova, Italy (Chapter - 4).

I extend my heartfelt appreciation to my wife Mrs. Kumkum Verma and daughter Ms. Ankita Chandra for their understanding, constant encouragement and patience during crucial moments, for their support and indulgence.

I have been able to contribute another title earlier, 'Invertebrates: Protozoa to Echinodermata' published by Narosa Publishing House, New Delhi. I am grateful to Narosa Publishing House, New Delhi once again for the excellent presentation of this book.

Ashok Verma

CONTENTS

Phylogeny; DNA Hybridization Technique; Comparison of DNA Sequence; Chromosome Painting; Molecular Clock; Importance of Molecular Phylogeny

Unrelated Characters; Character Displacement; Classification and Source of Data; Significance of Characters in Biodiversity Assessment

Chapter 1

HISTORICAL RESUME OF TAXONOMY

Ever, since the dawn of civilization, human beings have ever been anxious to unravel the mystery of nature, the flora and fauna, the sun and the star. In the beginning, taxonomy was based on the knowledge of local people about vast array of organisms. Later on, folk taxonomy gradually transformed in to formal system for naming and classifying organisms. In biological science, classification refers to the identification, naming and grouping of organisms into a system based on similarities in external and internal anatomy. In the scheme of classification, the vast assemblage of animal life on earth is organized in scientific manner. Taxonomy is that component of systematics which is focused on theory and practice of classification. Taxonomy lays the foundations for the Tree of Life, which is a requisite database for studies in ecology and conservation. Taxonomy is the discipline of science that helps in the identification, description and classification of the diversity of life which is amazing. The science of identifying, describing, naming and classifying all organisms is called 'taxonomy'. The study of nature has also involved the naming and classification of living organisms into groups. There have been different phases in the development of theory and principle of taxonomy. The historical aspect of taxonomy can be studied as mentioned herein as under:

THE VEDIC PERIOD

The earliest description of classification finds mention in Vedas and Upanishads (1500BC to 600BC). Several technical terms have been used to describe different plants and their parts. Eminent physicians and scholars like Charaka and Shushruta had also described plants having medicinal values. Early Greek scholars notably Hippocrates (460-377BC) and Democritus (465-370BC) described different animals; however, there is no conclusive evidence of useful classification.

ARISTOTLE' S CONTRIBUTION

Aristotle (384-322 BC) regarded as the father of biological science and taxonomy, was the first ever to develop the earliest known method of classifying organisms. Aristotle, student of Plato, was one of the most noted philosopher and scientist of the ancient world. He adopted his own method of enquiry, different from his teacher and took into consideration the morphological features, embryology, habits and ecology of different groups of animals. He emphasized that; animals may be characterized according to their habit and habitat, morphology and behaviour. He divided the animal kingdom into two major divisions.

Anaima

It included all invertebrates in which red blood cells were absent, i.e., animals without blood, e g., coelenterates, arthropods and molluscs.

Enaima

It included all vertebrates, where red blood cells are present, i.e., animals with blood. It is further subdivided into two categories:

Oviparous
Animals which lay eggs is known as oviparous e g., fish, amphibians, reptiles, birds and monotremes.

Viviparous
It included vertebrates, which give birth to young ones e g., mammals.

Aristotle also referred to the major groups of animals as birds, fishes, whales and insects. Further, in insects, emphasis was laid on the nature of mouthparts such as mandibulate and haustellate types and the state of wing viz., winged and wingless conditions. He also described insects belonging to present day orders like Coleoptera and Diptera. Interestingly enough, these orders have been recognized till today. Aristotle also established different categories using certain important characteristic features, viz., animals with blood versus animals without blood, two-footed versus four footed, hairy versus feathered conditions and animals with or without outer shell etc. Aristotle dissected hundreds of animals to know their anatomy and showed that dolphin is a mammal, not a fish. Thus, Aristotle's criteria of animal classification dominated the field of systematics for well over 2000 years.

According to Mayr (1982 a), Aristotle did not provide any consistent classification of animals. After death of Aristotle, people lost interest in the natural history and the study of animals.

In the year, 1172 Ibn Rushd (A Verroeas), who was Judge in Seville, translated and abridged Aristotle's book '*de Anima*' (On the soul) in to Arabic. His original commentary is now lost, but its Pre-Linnaean period translation into Latin by Michael Scot is available.

Theodore Gaza, a Greek refugee to Italy after the fall of Constantinople in 1453, translated the first Latin edition of Aristotle's '*De Animal bus*'.

THE PRE- LINNAEAN PERIOD

In late 15th century, several authors tried to arrange plants and animals according to the principles of logical division. Some of the important contributions deserve mention:

i. Guillaume Rondelet (1507); William Turner (1508) and Pierre Belon (1517-1564) extensively studied and catalogued birds into different groups on the basis of adaptations such as aquatic birds, wading birds, birds of prey, perching birds and terrestrial birds.

ii. A Swiss Scientist, Conrad Von Gesner (1516-1565) wrote *Historia Animalium* containing vivid description of many animals, never seen before by most Europeans.

iii. Andrea Cesalpino (1519-1603), an Italian botanist is considered to be the first taxonomist who classified 1500 plants based upon their habit viz., trees, the floral part, fruits and seed characters. He introduced the sophisticated hierarchy of class, section, genus and species and arranged the different plant species in graded order.

iv. Some authors like Gesner (1551) and Aldrovandi (1600) wrote encyclopedia on animals.

v. Marcello Malpighii (1583) expanded the hierarchical system of animal classification taking cue from Andrea Cesalpino.

vi. Joseph P. de Tourne fort (1627-1705) - a French Botanist developed description of 9000 plants and nearly 700 genera. He gave 'Genus' a distinct rank in the taxonomic hierarchy and introduced the practice of naming the plants according to the genus.

viii. John Ray (1627-1705) an English naturalist published his observations on plants, animals and natural theology, while giving due considerations to the variation among different individuals and importance of anatomical characters, e g., he used features, such as the shape and size of the beak to classify birds.

ix. Augustus Quirinus Rivinus (1652-1723) introduced the category of order while classifying plants on the basis of characters of flowers.

x. In the later part of the 16th century and the beginning of the 17th century, careful study of animals started. Anatomy was considered as the basis of classification and workers like Fabricius (1537-1619), Petrus Severinus (1580-1656), William Harvey (1578-1657) and Edward Tyson (1649-1708) brought this into practice. It is relevant to mention that lot of credit goes to Harvey for unraveling the concept of human anatomy.

xi.Certain advancements in the system of classification were made with the progressive improvement in the field of microscopy by workers like Marcello Malpighii (1628-1694), Jan Swanmerdam (1637-1680) and Robert Hooke (1635-1702).

xii. Lord Monboddo (1714-1799) did excellent work to unravel the mystery of species relationship, which helped in the establishment of the theory of evolution.

xiii. Workers like Adanson (1727); Lamarck (1744 -1829); Cuvier (1769 -1832); Leukart (1823-1898); Haeckel (1834-1919) and Ray Lankester (1897-1929) also contributed towards the general principles of classification. Haeckel revolutionized the concept of development. Later on, embryologists divided the animals into two subgroups:

Protostomia

In this group of animals, the mouth develops from the blastopore (**protostome**= first mouth). The coleom is schizocoelic and the embryo undergoes spiral cleavage, e g., members of phylum Annelida, Mollusca and Arthropoda.

Deuterostomia

In this case, the mouth develops from the second opening opposite blastopore (**deuterostome** = second mouth). The blastopore becomes the anus. The group includes members of phylum Echinodermata, Hemichordata and Chordata.

LINNAEUS' CONTRIBUTION

Around seventeenth and eighteenth century, animal taxonomy made little progress except the work of Willughby (1635-1672) on birds and that of Reaumur (1683-1757) on insects. However, Buffon (1707) and Linnaeus (1707-1778) dominated the natural history in the eighteenth century. It is relevant to mention that the Bauhins, who lived nearly two hundred years before Linnaeus, used the binomial nomenclature in various forms.

Swedish naturalist Carl Von Linne' (1707-1778) is regarded as the father of 'modern taxonomy', as he laid down the basic principles of **binomial nomenclature**. He Latinized his own name as Carolus Linnaeus and dedicated his life to the task of describing and classifying all the species. According to binomial nomenclature, every organism is given a scientific name, typically a two-word name in Latin, to distinguish it from similar organisms. This naming process creates a standard way for the scientists around the world to communicate about the same organisms by a single name. This standard of nomenclature minimizes confusion, particularly when common names are given to the organisms.

Linnaeus gave the concept of hierarchical system of classification, whereby he laid emphasis on the similarity in the organization of animals as well as on their differences. Linnaeus considered species to be the groups of individuals with mutual resemblances just as children resemble their parents. In his *Imperium Naturae*, Linnaeus established three kingdoms, viz., *Regenum Animale, Regnum Vegetable* and *Regenum Lapideum*.

Linnaeus produced two notable contribution viz., *Species Planterum* (1753) and *Systema Naturae* (1758). Linnaean taxonomy categorizes organism into a hierarchy of Kingdom, classes, orders, families, genera (singular: genus) and species. The category of phylum was added to the classification scheme later, as a hierarchical level just beneath kingdom. It may also be added that the basic hierarchy formulated by Linnaeus, is as follows:

- Imperium (Empire) - the phenomenal world.
- Regnum (Kingdom) - the three great divisions of nature at the time-animal, vegetable, and mineral.
- Classes (Class)-subdivisions of the above, in the animal kingdom six were recognized (mammals, birds, amphibians, fish, insects and worms).
- Ordo (Order)-further subdivisions of the above-the class Mammalia has eight.
- Genus - further subdivisions of the order, e g., in mammals- order Primates e g., *Homo*.
- Species-subdivisions of genus e g., *Homo sapiens.*
- Varietas (variety)- Species variants e g., *Homo sapiens europaeus*

However, certain suppositions of Linnaeus were by no means universal. For example, his classification of groups like birds, amphibians, and lower invertebrates (Vermes) were inferior to those of earlier authors.

THE POST LINNAEAN PERIOD

With further scientific development, artificial system of classification evolved into natural system of classification, where-in emphasis was laid on the degree of similarities between the animal groups and placing them in category.

Jean Baptist de Lamarck (1774-1829) published his famous *Philosophique Zoologique* in 1809. According to Lamarck, the living beings evolve from lower to higher stages. He also published *Flora Francoise* in 1778; and laid down the principles of his concept of natural classification. He refined and modified Linnaean scheme of classifying invertebrates into two classes only viz., insects and worms and distinguished molluscs and other classes.

Cuvier (1769-1832) the founder of the comparative anatomy which formed a sound basis of systematics, outright, rejected Lamarck's theory of evolution. He divided animals into four groups- animaux, vertebrates, molluscs, articules and rajonnes. However, Cuvier scheme of classification failed to get support, as it did not gave any emphasis on phylogeny, though Cuvier stressed to include extinct fossil groups as well in his scheme of classification.

DARWIN'S CONTRIBUTION

Charles Robert Darwin (1809-1882) during his voyage on *HMS Beagle* under the command of Captain Robert Fitzroy to survey different coast areas of South America, observed different groups of animals and plants and studied the variation, structural differences and adaptations. On the basis careful observation, Darwin put forward the theory of evolution due to natural selection and published the famous *Origin of Species*. Darwin also provided a set of criteria, which further helped in the establishment of a scientific basis of classification. Darwin suggested that classification systems should reflect the history of life i.e., the species should show relationship based on their shared ancestry. A phylogenetic approach to the classification was developed and Darwin laid emphasis on the characters, which were useful in classifying different groups of animals. At the same time, Alfred Russell Wallace (1823-1913) working in East Indies had developed similar theory. Both the theories were published in the bulletin of Linnaean Society.

Haeckel (1834 -1919) proposed the famous 'biogenetic law' and introduced the concept of phylogeny represented with the help of phylogenetic tree, which showed the origin of different groups of animals from an ancestral form. It also helped to deduce the phenomenon of **divergence** and **convergence** along with the **missing link** between different groups of animals. Haeckel placed **bacteria** within kingdom **Protista** in a separate group-**Monera**.

It is relevant to mention that the development and use of microscopes in the late 16th century revolutionized the pattern of classification. Initially scientists had relied on a two kingdom classification:

i. Kingdom Plantae included single celled organisms capable of carrying out photosynthesis.

ii.Kingdom Animalia included organisms that ingested foods.

By the end of 19th century, a wide variety of microscopic organisms with diverse cell anatomies and specialized internal structure were identified. Haeckel placed these unicellular organisms in a separate group Monera under kingdom Protista, as these organisms differed from all other cells because they lacked nuclei,

In 1930s, French marine biologist Edouard Chatton made crucial distinction between prokaryotes (organism such as bacteria as they lack nuclei) and eukaryotes, (more complex organism having nuclei).

In 1928, American biologist Herbert Copeland also showed a clear cut distinction between prokaryotes and eukaryotes; as such prokaryotes were kept in fourth kingdom known as Monera.

In 1950s, American biologist Robert H. Whittaker proposed a fifth kingdom Fungi.

By the 1970s advances in molecular systematics provided a new insight which led Carl Woese to propose a six kingdom classification system in which he separated prokaryotic organisms into two kingdoms, i.e., Archaebacteria and Eubacteria and the eukaryotic organism into the kingdoms Plantae, Animalia, Fungi and Protista. He later advocated the use of a new category called the domain. In his new system of classification all life forms have been grouped into three domains: bacteria, archaea and eukarya.

Cavalier-Smith (1998) presented a revised six-kingdom system of life down to the level of infraphylum in which Bacteria are treated as a single kingdom, and eukaryotes are divided into only five kingdoms: Protozoa, Animalia, Fungi, Plantae and Chromista. Intermediate high level categories (super kingdom, sub kingdom, branch, infrakingdom, super phylum, subphylum and infraphylum) are extensively used to avoid splitting organisms into an excessive number of kingdoms and phyla (60 only being recognized).The different scheme has been given in Table 1.

POPULATION SYSTEMATICS

The population systematics laid emphasis in studying the variations in the intraspecific populations. Initially, species were considered to be invariant units, later on, when population samples from different geographical range of species were considered, they showed quite differences. This led to the development of typological concept of species being considered to be 'polytypic' having many subspecies consisting of different populations. In this case, emphasis has been laid on the biological aspect of species taking into consideration the ecological, geographical, genetic and other factors.

English statistician, Ronald A. Fisher (1890-1962), British geneticist JBS Haldane (1892-1964) and American geneticist Sewell-Wright (1889-1980) analyzed the genetic problems by using mathematical models. According to them, mutation is the sole cause of evolution and the direction and speed of evolution is determined during the process of natural selection. Huxley (1940) introduced the term 'new systematics' which formed the basis for re-evaluation of the status of 'species'. At the same time Camp and Gilley (1943) introduced the term **biosystematics**.

According to the concept of population systematics, all organisms occur in nature as members of the population and they can best be studied and classified by being treated as samples of natural populations. This prompted taxonomists to undertake the field study of natural populations by taking into account significant characters like their behavioral, ecological and overall biological features. Thus, the classical taxonomy transformed into biological taxonomy. This prompted Mayr (1942) to consider species as a 'group of interbreeding population'.

PHYLOGENETIC SYSTEMATICS

German entomologist Hennig (1966) proposed that systematics should reflect evolutionary history of lineages as closely as possible and he referred this as 'phylogenetic systematics'. The relationship established by phylogenetic systematics describes the evolutionary history of species and hence the phylogeny i.e., the historical relationship among different groups of organisms.

Taxonomy - the science of discovering, naming, and understanding the immense biodiversity of earth has enabled to understand the fascinating knowledge on features of life and has also contributed in many ways to the sustainability of our planet. The benefit and impact of taxonomy is immense. It would be relevant to quote Moore (1993), "It's not a hyperbole for us to say that all of biology is footnote to Aristotle. He defined the field, outlined the major problems, and accumulated data to provide answers - he set the course. "

Table 1 Different Classificatory Scheme

Linnaeus, 1735	Haeckel, 1866	Chatton, 1925	Copeland, 1938	Whittaker 1969	Woese et al., 1977	Woese et al., 1990	Cavalier-Smith 1993	Cavalier-Smith 1998
2 kingdoms	3 kingdoms	2 empires	4 kingdoms	5 kingdoms	6 kingdoms	3 domains	8 kingdoms	6 kingdoms
Not treated		Prokaryotes	Monera	Monera	Eubacteria	Bacteria	Eubacteria	Bacteria
					Archaebacteria	Archaea	Archaebacteria	
	Protista	Eukaryota	Protista	Protista	Protista	Eukarya	Archezoa	Protozoa
				Fungi	Fungi		Protozoa	
Vegetabilia	Plantae		Plantae	Plantae	Plantae		Chromista	Chromista
Animalia	Animalia		Animalia	Animalia	Animalia		Plantae	Plantae
							Fungi	Fungi
							Animalia	Animalia

Chapter 2

BASIC CONCEPTS OF ANIMAL TAXONOMY

The word **taxonomy** is derived from a Greek word (Greek. taxis-'arrangement/order' and nomos-'law'). It was coined by French Botanist APde Candolle. Taxonomy is the science of identification, describing, naming and classifying organisms and includes all flora and fauna and the microorganisms of the world. In other words, taxonomy is the branch of biology which focuses on the description of the taxa and the organization of these into the classificatory schemes. In general, taxonomy is the theoretical study of classification, including its bases, principle, procedures and rules. Taxonomy has a well-defined role in classifying the organisms. Taxonomy is one aspect of much larger field called systematics. Taxonomy is the science that interprets biological data to test hypotheses of what makes up a taxon. Taxonomy is the science of the description and classification of organisms, essential to the inventory of life on earth (Lincoln et al., 1998; Wägele, 2005). Godfray (2002a) indicated that taxonomy, the classification of living things, has its origins in ancient Greece (with the first basic classification of Aristotle) and in its modern form dates back nearly 250 years, to when Linnaeus introduced the binomial classification still used today. Specific rules have been established for recognizing, naming and classifying species to avoid redundant descriptions or the use of the same name for more than one species. These rules were introduced in the late 19th century and are continuously monitored by international commission of scientists (Tautz et al., 2002).

Different workers have tried to define taxonomy from their own viewpoint. Some of the accepted definitions of taxonomy are given as under.

- According to Mason (1950) taxonomy is the synthesis of all the facts about the organisms into a concept and expression of the interrelationship of organisms.

- Harrison (1959) defined taxonomy as the study of principles and practices of classification, in particular the methods, the principles and even in part, the result of biological classification.

- According to Simpson (1961) taxonomy is the theoretical study of classification, including its bases, principles, procedures and rules.
- Heywoods (1967) defined taxonomy as the way of arranging and interpreting informations.
- According to Blackwelder (1967) taxonomy is day-to-day practice of handling different kinds of organisms. It includes collection and identification of specimens, the publication of data, study of literature and the analysis of variations shown by the specimens.
- According to Johnson (1979), taxonomy is the science of placing biological form in order.

SYSTEMATICS

The word systematics is derived from Latin word *Systema* and can be defined as the classification of living organisms into hierarchical series of groups. In fact systematics is the science which deals with the diversity of life and relationships among the living organisms through time. There is vast assemblage of animals on earth which show a great degree of variation and it is very difficult to tackle this diversity. This is achieved by means of systematics. In other words, systematics is the scientific study of diversity in living organisms or science of working out their relationships to each other. The relationships are visualized as evolutionary trees (synonyms: cladogram, phylogenetic trees, and phylogenies). Phylogenies have two components namely, branching order (showing group relationships) and branch length (showing the extent of evolution). Thus, systematics is used to understand the evolutionary history of life on earth.

According to Blackwelder and Boyden (1952), "systematics is the entire field dealing with the kind of animals, their distinctions, classification and evolution".

Simpson (1961) defined systematics as "the scientific study of the kinds and diversity of organisms and of any and all relationships among them". It is relevant to mention that the relationship does not literally mean phylogenetic affinities but on the other hand includes all associations of contiguity and of similarity.

According to Padian (1999), systematics can be seen as the philosophy of organization nature, taxonomy as the use of sets of organic data guided by systematic principles, and classification as the tabular or hierarchical end result of this activity.

Kapoor (1998) considered that the relationship of taxonomy to systematics is somewhat like that of theoretical physics to the whole field of physics. In this sense, taxonomy would be just a part of systematics; taxonomy includes classification, but depends a lot on systematics for its concepts, and systematics includes both, taxonomy and phylogeny. Wägele (2005) is of the opinion that, although theoretically the terms taxonomy and systematics could be synonyms, in practice, however, differences in usage are obvious and a systematist and a taxonomist can conduct different analyses.

The discipline of taxonomy traditionally covers three areas of stages: alpha (analytical phase), beta (synthetic phase) and gamma (biological phase) taxonomy (Kapoor, 1998 and Disney, 2000). Alpha taxonomy is the level at which the species are recognized and described; beta taxonomy refers to the arrangements of the species into a natural system of lower and higher categories and gamma taxonomy is the analysis of intraspecific variations, ecotypes, polymorphisms, etc.

It is also relevant to mention here that systematics includes taxonomy, which is concerned with the identification and nomenclature of present and past animals and the process of evolution describing the manner by which present day complex individuals have been derived from the simpler ones. Thus, systematics is fundamental to biology because it shows the way how organisms are related to each other.

WHY CLASSIFY ORGANISMS?

In the living world, the diversity of animals is amazing. For the sake of convenience, the animals can be grouped on the basis of their eating habits-into herbivores or carnivores or on the basis of their habitat -into terrestrial or aquatic ones. This process requires that closely related individuals on the basis of similarities be classified together. Thus, **classification** (Latin classis-'a class', facere-'to make') brings an orderly sequence of individuals amongst wide array of populations. Classification involves practical application of taxonomic principles, which helps in the correct identification of species and assigning it a suitable place in the **hierarchical** system. In order to stand how organisms are related they are arranged in groups on the basis of similarities. This helps to manage the vast array of animal populations for further studies.

FEATURES OF CLASSIFICATION

The classification should be meaningful. The essential features of classification are detailed herein as under:

- Grouping of organisms together should be on the basis of biological similarities or relationships as for as possible.
- To organize the diversity among different groups of organisms.

• The classification should be consistent with the relationship, which forms its basis.

• It would be observed that evolution is considered to be the basis of classification. With the addition of many new genera and ever-increasing advancement in the field of taxonomy, there is possibility of change in the status of genera described earlier. It implies therefore, that the classification must not be constant on the basis of some relationship deduced in the past, instead may be plastic, so as to accommodate new characters. Thus, a balance has to be maintained, otherwise when classification and nomenclature vary from time to time, it may create quite confusion.

• A system of classification must be practical and purposeful. It must be able to incorporate characters of newly discovered genera and at the same time should retain its originality.

HIERARCHICAL CLASSIFICATION

Carolus von Linnaeus created a hierarchical classification system using only six taxonomic categories: kingdom, class, order, genus, species and variety. Since the kingdom was used by Linnaeus to separate plants from animals, he used only three higher categories: Class, order and genus. (kingdom, phylum, class, order, family, genus and species). These categories are based on shared physical characteristics or phenotypes within each group and the differences. Linnaeus presented a system for organizing different groups of animals into a series of nested groups based on their similarities and differences. This hierarchical Linnaean system uses clearly defined shared characteristics to classify organisms into each group represented by these different levels. It is interesting to mention that organisms are not classified as individual but as groups of organisms.

The hierarchy is a systematic framework for zoological classification. It involves placing the animals based on similar features into groups i.e., **phena**, and arranging them in definite named taxonomic categories, i.e., **taxa**. Thus, a hierarchical system of classification is the one where the largest taxa are subdivided into successively smaller taxa. Thus, each taxon (rank) has a particular level within the system.

In a scheme of classification, sequence of steps is needed to classify a particular individual. The different features of an individual or phenotype occurring within single population are also considered. The important steps include establishing phenon, taxon, and other categories like species, genus, family, order and species. Terms like species, genus, family and order constitute category. Thus, categories designate a given rank or level in a hierarchal classification. Certain additional categories are also developed with the use of prefix-super and - sub. For example, in insect taxonomy, order

Hymenoptera includes superfamily -Vespoidea and subfamily-Vespinae. Before giving an account of hierarchical classification, an idea about categories, rank and phena seems desire:

Categories

A taxonomic category designates a rank or level in a hierarchical classification, in other words, taxa of a given rank are known as **category**. Groups of same rank belong to the same taxonomic categories, to which a particular name is given. The category for a given taxa shows its rank in the hierarchy. The categories are based on concepts. The organisms placed in these categories are concrete zoological objects. It must be made clear that the category designates a rank in hierarchy and taxa designates named grouping of organisms. The categories may be lower or higher. The lower category includes **species** and **subspecies**, while the higher categories include groups above species level. Genus is the lowest higher category. The highest category in the animal kingdom is the phylum and the species constitute the lowest category. A Linnaean hierarchy consists of taxon of different ranks. Rank shows the degree of similarity and period of common origin. One interesting fact emerges from this *viz.*, the lower the rank of a taxon; there is every possibility of grouping individuals showing common attributes. An acronym mnemonic for remembering the above scheme in sequence is **K**ing **P**hillip **C**alled **O**ut **F**or **G**ood **S**oup. Accordingly, the concept of Linnaean hierarchy can be illustrated as under:

Like	individuals	make	Species
Like	species	constitute	Genera
Like	genera	constitute	Families
Like	families	constitute	Orders
Like	orders	constitute	Classes
Like	classes	constitute	Phyla
Like	phyla	constitute	Kingdoms

Due to the discovery of more and more species, it became imperative to establish the exact similarities between the species and assign its correct position in the Linnaean hierarchy. Thus, additional categories were created and for this, prefixes like -super and -sub were used. Thus, subphylum, subclass, super order, sub order and subfamily were created to meet the specific need. In insects, a new category known as **tribe** has been added in between genus and family. Still other categorical designations include cohort, phalanx, series, section and division. Simpson (1945) used twenty-one levels of categories in his classification of mammals as compared to six given by Linnaeus (1758) in *Systema Naturae* (For detail, please see Chapter 15).

The animal kingdom includes vast assemblage of animals that occur as distinct species. Thus, the species is an assemblage of animals that differ significantly from any other species in the features it possess or inherits. Once, species are identified and known to possess many characters in common, they are grouped into a larger assemblage known as **genus**. Similar genera constitute families, followed by order, class and finally phylum.

Taxa

A taxon is a group of related organisms recognized as a formal unit at any level of hierarchical classification. Thus, the unit of classification is known as taxon (pleural-taxa). It is a name designating an organism or groups of organisms. A taxon is assigned a rank and can be placed at a particular level in a systematic hierarchy showing evolutionary relationship. It may include species, genus or order, for e g., Arthropoda (a phylum), insecta (a class), Coleoptera (an order) Chrysomelidae (a family), *Dysdercus* (a genus), *koenigii* (a species). Further, there are two important features of taxon:

• Members of lower level taxa e g., species are more similar to each other than are members of higher level taxa e g., kingdoms.

• Members of definite taxa are more similar to each other than any two members of different specific taxa found at the same hierarchical level, e g., humans are more similar to apes than to kangaroo.

Rank

It is the level i.e., the relative position in a given hierarchical sequence for nomenclatural purpose of a taxon in a taxonomic hierarchy. Rank is designated on the basis of degree of differences among taxa. If a particular taxon differs at a higher rank, the ranking constitutes one of the major components of Linnaean hierarchy. For example, species, genus, family and class are different ranks. The ranks of the family group, the genus group and the species group at which nominal taxa may be established are stated in Articles 10.3, 10.4, 35.1, 42.1 and 45.1 of the Code.

Species

A species is an assemblage of animals that differ significantly from any other individual groups in the characteristics it inherits. It is the fundamental taxonomic unit. The scheme of classification begins with the concept of species. It is the unit category of taxonomy. It is relevant to note that every higher taxon arises by a process of speciation and the species are the units of evolution.

Genus

It is the lowest higher category containing a single species or a monophyletic group of species, which is separated from other taxa of the same rank by a decided gap (like behavior, morphology or some other characters). Genera are made up of species that share one or more characters.

Family

Classes are divided into families. A taxonomic category includes one genus or group of genera having common phylogenetic origin which is separated from other families by a decided gap. A family usually contains more than one genus and each genus usually includes more than one species.

Order

It consists of one or several similar or closely related families. In other words, each class is divided into small groups.

Class

It consists of one or several similar or closely related orders. The phylum is divided into even smaller groups known as classes.

Phylum

A phylum is a major category or taxon of organisms with a common design or organization. The animal kingdom is divided into nearly 40 smaller groups known as phylum. Here the animals are grouped by their main features.

Kingdom

It is one of the highest primary divisions into which all objects are placed. Each kingdom is divided into smaller units known as phyla (plural, phylum). For example, animals having notochord are classified under phylum Chordata. The chordates are further divided into classes such as Amphibia, Pisces, Reptilia, Aves and Mammalia, on the basis of certain special features which they posses. In the 1960s a rank was introduced above kingdom, namely domain (or empire), so that kingdom is no longer the highest rank. Prefixes can be added to make **subkingdom** and **infrakingdom** which are the two ranks immediately below kingdom. Superkingdom may be considered as an equivalent of domain or empire or as an independent rank between kingdom and domain or subdomain. In some classification systems the additional rank **branch** (Latin: *ramus*) can be inserted between subkingdom and infrakingdom (e g., Protostomia and Deuterostomia in the classification of Cavalier-Smith). Because of its position, branch can be considered as a minor rank of the kingdom group even if it is not etymologically derived from it.

TAXONOMIC CHARACTERS

Character is an attribute or feature which enables to distinguish different taxa. In classification of animals, species sharing a common character are grouped into taxa, which can be defined as group of populations (taxa) that can easily be identified by having common characters. A taxon must possess certain characters of its own to establish itself as a separate category having scientific name for it. Many taxon can be differentiated based on just one character like presence of notochord in the chordates. It must be emphasized that almost any character of an organism may be of great taxonomic importance, if it subserves the function of differentiating it from the other taxon. It may be mentioned that the vast majority of classifications are based on morphological characters, which can be easily observed and help to evaluate variation.

ATTRIBUTES FOR CLASSIFYING ORGANISMS

While classifying life forms, the taxonomist encounters the problem of identification of individuals. In the initial stage of identification, certain specific characters serve as an excellent parameter to distinguish a particular individual from the other. For example, the members of insect order viz., Orthoptera and Coleoptera can be identified easily with the help of following characters:
 • Wings straight, fore wings narrow, leathery and toughened, hind pair membranous.
Based on the above characters, the insect in question belongs to order Orthoptera.
 • Forewings modified into elytra and cover the hind wings.
Based on the above features, the given insect belongs to order Coleoptera.
  To be more precise, an ideal classification is based on homology; i.e., the shared features which have been inherited from a common ancestor. Generally, the hierarchy and classes are difficult to differentiate in a system of classification. However, following is the broad line distinction between hierarchy and class. In simpler terms, hierarchy denotes that both the groups of insects are hoppers but the keys serve as guide to point out the exact differences between them.

Table 2.1 Difference between hierarchy and class

Hierarchy	Class
Each class belongs to a single class at the next higher level.	Each class belongs to two or more classes at the next Higher level.

Family	Acrididae		Tettigonidae
Key	Ovipositor sword shaped		Ovipositor sickle shaped

Fossils

Discovery of certain fossils fill the gap between evolutionary sequences of certain groups of animals, e g., *Archaeopteryx* serves as missing link between reptiles and birds. Similarly, *Seymouria* is a missing link between amphibians and reptiles and *Ichthyostega* is a missing link between fishes and amphibians. Moreover, to mention that just one fossil record may not be enough to trace the evolutionary status of vast assemblage of different groups of animals.

THE TASK OF SYSTEMATIST

We know that the earth evolved over 3.5×10^9 years ago. The biosphere consists of around 1.8 million described species. It is very difficult to address such vast number of species unless they are arranged in some systematic order based on similarities or differences, categorized and finally classified. The scheme of classification show **'phylogenetic relationship'**. The basic task of the systematist is to estimate the diversity of species investigated and assign them appropriate categorical rank. Systematists search for a phylogenetic system, but they do not necessarily have to acquire special knowledge on the distinction, validity of proper name and the numbers of known species. Many systematist study the phylogeny of supraspecific taxa but are not able to identify a new species.

According to Wägele (2005), the systematist can, but must not necessarily know, the rules of taxonomy. On the other hand, the taxonomist should know the logics of phylogenetic systematics in order to be able to systematize new species correctly. In practice, however, it is also possible to describe species without knowledge of the theory of phylogenetics. Once an organism or group of individuals is encountered in nature, some of the specimens are collected and studied. The sample collected may show variability, then the problem arises as how to characterize a species and how to categories species so collected, into different categories. Depending on the taxon investigated, a systematist may proceed through following steps:

- Data about the collected specimen is recorded i.e., various attributes of the taxon in search of diagnostic characters as well as characters it shares with related members of its group are studied. New taxa may be discovered during new collections and through taxonomic processes (i.e., taxon based reviews and revisions, and regional reviews), the identification is done (i.e., using existing classifications and keys).

- Scientists classify organisms using a series of hierarchical categories known as taxa (taxon, singular). This hierarchical system moves upward from a base containing large number of organisms with specific characteristics. This base taxon is a part of larger taxa which in turn becomes part of an even larger taxon. Each successive taxon is distinguished by a broader set of characteristics. The case level in the taxonomic hierarchy is the species.

- The individual is categorized into taxonomic units like species or sub species and the variation between the species is worked out.

- The observed characteristic features of different species are compared and subjected to statistical analysis.

- The relationship based on comparisons drawn out from homology, parallelism, convergence, primitiveness and specialization is worked out.

- Based on affinities and **divergence**, different group of organisms are grouped into taxa of various ranks. The systematist groups individuals and taxa by looking at similarities arising from a common genetic origin. When two species have similar characters inherited from a common ancestor, these features are said to be **homologous**. Using homologies, systematists discover pairs of taxa, or sister taxa, that share a common ancestor. Individuals or taxa that do not conform to a group are excluded by default. The phylogenetic relationship is then represented with the help of **dendrograms.**

- The taxa so discovered are given names applicable to recognized taxa and if no applicable published name is available, a new name is assigned to the specimen based on provisions of International Code of Zoological Nomenclature. Naming a taxon is more than simply giving it a name. Before a name is accepted (i.e., published) it has to comply with the rules of the Code. Nomenclature is a system of names and the provisions for their formation and use. The starting point of Zoological Nomenclature is set to 1st January 1758. This is the date of publication of *Systema Naturae* (10th edition) by Swedish botanist Carl Linnaeus.

- When taxa are described and named, the next step is to investigate the relationships between them. Systematists attempt to reconstruct phylogenetic trees and unravel the patterns that led to the distribution and diversity of life.

- Of late, the techniques of molecular biology are being employed by the systematists to compare the sequence of genes among organism. Individuals sharing similar DNA structure may be closely related. Thus, molecular systematics provides valuable insight into the classification.

It is also important to mention that taxonomy aims at identifying species not the specimens. A species is a group of individuals (specimens) that may, to a greater or lesser degree, show the ever present intra-population variability. Knowledge of polymorphism is essential for the species' circumscription. Furhter, it is always recommended that the specimens should be identified correctly. In brief, the chief task of the taxonomist is to know the variability in the population and group them separately on the basis of similarities and differences.

TAXONOMY- MYTHS AND CONCEPTIONS

Taxonomy in general appears to loose ground, because it is tedious and requires lot of field work. It further suffers due to the lack of sufficient information and detail study. However, taxonomy has been the foundation of evolution and recapitulation and has also provided the basis for the development of population concept. According to Wheeler and Valdecasas (2007) there are different opinions about taxonomy detailed herein as under:

- Taxonomy is descriptive, yet much of its descriptions like characters, species and clades are based on tested hypotheses about similarity and group membership.
- Mayr (1942, 1969) perpetuated the myth that taxonomy is based upon typological thinking but this is neither historical nor contemporaneous fact.
- Species in taxonomy are subjective, thus species can be tested on the basis of scientific hypotheses. Species based on genetic distances are merely arbitrary constructs of convenience that only occasionally correspond to truly evolutionarily meaningful entities.
- There is no single or necessary species concept. There are dozens of species concept and in several cases there are distinct distinction between concept and criteria. According to Platnick (1979) the observable attributes which are inherited in their original or some modified form are important.
- It has been argued that many traits that vary within populations and species often create irrelevant, uninformative and misleading informations. However, when the detailed investigations are made to find out the different processes that result in speciation e g., allopatry, sexual selection etc., other than pattern of character distribution, it is possible to recognize species as the result of any or all micro-evolutionary causes.
- According to Sokal and Rohlf (1966); Rohlf and Sokal (1967); species are identified just on the basis of matching **morphometric** characters or arbitrarily chosen distances, thus taxonomy is regarded as a mindless, mechanistic procedure.
- Taxonomy is under practice for centuries still utilizing the binomial nomenclature and hierarchical grouping of individuals as advocated by Linnaeus.

However, Valdecasas et al., (2007) has treated taxonomy somewhat old and out dated.

• Many biologist challenge taxonomy being a historical science not based on any experimentation. However, due to nature of evolutionary novelties and their inheritance by descendant species precise predictions about character distributions make taxonomic assertions quite scientific (Platnick, 1979; Nelson and Platnick, 1981).

IMPORTANCE OF TAXONOMIC STUDY

Taxonomy is the oldest discipline in biology, with Carl Linnaeus (1707-1778) bringing a hierarchical order to natural history and establishing the basis for biological nomenclature and classification that guides the science of life to this day (Blunt, 2001, Warne, 2007). Taxonomy usually refers to the theory and practice of describing, naming and classifying living things. Such work is essential for the fundamental understanding of biodiversity and its conservation. Taxonomy provides detailed account of existing diversity of life on earth and helps to reconstruct the phylogeny. Numerous evolutionary phenomena can be explained with the help of taxonomy. The purpose of taxonomy is to recognize categories and identify the vast array of individuals. The species are grouped into higher-level taxa based primarily on apparent similarity or by the possession of shared traits. Later on, with the development of theory of evolution, the species are grouped in accordance with their evolutionary history.

Taxonomy with evolution and speciation perspectives is known as 'systematics', the foundations of biology, fundamental to all branches of biological and environmental sciences since the New Systematics (Huxley, 1940). Modern systematics is based on Darwin's theory that evolution is a process of descent with modification. A more appropriate name for systematics would therefore be the phylogenetic (evolutionary) systematics. One question often comes in an intelligent mind, that where did we come from and what is our future? Answer to this question is not possible without systematics. In science, systematics forms the basis of all other fields of comparative biology. The systematic biology includes the study of each species across all levels of biological organizations and the diversity of all species of **biosphere**.

Systematics is now in a stage of technology driven revolution. It provides the framework (i.e., classification) and language (i.e., nomenclature) by which biologists communicate information about organisms. Further, through phylogenetic hypotheses, the basis for evolutionary interpretations is worked out (e g., mode of speciation, historical biogeography and ecology). Phylogenetic hypotheses and corresponding classifications predict properties of newly discovered or poorly known organisms (e g., fossils). Thus, taxonomic classifications are of great help in unraveling the mystery of biology and paleontology.

Due to ongoing anthropogenic development, there has been a massive degradation of several populations which is most commonly measured by changes in the distribution and abundance of species. The loss of individual populations of species may affect the genetic diversity of a species and thus impact the survival of the species itself. The loss of ecosystem diversity restricts the habitat available for a species and may also affect its survival.

Taxonomy is therefore an integral part of biology and science as a whole reaching out into other fields almost unnoticed, but most especially medicine. As a subject, it is fundamental to our understanding of the natural world and critical to future research. We must know what species we have and what they are to be able to continue to study everything from infectious diseases to pollution and the history of life. If we are going to save our planet, we have to know what is on there to start with. Nevertheless, what really matters is that without taxonomy it is impossible to know which species lived yesterday, are living today, and will have the chance to be alive tomorrow in a given area; what type of balance exists within a community that occupies an area, and why that balance dominates; what is the cost of the biodiversity of a certain area; what will happen to the biological balance of a given area if the dominant environment conditions change, etc. In conclusion, none of the above will become reality if there is neither taxonomy nor taxonomists.

Chapter 3

TYPES OF CLASSIFICATION

Taxonomy deals with the identification and nomenclature of the animals and is part of a larger division of biology known as systematics. It deals with the identification of organisms and assigning them into groups, based on similarities and differences. The arrangement of organisms into groups and finally considering how evolution has resulted in the formation of these groups is the essence of classification. Classification is based upon the recognition of homologous structures (e g., relative position or origin during embryonic development etc.). Thus, on the basis of system of classification, one can identify particular genera easily and can assign its systematic position. The classification is not static but may change in the light of new information and addition of new species. These changes are apparent at the level of species and genera.

PRINCIPLES OF ZOOLOGICAL CLASSIFICATION

Classification is based on the relationships among organisms which may vary. Initially, animals were grouped on the basis of certain relationship, later on similarities and differences formed the basis of animal classification. Before attempting to classify animals, following facts should be considered:

Differences

It refers to the features on the basis of which a species can be differentiated from the other species.

Accident

It refers to the characters which may be present in some species but not in all.

Property

The characters which are present in all the members of the species but may not serve as a basis for differentiation.

FEATURES OF CLASSIFICATION

Zoological classification involves the ordering of animals into group based on their relationships, i.e., of associations by contiguity, similarity or both. There are two distinct processes associated with the classification viz., identifying different animals and arranging them into groups based on certain common features i.e., finding out similarities.

In classification, vast assemblage of animals are categorised into clearly defined groups, based on similarities for their proper understanding as well as to have an idea of interrelationships of diverse groups. Accordingly, classification is grouping of like and unlike individuals in to different groups. This grouping does reflect certain relationship as well as degree of differences and similarities. Thus, classification itself is an art. The most important basic concept of systematics is the naming and classification of animals. Taxonomy, thus is the framework on which systematics is based.

Once, it was taken for granted that the evolution is the only mechanism of biological diversity and speciation, it became very clear that the classification of animals should reflect the phylogeny of organisms, where each taxon should have a single ancestral form. To be more precise, an ideal classification should be based on homology; i.e., the shared features, which have been inherited from a common ancestor.

The taxonomists work with the unit referred as **taxon** (plural- taxa), which represents the grouping of closely related entities, may be individuals (i.e., a **subspecies** or a **species**), several species (i.e., a **genus**) or several genera (i.e., a family), etc. Therefore, the lowest taxon includes individuals and higher taxa include subordinate taxa and individuals. Once discovered, taxa are described using characters. There are many attributes of a taxon that can be used for recognizing, defining or differentiating it from other taxa. The characters that are used for this purpose may be morphological, anatomical, behavioural, ecological, physiological, and molecular as well.

The binomial nomenclature, a combination of genus name and a specific name forms the basis of identifying different flora and fauna. Rules governing the nomenclature and classification of different animals are contained in the International Code of Zoological Nomenclature. Similar code exists for plants and bacteria as well.

TYPES OF CLASSIFICATION

The grouping of individuals has been done in different manner. Classifications have been made on different types of relationships discussed herein as under:

Ecological Classification

It describes groups of population on the basis of habitat, e g., aquatic **biota**, grassland ecosystem, soil micro-organisms; or on the basis of environmental factors for e g., fish population of aquatic or marine habitat, insects living in aquatic or terrestrial habitat, **thermophilic** bacteria, cave dwellers or flying animals etc. This type of classification gives a brief idea of different groups of organisms. However, with the advancement in the field of taxonomy, ecological classification has to be verified after correct identification and nomenclature of organism based on other significant characters.

Teleological Classification

In this scheme of classification, animals have been classified on the basis of their utility. For e g., beast of burden for carriage like-ox, horses, bulls, domesticated animals, pets that serve people, edible fishes and poultry provide eggs and meat for human consumptions. In case of plants, the classification was based on the basis of products they yield like, medicinal herbs and spices.

Artificial Classification

An artificial classification takes into account only a few characters. The animals can be classified in to different category like, flying animals, running animals and swimming animals. However, this would result in bracketing together certain types like insects, birds and bats in the first category, some other insects, non flying birds and mammals in the second category and certain protozoan, fishes and whales in the third group. Thus, when organisms are grouped arbitrarily based on single characters, the classification is said to be the artificial one.

Such classifications are more or less logical and sometimes useful, if only certain practical aspect of classification is considered. It is relevant to mention that classifications based on similar adaptations or modes of living have been quite popular. Initially, animals were classified as animals living on land, water and air by the Roman naturalists in *Historia Naturalis*. Birds with webbed feet were classed together and so were with those having long legs. The rodents and lagomorphs were placed in a single order due to an adaptive similarity of the **incisors**. The artiodactyls and perissodactyls were classified as ungulates due to similarities in feeding habits, structure of foot and general body built up.

Merits of Artificial Classification

Artificial classification serves as basis of identifying animals depending on their habitat. It is practical for the preparation of synoptic keys. Artificial classifications are designed in such manner that organisms belonging to different taxa within the system can be separated just on the basis of one character.

Demerits of Artificial Classification

Early taxonomists used morphological characters almost exclusively to distinguish taxonomic categories as a basis for classification. For example, animals grazing on leaves and grasses were classified as **herbivores** and those preying upon other animals were referred as **carnivores**. Further, consider two species of grasshoppers, one belonging to super family **Catentopinae** and the other to family **Acridinae**. In Catentopinae, the **prosternum** is armed and is provided with spines, whereas, in Acridinae, the prosternum is unarmed. However, in classification the importance of spines is not taken into consideration, instead shape and degree of approximation of **metasternal plates** is taken into account. This is one of the greatest drawbacks of artificial classification because only one character is measured at a time for one group and for another group, different character is considered. Thus, the artificial classification based on few characters is not accepted. Further, this classification does not provide any indication of the 'true' or 'natural' relationships of the species considered.

Natural Classification

A classification which is based on maximum number of important attributes is referred as natural classification. A natural system of classification provides a basis whereby the true phylogenetic relationships of the group are expressed. A natural classification helps to provide steps during evolution. The sequence of events may not be clear due to the lack of fossil evidence and also due to parallel and convergent evolution. However, the gradually increasing complexity observed in taxonomic scheme is in consonant with Darwins' hypothesis *descent with modifications*. This holds true, so far as the relationships of the groups within the phylum are concerned but the relations between the phyla are difficult to interpret. Although, the phyla may have little apparent relation to one another, it is possible to arrange them in order of degree of complexity of organization. Further, certain phyla can be arranged into still larger groups such as sub-kingdoms and branches, which serve to emphasize that gradation between phyla, are not of equal magnitude and member of some phyla have similar features, which indicate their ancestry from a common stock.

Many systematists feel that a natural classification provides a basis whereby the true phylogenetic relationships of the group are expressed; however this is far from true. A natural classification at best is said to provide an indication of evolutionary status of the group. One can have an extensive idea about the trends of evolution but due to the lack of fossil records and also following parallelism and convergent evolution, the details are often not available.

It must be made clear that considerations in taxonomy do take into account the progress of organic evolution, but the relationships based on phylogeny do not indicate true phylogenetic relationships. In other words, it can be made clear that the natural classification indicates the affinity i.e., degree of similarity between organisms. Within the classification scheme, organisms placed in the same taxon showing the greatest affinity said to form a natural group. Almost all the modern accepted classifications are natural, i.e., they indicate the degree of similarity between the organisms. A natural classification is based on following:

Morphology

Initially only morphology was considered as the basis of classification. However, it was observed later on that structure of one character or a set of character alone will not yield the proper information for classification, so it will be practically impossible to have classification based on morphological characters alone. Thus, modern taxonomy is based on characters drawn from different sources.

Homology

In natural classification, homology serves as the main criteria for creating taxonomic groups. However, homology should not be used as the sole basis for classification because it creates only phyletic relationship. At the intraspecific level, it is difficult to ascertain the **homologous** characters. It becomes quite difficult to decide the homology of many characters, which the animals under question possess.

The basic concepts of homology are often confusing, because it may be simply due to the **cladistic** relationship between two closely related groups through recovery of common ancestry without taking into consideration the number of characters. Thus, homology alone cannot serve as the basis of natural classification, because it may result in the development of some similar characters in different groups of animals due to **parallelism** or **convergence**. Thus, the basis of natural classifications is the establishment of **homology**. Once homology is established, then it has to be seen whether the characters are advanced (derived) or primitive (ancestral) i.e., **apomorphic** or **plesiomorphic**. Comparative morphology of extinct forms and the fossil records help in such analysis.

Population

One of the characteristics features of natural classification is that, while making natural system into consideration the populations are accounted, which are related to each other through common ancestry. In Fig 3.1 it has been shown that T is a common ancestor of subgenera P and Q.

Fig. 3.1 T represent common ancestor to subgenus P and Q

In Fig.3.2, the population represented by A, B, C, and D, E, F is related to each other through common ancestry and it also reflects certain relationship. Further, population represented by A, B and C may be terrestrial and those represented by

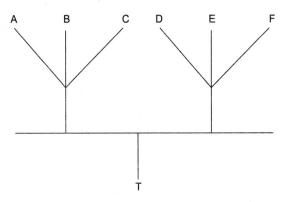

Fig. 3.2 Two population P and Q arising form a common ancestor T. A, B and C represent different population arising from subgenus P and Q

D, E and F may be aquatic. This indicates the true nature of population and its objective state that help in making the framework of natural classification.

Natural classification can be categorized under following three groups:

Cladistics

The cladistic line of approach relies upon the fact that the animal groups have evolved by descent with modification (Hennig, 1957). In other words, the evolutionary relationship is determined based on shared derived characters. Shared primitive characters are not used as a basis of classification, since these may be lost or modified during the course of evolutionary process. Further, the cladistics scheme is not based on the extent of evolution after the branching of evolutionary line. For example, chimpanzees and orangutans are more similar to one another than to humans; however, cladistics taxonomy places human and chimpanzee together, since they share a more recent common ancestor than orangutan. The significance of cladistics taxonomy may be summarized as under:

• Cladistic analysis is useful for creating a system of classification, as it provides hypothesis about the relationship of individuals.

• Cladistic helps to elucidate the way in which the characters change within group over time and the direction of the change, as well as the relative frequency with which the characters change.

Phenetics

The phenetists (initially referred as numerical taxonomists) led by Sokal and Sneath (1963); Sneath and Sokal, (1973); relied upon number of characters common among organisms regardless of the fact whether these were primitive or advanced. This is based on following:

• More and more characters should be studied

• All characters should be given equal importance

• The greater the similarities between the groups, closer would be the relationship between them. In other words, the groups sharing many characters in common are closely related. The phenetic study is generally represented in the form of phenograms (For detail, please see Chapter 10).

Phyletics

This school of thought has been supported by Simpson (1961) and Mayr (1969, 1981, 1982 a). It incorporates both cladistic and phenetic approach in classification.

In this case, the taxa are grouped together based on the number of similarities or differences depending on the numerical coefficient employed. Traits are measured and either converted into integers or input directly as numerical data. These data are then mathematically processed using an **algorithm** that generates a similarity (or distance as the case may be) matrix.

Merits of Natural Classification

A natural classification is based on overall similarity of population and probable evolutionary line. The natural classification has more and more attributes because it draws information from a greater source. The more natural the information is available, the more natural the classification becomes. The natural classification has large number of tools, because it is based on a several characters and provides information based on recent trends in taxonomy. The natural classification shows following essential features:

- It relies on phylogeny.
- It groups animals with maximum possible characters.
- It conforms to the available knowledge of genetic traits.
- It reflects objectively ascertainable state of things.
- It reflects abstract things.
- It is meant for natural purposes.
- It enables to make maximum number of prophecy and deductions.
- It truly reflects the nature of a character.
- It is based on probable evolutionary relationship.
- In this case, the organisms resemble each other in multitude of characters and appear to be grouped by nature.
- It shows that the groups are related to each other by common ancestry.

From the foregoing discussions it can be concluded that, "a natural classification is one, in which the groups are recognized by having maximum number of characters common with their limits set by discontinuities in the diversity and capable of yielding maximum number of correct deductions."

Biological Classification

Biological classification involves grouping and placement of organisms by biological type such as **genus** or **species** based on scientific criteria. Biological classification was established by Linnaeus (1758), who grouped animals according to shared physical features.

Evolutionary Classification

According to Darwin (1859), a classification must be 'genealogical' showing common ancestry of members of each taxon. Darwin also emphasized to consider divergence of descendant groups. Evolutionary classification relies much upon evolutionary relationships and biological attributes rather to restrict morphological relationship. Evolutionary classification entails the genealogical relationships between the groups and the degree of evolutionary change under gone by them.

There is much debate, as how to measure the evolutionary distance and how to correlate the degree of evolutionary change taken place in the group under study. Further, when two factors are considered at a time, it creates confusion and difficulty. Moreover, there is no perfect methodology, which may help to consider similarities and descent with modification. As such these two factors should be considered sequentially. Cladistic school of thought may first determine descent and then assign the holotypic taxa its correct position in the Linnaean hierarchy. Evolutionary taxonomist like Simpson (1961) and Mayr (1968) generally preferred to establish reasonably homogenous taxa and then tested them for monophyly and discarded other non-significant features. Darwin (1859) brilliantly illustrated the *Origin of Species* and *struggle for existence* through natural selection.

Phylogenetic Classification

It is based on phylogeny and refers to the evolutionary history of living animals. It draws inferences from paleontology. In choosing phylogeny as the basis of classification, vertical hypotheses are made and it is likelihood that one of the vertical hypotheses may be based on wrong assumptions? However, it may be mentioned that Mayr (1953) gave two statement in support of phylogeny. The phylogenetic system is the only known system having similar basis of classification. It has practical advantage of combining forms. There are some phylogenetic considerations, which represent a very complex history. Since fossil records of most groups are incomplete and there are certain gaps, so phylogeny can not be taken as the basis for natural classification. In groups with adequate fossil records such as horses, none of the proposed phylogeny has stood to the test of time for a longer period.

Each taxon, regardless of the rank will have a sister group its closest relative so that the development of classifications and evolutionary history (phylogeny) can be illustrated with the help of a branching diagram popularly referred as **phylogenetic tree**. An ancestor and all of its descendants form a **monophyletic** group, whereas, when some of the descendants are lacking the remaining descendants said to constitute **paraphyletic** group. Groups derived from more than one ancestor are said to be **polyphyletic** (For detail, please see Chapter 7).

As discussed in the preceding paragraphs, Darwin had advocated the classification based on phylogeny and Haeckel (1866) gave the famous **recapitulation hypothesis.** However, determination of phylogenetic relationship is often difficult. Since the common ancestors of different groups of organisms are usually long extinct and the fossil records are poor, this puts a great deal of emphasis on constructing phylogenies by comparison between known organisms whether existing or in the form of fossil.

In general, the more a group of species shares common inherited attributes, there is more possibility of their descent from a common ancestor. Thus, all the available heritable characters are considered while making comparisons between species. The more attributes they share, there is more likely hood of their common ancestry. Greatest blow to Haeckels' theory came from the objections that it is difficult to trace the phylogeny. However, according to Mayr (1969) neither the phylogeny is based on classification nor classification is based on phylogeny. In fact, both are based on a study of natural groups found in nature. Groups having common characters reveal that they are the descendants of common ancestor. It is based on the comparisons of organisms and their characteristics and on a careful evaluation of the established similarities and differences.

Genetic Basis of Classification

Genetic relationship between animals can serve as a basis in framing the natural classification. Classification is a basic and vital activity. All biological classification systems are designed to serve particular purposes and must be judged with reference to their specific goals. Such systems can reflect similarities in the measurable traits of organisms (phenetic classifications), their phylogenetic relationships (cladistic classifications), or some combination of the two (orthodox classifications). Mixed systems are common because often people wish to accommodate more than one goal within a system. The different biological classification systems all are designed to express relationships among organisms, but they differ with respect to the kinds of relationships they attempt to express. Classifications are based on features selected according to the goals of the system and help to bring disorder into order on the basis of similarities and differences. It determines methods for organizing immense diversity of life on earth. It is a dynamic process that reflects the very nature of living organisms, which is subject to modification and change over many generations during the process of evolution.

Chapter 4

ZOOLOGICAL NOMENCLATURE

Taxonomy deals with the identification and naming of organisms into formal system based on similarities, such as internal and external anatomy, physiological functions and genetic make up along with evolutionary history. With an estimated 2 to 50 million species on earth, it is quite difficult to categorize and name the vast variety of animal groups. As the diversity of life is enormous, classification provides the method for organizing this vast array of life systematically. As we know that life appeared some 3.5 million years ago and new organisms have evolved during the course of grand evolutionary march, many of these organisms have become extinct, while some have developed into present day flora and fauna. Thus, extinction and diversification continues endlessly and scientists are frequently encountering fluctuations, which may affect the way this vast assemblage of animals and plants have been classified and arranged. Consequently, the strength of Linnaean taxonomy lies in the fact that the binomial nomenclature represents the genus and species name. On the other hand, as explained by Dubois (2005 c), taxonomy and nomenclature are different disciplines. Taxonomy recognizes classificatory units or taxa, whereas nomenclature attaches a given scientific name to each of these units. Taxonomy is a scientific discipline, whereas nomenclature is a technique.

EARLY SYSTEM OF NAMING

Before Linnaeus, biological nomenclature existed in many forms. It was Aristotle who classified all living forms based on their mode of movement i.e., air, water or land. Some of the contributions of naturalist in this field are being given here in as under:

- Ibn Rush (1172) translated and abridged Aristotle's book *de Anima* (On the Soul) into Arabic.

- Conrad von Gesner (1516-1565) made a significant contribution on the form of life known at that time.

- An Italian philosopher, physician and botanist Andrea Caesalpino (1583) in *de Plantis libri* XVI (1583) proposed the first methodological arrangement of plants. He divided the plants into 'fifteen higher genera' based on the structure of trunk and formation of fruity bodies.

- John Ray (1627-1705) an English naturalist classified plants in *Historia Plantarum*.

- Augustus Quirinus Rivinus (1652-1723) while classifying plants introduced the concept of category.

- Joseph Pitton de Turnforte (1656-1758) introduced the hierarchy of class, section, genus and species. He for the first time used uniformly composed species names, which consisted a generic name and a diagnostic phrase e g., *differentia specifica*.

BINOMIAL NOMENCLATURE

The Swedish naturalist Carolus Linnaeus (1707-1778) developed a system of naming plants and animals. Linnaeus (1758) created a system of binomial nomenclature which is a recognized method of naming individuals. As the word 'binomial' suggests, the scientific name of a species consists of combination of two terms: the genus name and the species name. For example, the binomial name of human beings is *Homo sapiens* and no other species of animals can have this binomial name. In this manner, every species has a unique and stable name compared with the common name, which are neither unique nor constant and may vary.

Binomial nomenclature is also referred as 'Binomial Classification System'. In formal usage, the genus name is a noun and always written in capital letter, while the species name is an adjective, which should agree in gender with the genus, and is written in lower case. A genus (plural-genera) is a higher level category and it may contain single or more than one species. The species is the smallest unit of classification and each species (**taxon**) is nested with a higher category. Unlike common names which change from place to place, scientific names are unique and remain constant. Linnaeus' excellent work drew a line between taxonomy and nomenclature. The systematists now accept the uniqueness and stability of biological nomenclature across the world over.

Salient Features of Binomial Nomenclature

Binomial nomenclature is quite significant due to its brevity, widespread use and stability of names. Binomial consists of the name of a genus followed by the name of the species. It has following features:

• The first name i.e., name of the genus (generic name) is written first and always begins with capital letter.

• The species name is written after the name of the genus and the first letter of species is always written in lower case.

• The name of genus and species are always either printed in italics, such as *Homo sapiens* or underlined separately when hand written.

• The scientific name is written in full, when it is used for the first time or when several species from the same genus are being listed or discussed in the same paper or report.

• The name of the genus must be written full without abbreviation during its first usage e g., *Passer domesticus*, it may then be abbreviated in subsequent usage in the same text and can be written as *P. domesticus*. After genus and species, the name of the author (the authority) is written who first described them in publication. This is never italicized or underlined, e g., *Dysdercus koenigii* Fabricius or *Papilio atalanta* Linnaeus, 1758 and *Balaena mysticetus* Linnaeus, 1758. Since Carolus Linnaeus was the first person to name many plants, therefore after the species name of many individuals the author Linnaeus is written.

• Date of official descriptions can also be included with scientific names to further clarify the period when the description about the individual was actually published. For example, *Rana tigrina* Linnaeus, 1758. This means that Linnaeus had described the species in the genus *Rana* in the year 1758.

• If the authors name appears in parenthesis after the name of the animal, it means that this species was described by someone else but the species has been moved to correct genus by the author whose name appears in parenthesis. For example, *Anser albifrons* (Scopoli, 1769) the white-fronted goose was first described (by Giovanni Antonio Scopoli), as *Branta albifrons* Scopoli, 1769. It is currently placed in the genus *Anser*, so author and year are set in parenthesis. The taxonomist who first placed the species in *Anser* is not recorded (and much less cited).

• A subgenus is sometimes given in parenthesis after the genus, thus *Bison* (*Bison*) *bison bison* (Linne, 1758).

• In some cases authors may use a scientific name differently than the person (author) who originally described the species. In such cases the scientific name, as listed in catalogs and other writings, is separated from users name by a colon e g., *Phytocoris marmoratus* Blanchard; *Phytocoris marmoratus* : Stonedahl.

- The abbreviation 'sp.' (or spec.) is used, when the actual specific name is not known, for example, *Passer* sp. denotes a species of the genus *Passer*. *Dysdercus* sp. means the specimen under study belongs to genus *Dysdercus*, however, it may be *D. cingulatus* or *D. koenigii*. Further, when a specimen is named using 'sp' after a genus name, this indicates that it is identified well enough to be assigned to a genus.
- The abbreviation 'spp.' (plural) indicates 'several unknown species'.
- The abbreviation 'cf' is used when the identification is not confirmed. For example, *Corvus cf. splendus* indicates that the species is in doubt.

Some examples of scientific names are as under:

Canis familiaris	-	Dog
Felis domesticus	-	Cat
Homo sapiens	-	Human beings

Linnaean Hierarchy

After giving binomial nomenclature, Linnaeus used five ranks: class, order, genus, species and variety. The Linnaean system of grouping of individuals is based on similarities and differences. It reflected a nested set of groups within groups. The biologists later added larger and higher level group names known as taxon from family to kingdom arranged in hierarchical order. Human beings, for example, can be assigned systematic position in Linnaean hierarchy as under:

Sub species	*sapiens*
Species	*sapiens*
Genus	*Homo*
Subfamily	Homininae
Family	Hominidae
Superfamily	Homonoidea
Suborder	Anthropoidea
Order	Primates
Infraclass	Eutheria
Subclass	Theria
Class	Mammalia
Superclass	Gnathostomata
Phylum	Chordata
Subkingdom	Metazoa
Kingdom	Animalia

- It is relevant to mention that the referred categories are completely arbitrary. The use of expanded hierarchy depends entirely upon needs and preferences of the taxonomist studying a particular species and therefore any combination is possible.

The same name of animal under study is used all over the world in all languages, this avoids difficulties of translation.

• The zoological nomenclature is somewhat stable. However, when species are transferred between genera, the species descriptor is kept the same. Similarly, if the species that were previously thought to be distinct are demoted from species to lower rank, former species names may be retained as infra species descriptors.

• Despite the rules favoring stability and uniqueness in practice several species may have several scientific names in circulation, depending largely on taxonomic point of view (synonymy).

• When considering micro-organism species, a category (not usually considered a taxonomic one) found below the level of species is **strain**. A strain in some way is equivalent to a breed or a subspecies among plants or animals. By convention, names of some categories always end as under:

Superfamilies-oidea; Families-idae; Subfamilies-inae; Tribes-in i; Subtribe - ina.

Trinomial Nomenclature In late 1740's, Linnaeus began to use a parallel system of naming species with *nomina trivilia* or *Nomen triviale*, a trivial name having a single or two word epithet placed on the margin of the page next to the many worded 'scientific name'. Linnaeus followed a rule that the trivial names should be short and unique within a given genus. He consistently applied *nomina trivilia* to the species of plants in *Species Plantarum* (Ist edition 1753) and to the species of animals in the 10th edition of *Systema Naturae* (1758).

A trinomen is a name consisting of three names: generic name, specific name and sub-specific name. All the three names are type set in italics and only the generic name is capitalized. *Butea jamaicensis borealis* is one of the subspecies of red tainted hawk *Butea jamaicensis*. Further, Great Comorant (*Phalacrocorax carbo*) found in New Zealand differs slightly from those found elsewhere and are classified as the sub-species *Phalacrocorax carbo novaehollandiae*. Similarly, *Diabrotica undecimpunctata undecimpunctata* Mannerheum indicates that the third name denotes a subspecies, a group of individuals of the same species that have differences in body form or colour and geographical differences.

Following is the order in which different categories are outlined.

Kingdom
Phylum
Sub phylum
Super class
Class
Subclass
Cohort
Superorder

Order
Suborder
Superfamily
Family
Subfamily
Tribe
Subtribe
Genus
Subgenus
Species
Subspecies

No indicator of rank is included in zoology; subspecies is the only rank below the species, only the single infraspecific rank is used, so no additional indication of rank is required, with the third position sufficient to indicate that it is a subspecies.

TYPES OF NOMENCLATURE

Since classification relies on phylogeny based on entire wealth of ever-growing knowledge, the process goes on changing. If there is just one type of classification, it may be insufficient to describe all the available species or the one described later.A constant classification scheme may not serve the purpose. The nomenclature must support ever changing classification, which implies that a name can only be associated with just one attribute of a taxon. Based on the features with which a name is associated, several fundamentally different types of nomenclature are described herein as under:

Rank Based

In this type of nomenclature, a name is associated with certain rank of a taxon and is subject to change whenever the rank changes, but remains the same when other attributes such as circumscription or position change. The ranking nomenclature still plays a major role in taxonomy, because all International Codes including ICZN are based on this principle. However, the ranking nomenclature has significant shortcomings.

Circumscription Based

Under this scheme, a name is associated with a certain circumscription of a taxon without regard of its rank or position. Circumscription naming is applied to higher taxa.

Description Based

In this type of nomenclature, a name is associated with the definition of taxon and changes accordingly, whenever the taxon is redefined. For example, different authors referred to the same taxon (springtails) as Collembola (derived from collophor or embolium, the sticky ventral tube).

Phylogeny Based

In this case, the name is associated with a common ancestor of the taxon, which means a name may only refer to a holophytic taxon evolved from such ancestral species. This nomenclature is of little practical value. A phylogenetic name may become available only, if its publication refers to a common ancestor of the taxon.

Hierarchy Based

Under this case, the name is associated with the taxon's placement within hierarchical classification and does not depend on rank. This system is based on recently enacted International Code of Zoological Nomenclature, yet overcomes some important flaws of the ICZN's ranking principle.

NECESSITY FOR RULES OF NOMENCLATURE

There are millions of animal species and supra specific taxa and individuals occurring in different parts of the world. Further, all the hundreds of previously unknown species are described, thousands of already described species are redefined, thousands of new genera and subgenera and many new families may replace the older ones. This is likely to create ambiguity in nomenclature. Therefore, certain set of rules have been framed. The most accepted rules include the International Code of Zoological Nomenclature, International Code of Botanical Nomenclature (ICBN) and International Code of Nomenclature for Bacteria and some other Codes which govern ranking names of species and other biological taxa. Competent International authority enacts these Codes and the provisions of the Codes are binding on all the biologists. In rarest of rare case, if the rules fail to provide solution to the name related issue, the commission makes a special ruling. The rules in the Code determine which names are potentially valid for any taxon including the ranks of subspecies and super families. The provisions can be waived or modified, when applied to a particular case, in situation where strict adherence would cause confusion.

With the recent emphasis on phylogeny based classification, everyday new information is being added. Thus, to regulate the usage of taxonomic names, there should be universal rules of nomenclature which would help in naming and assigning correct position of zoological taxa. The International Commission on Zoological Nomenclature (ICZN) acting on behalf of the zoologists from across the world, which can make such exceptions. A proposal to this effect is to be submitted to the commission.

NOMENCLATURE CODE

The International Code of Zoological Nomenclature is a set of rules adopted in Zoology which provides the maximum universality in naming of the animals according to taxonomic judgment. The nomenclature Code is the rule book that govern biological nomenclature i.e., the name of the animals (International Code of Zoological Nomenclature), plants (including Fungi, Cyanobacteria) (ICBN) and the viruses. The rules for nomenclature are similar, but are governed by separate Codes, for zoology, botany and bacteria. The botanical code (now renamed the *International Code of Nomenclature* for algae, fungi, and plants (ICN)) changed its rules on publication earlier this year as well, allowing e-publication without any additional requirements for archiving or registration. Under the ICB, legitimate publication of new plant names requires only generation of a PDF document published in a journal with ISSN or a book with ISBN. The date allowing this transition was 1st January, 2012, which is now, with the current ruling, matched by the amendment to the zoological code. The scale of the nomenclatural challenge is considerably smaller for plants; however, as to date only about 260,000 plant and 100,000 fungi species have been described.

The bacterial code (*International Code* of *Nomenclature* of *Bacteria* (ICNB)) has very different requirements, as all new nomenclatural acts need to be published in one of two journals, tightly linked to their Code. Moreover, the bacteriologists decided to essentially amputate the historical legacy of research with most work before 1980 being deemed unrecoverable. Only about 9260 bacterial and archaebacterial names are regulated by the ICNB, as there is also a requirement that named taxa have a type culture. Much of bacterial diversity can not be cultured, thus can not have a type culture and therefore remain beyond the bacterial nomenclatural system.

Work is now on-going to bring together the different nomenclatural codes, primarily through development of tools in common. The development of

ZooBank and a registration system for fungi has become mandatory since January 2013. This and other projects to unify the Codes in practice, is being helped by an International Committee on Bionomenclature (ICB). These codes differ from each other, for example:

• The plant code (ICBN) does not allow **tautonyms**, whereas, there is scope for tautonymy in ICZN.

• The starting point i.e., the time from which these Codes are brought in to effect is 1758 for zoological nomenclature whereas; it is 1753 in case of botanical nomenclature.

• In case of bacteria, the starting year of the Code is 1980.

• The Draft BioCode (DBC) is the result of an attempt at unifying the nomenclatural Rules currently in force in different taxonomic domains (mostly zoology and botany), which are the result of a long historical process during which they have widely diverged in several important respects (Dubios, 2011).

• International Code of Phylogenetic Nomenclature, known as the **PhyloCode** is a developing draft for a formal set of rules governing phylogenetic nomenclature. Its current version is specifically designed to regulate the naming of **clades**, leaving the governance of species names up to the rank-based Nomenclature Codes (ICN, ICZN, ICNB). This has been developed to name clades of phylogenetic tree but not of paraphyletic or polyphyletic groups. Instead of being grouped into ranks, such as genus, family and order, the organisms are grouped into clades defined as any set of organisms with a common ancestor. Once implemented, it will be associated with a registration database, known as RegNum, which will store clade names and definitions. The PhyloCode will cover the naming of clades and species. It is believed that this will provide a usable tool for associating clade name with definitions, which could then be associated with sets of subtaxa or specimens through phylogenetic tree databases (such as TreeBASE). The first International Phylogenetic Nomenclature Meeting, which took place in Paris in July 2004, was one of the final steps towards the proposed implementation of the PhyloCode. As with other nomenclatural Codes, the rules of the PhyloCode are organized as articles, which in turn are organized as chapters. Each article may also contain notes, examples, and recommendations.

In 1842, the British Association for the Advancement of Science adopted a general set of rules regarding procedure in nomenclature. The American Association developed a Code of its own in 1821, followed by French in 1881 and the German in 1894. American Ornithologists Union (A.O.U.) devised a code in 1885 for naming the birds.

THE GENERAL PRINCIPLES OF NOMENCLATURE

There are certain set rules outlined in the Code of Zoological Nomenclature being followed during biological nomenclature, discussed herein as under:

Principle of Binomial Nomenclature

The scientific name of a species and not of a taxon of any other rank is a combination of two names (a binomen), the first being the generic name and the second is the specific name. The principle applies to the names of species and the fact that family-group and genus-group names are uninomial and subspecies names trinomial, is not consistent with the principle.

Principle of Name bearing types

The principle that each nominal taxon had actually or potentially its name bearing type that provides the objective standard of reference by which the application of the name is determined.

For example, if it comes in the knowledge, that the same specific name is being applied to more than one species, the original name is applied to the animal conspecific with the type specimen. Sometimes, an original description establishing a species name mentions only characters shared by several species; in this case the type specimens should be re-examined to decide which of these species will keep the name (For details, please see Chapter 14).

Principle of Priority

It refers to the priority of publication of names given to categorical levels of species, subspecies, genus and family. If two names are given to the same taxon, the rule of priority comes into operation. It is quite simple; the first name applied to a taxon is the name that will be used. Priority thus relates to the date of publication or mailing date. According to this, the oldest published name is considered to be the valid name of taxon, provided that the name is not invalidated by any provision of the Code or by any ruling of the commission. This is true, even if there is some error in the original name. The valid name of the taxon is one, published first in accordance with all applicable norms of ICZN and takes the priority; later names for that organisms or groups are junior synonyms and are not considered valid.

The principle is applied only to the names published after 01.01.1758, the cut-off date from which the Code was brought into the effect. Any works published in 1758 or after are considered published. The Law of Priority (Article 25) mentions that if a genus or species has been accidentally given two names, then the name given earlier is held valid and the later name becomes a 'junior synonym'. Baseline priority for botanical nomenclature dates to Linnaeus' *Species Plantarum* (1753).

The principle of priority also provides that every name must have its authorship, since priority does not apply to anonymously published names. Authorship refers to the particular paper in which the taxon was first described with exact date of publication. It is relevant to mention that the first published name takes priority; later uses of a name spelled the same but used to refer to different organisms that are junior **homonyms** and must be given replacement names. It is also added that the first published description of a species fixes the species epithet (a word used to characterize the species); if the species is later moved to another genus, it retains the first published epithet unless that would create a homonym. There are approximately 2-3 million cases of this kind for which this principle is applied in zoology. Following are some example:

- George Cuvier (1797) proposed the Genus *Echidna* for spiny anteater. However, Johann Reinhold Forster (1777) had published the name *Echidna* for a genus of many eels. Forster's use thus has priority with Cuviers', being a junior homonym.

- Johann Fridrich Blumenbach (1799) named common chimpanzee as *Simia troglodytes*. Lorenz Oken moved it to the new genus *Pan* in 1816, so that the valid name is now *Pan troglodytes*.

- Johann Karl Wilhelm Iliger (1811) published the name *Tachyglossus* and this is considered to be the valid replacement or *nomen novum.*

- Nunneley (1837) established *Limax maculatus* (Gastropoda), Wiktor 2001 classified it as a junior synonym of *Limax maximus* Linnæus 1758 from Southern and Western Europe. *Limax maximus* was established first, so if Wiktor's 2001 classification is accepted, *Limax maximus* takes precedence over *Limax maculatus* and must be used for the species

- Andre Marie Constant Dumeril and Gabriel Bibron (1844) gave species epithet *madagascariensis* to two species of Madagascar snake viz., *Pelophilus madagascariensis* and *Xiphosoma madagascariensis.*

- John Edward Gray published the name *Antilocapra anteflexa* in 1855 for a species of pronghorn, based on a pair of horns. However, it is now thought that his specimen was an unusual individual of the species. George Ord (1815) published *Antilocapra americana*, which takes priority with *Antilocapra anteflexa* being a junior synonym.

- Johann Jakob published the name *Leptocephalus brevirostris* in 1856 for a species of eel. However, in 1895 it was realized that the organisms described by Kaup was in fact the juvenile form of European eel, which was named *Muraena anguilla* by Carolus Linnaeus in 1758 and moved to the genus *Anguilla* by Franz

Paula Von Schrank in 1798. So *Anguilla anguilla* is now the valid name for the species and *Leptocephalus brevirostris* is considered to be a junior synonym.

- American paleontologist Marsh (1879) described Jurassic dinosaur *Brontosaurus* (thunder bolt). However, it was found that it was the same animal as *Apatosaurus* (deceptive liazard) described by Marsh in 1877. Therefore, *Apatosaurus* is the accepted name.

- George Albert Boulenger (1893) moved the genus *Pelophilus* to the genus *Boa madagascariensis*. This meant that when Arnold G. Klug moved *Xiphosoma madagascariensis* to the genus *Boa* in 1991, the name *Boa madagascariensis* was invalid - a junior secondary homonym. Klug gave the species the replacement name *Boa manditra*. As mentioned earlier, the oldest valid published name is the one that gets priority in usage, even if there is some error in the original name.

Limitations of Rules of Priority

If the species does not refer to the valid name, it is difficult to ascertain its priority. When a new species is described, its author just mentions its features, on the basis of which this species is considered to be different from the one described earlier. Some times, the author fails to indicate all characters which distinguish it from the forms not described by that time. So when a new species is described, the author is advised to refer to the original description. In cases, where strict enforcement of principle of priority may result inconvenience or is practically difficult, special ruling is made to have an overriding effect on principles of priority. The International Commission on Zoological Nomenclature issues such rulings. Under the current Code, the principle of priority firmly applies only to those cases, where a valid name is being chosen among names published on different dates, competing names published in the same paper are subject to the principle of the first reviser. The first reviser is the first worker in the field to establish synonymy and to choose the valid name among competing ones. Thus, the principle of priority as applied to the genus group works in such way that whenever a genus gets divided into different genera, one of the subgenera is always assigned the same name as the genus to which it belongs, such subgenera is called nominal.

Authorship

In a taxonomic publication, a name is incomplete without an author citation and publication details. The citation indicates who published the name; in what publication; with the date of the publication. In taxonomy, it is generally conventional for the species to have the name of the person as a suffix, who first described the species. In citing the name of an author, the surname is given in full, not abbreviated, with any mention of the first name. The date (year) of publication in which the name was established is added, if desired with a comma between the author and date (the comma is not prescribed under the Code, it contains no additional information, however it is included in examples therein and also in the ICZN Official Lists and Indexes). Some cases are cited herein as under:

• *Balaena mysticetus* Linnaeus, 1758 the Bowhead whale was described and named by Linnaeus in his *Systema Naturae*.

• *Dysdercus koenigii*, Fabricius; means author Fabricius had described the species *koenigii* under genus *Dysdercus*. Author's name without parenthesis means that the species remains in the genus originally used by the author who described it for the first time.

• The parenthesis around author's name indicates that when the author originally described the species, he placed it in another genus. For example: *Anser albifrons* (Scopoli, 1769) the White-fronted Goose was first described (by Giovanni Antonio Scopoli), as *Branta albifrons* Scopoli, 1769. It is currently placed in the genus *Anser*, so author and year are set in parentheses. The taxonomist who first placed the species in *Anser* is not recorded (and much less cited), the two different genus-species combinations are not regarded as synonyms.

• For example, *Spilostethus pandurus* (Scopoli) indicates that when Scopoli originally described the species, he placed it in another genus, but some one working later moved the species to a new genus. Thus, if a name is later changed (e g., moved to a new genus), the name of the original author is given in parenthesis.

• The principles of priority prevent arbitrary renaming of taxon. Anyone can make up a few names for any known taxon and publish such name meeting all the available criteria; but principles of priority prevents such names from becoming valid, since the name of the taxon given earlier is valid. The principles of priority discourage unnecessary name creation and permits zoologists to reach consensus on which one of the existing names should be used. It is also significant to mention that to maintain stability in nomenclature, the rule of priority can be reversed if a junior name has been used very widely and for a long period. For example:

• Carolus Linnaeus named domestic cat *Felis catus* in 1758 and Johann Christian Daniel von Schreber named the wild cat *Felis silvestris* in 1775. Thus, if these two cat varieties are considered to be a single species, rule of priority being applied indicates that the species ought to be named *Felis catus*, but it is common practice that almost all biologists have used *F. silvestris* for the wild cat.

• In opinion 2027 (published in Volume 60, Part 1 of the Bulletin of Zoological Nomenclature, 31st March 2003 (1) the Commission "conserved the usage of 17 specific names based on wild species, which are predated by contemporary with those based on domestic forms", confirming *F. silvestris* for the wild cat, as *nomen protectum*. Taxonomists who consider the domestic cat a subspecies of the wild cat should use *F. silvestris catus*; the name *F. catus* remains available for the domestic cat where, it is considered to be a separate species. To make it more clear, it is stated that if these wild animals and their domesticated derivatives are regarded as one species, then the scientific name of that species is the scientific name of the wild animal.

Principles of Co-ordination

According to this, within the family group, genus group and species group, a name established for a taxon at any rank in the group is simultaneously established with the same author and date for taxa based on the same name-bearing type at other ranks in the corresponding group. In other words, publishing a new zoological name automatically and simultaneously establishes all corresponding names in the relevant other ranks with the same type. The International Code of Zoological Nomenclature rules for each group being different but every group is subject to co-ordination. The principle of co-ordination is described as under:

Within the species group (i.e., among species and subspecies), a species name is automatically available with its authorship and date, similarly a subspecies name also gets available as well. For example, publishing a species name (the binomen) *Giraffa camelopardalis* Linnaeus, 1758 also establishes the subspecies name (the trinomen) *Giraffa camelopardalis camelopardalis* Linnaeus, 1758. The same applies to the name of a subspecies; this establishes the corresponding species name.

Within genus group (i.e., among the names of genera and subgenera) coordination works in the same way; a generic name is automatically available as a subgeneric one and vice-versa. In both the cases, names keep their original authorship and date. Thus, the oldest valid name of a genus or subgenus is chosen among both generic and subgeneric names and not only among names of the same rank. Thus, publishing the name of a genus also establishes the corresponding name of a subgenus (or vice versa): genus *Giraffa* Linnaeus, 1758 and subgenus *Giraffa (Giraffa)* Linnaeus, 1758.

In the family-group, publication of the name of a family, subfamily, superfamily (or any other such rank) also establishes the names in all the other ranks in the family group (family Giraffidae, superfamily Giraffoidea, subfamily: Giraffinae). Co-ordination within the family group, under International Code of Zoological Nomenclature works without regard to the genus group names. Family group names are typified and each is derived from a genus group name by replacing its ending with a standardized rank suffix ending -oidea for superfamily, -idae for family, -inae -or subfamily, -ini for tribe, -ina for subtribe etc. So, unlike genus group names i.e., generic and sub generic, the names of family group taxa feature distinct endings for each rank.

Availability and Validity of Names

The name given to a particular taxon must be available with reference to the provisions of the Code. The taxonomic name which should be used is called valid name (accepted, correct) which is chosen, subject to the rules of nomenclature, among the names considered to be available for the purposes of the same Code. Thus, an available name can be valid or not, while an unavailable name is never valid. Availability criteria vary depending on nomenclature (see below), yet the following points are shared by all Codes. Availability of a name is established by a set of availability criteria. It should meet the following criteria:

- The name should appear in a work published after the year 1758 for animals.
- It should meet the criteria for publication designated by the Code.
- A binomial name should always be given.
- It must have certain format e g., Latin or Latinized in taxonomy i.e. written in Latin letters only and treated in accordance with the rules of Latin grammar (now written in English). Words of any derivation (Latin, Greek, or other languages) or any combinations of letters may be used as zoological names.
- The text of the name's first publication must meet some requirements (they vary depending on Code and name category). Only names meeting all availability criteria are considered available for purposes of a given nomenclature, so within a nomenclature many different available names can be created for a taxon, while available names for different taxa may be identical.

Publication of Names

A name becomes available only after its publication in scientific literature; unpublished names are unavailable. So the author publishing a scientific name has to be fully aware whether the name has already been used or is a new one. For a name already used, one should make sure that it is available (in such case publishing this name will not affect the nomenclature in any way). In case the intent is to publish a new name, the author must analyze everything in the Code applicable to that case and check all availability criteria to make sure that the suggested name will become available. Once the name is given to any taxon, it has to be published; it should meet the following criteria:

- It must be made public for the purpose of providing a permanent record.
- It must be obtainable, when first issued, free of charge or by purchase.
- It must be published in a journal having vide circulation
- Availability, authorship and date of publication are effective throughout the genus group, not just among names of the same rank. Author citations for such names (for example a subgenus) are the same as for the name actually published (for

example a genus). It is immaterial if there is an actual taxon to which the automatically established name applies; if ever such a taxon is recognized, there is a name available for it.

Further, authorship and date of any name in the family group (superfamily, tribe, subtribe or any other) automatically applies to all family group names derived from the same generic name. For example, the name *Heptagenia* Walsh 1863; was originally established for the genus *Heptagenia*. Subsequently, this genus was divided into subgenera. One of these subgenera included *H. flavescence*, the type species of the generic name *Heptagenia*. The correct name of subgenus should be such available genus group name (whether generic or sub generic) which is oldest one among all names, whose type species are members of the entire genus not the subgenus.

First Reviser Principle

According to this principle, in cases of conflicts between simultaneously published divergent acts, the first subsequent author can decide which has precedence. It supplements the principle of priority, which states that the first published name takes precedence. The Principle of the First Reviser deals with situations that cannot be resolved by priority. These items may be two or more different names for the same taxon, two or more names with the same spelling used for different taxa, two or more different spellings of a particular name, etc. In such cases, the first subsequent author who deals with the matter choose and publishes the decision in the required manner is the First Reviser, and is to be followed [Art. 24.2].

For example: Linnaeus 1758 established *Strix scandiaca* and *Strix noctua* (Aves), for which he gave different descriptions and referred to different types, but both taxa later turned out to refer to the same species, the snowy owl. The two names are subjective synonyms. Lonnberg 1931 acted as First Reviser, cited both names and selected *Strix scandiaca* to have precedence. Sometimes the First Reviser is unknown. For example, for the sperm whale, Linnæus 1758 established three subjective synonyms, *Physeter macrocephalus*, *Physeter catodon*, and *Physeter microps*. The First Reviser remains unknown; currently both *P. macrocephalus* and *P. catodon* are used.

The first reviser principle can help to stabilize the nomenclature. If names are published simultaneously (See Article 24 a, and Recommendation 24), the reviser can select the name that is well known. If a new species is spelled in more than one way in the original publication, Article 32 a, allows the first reviser to accept the spelling which is most commonly used. If no type species is fixed in the publication in which a new genus was described prior to 1931, the first reviser can select the species whose designation is in the best interest of stability (Article 69).

DESCRIBING NEW SPECIES

It is usual to identify the animals one comes across, with its name, but due to differences in language, region and knowledge, the same species may be given different common name or the same name may be used to refer to several different species. In order to have a record of all the species that have been discovered and to make it possible to precisely communicate about them, a formal scientific process for naming and describing species has been developed over the course of history, as discussed earlier.

In order to describe a new species, scientists must carry out a careful investigation to make sure that the species has, in fact, not yet been described. This often involves consulting with other experts on the particular taxon, visiting museums and collections to examine voucher specimens, reviewing the historical literature and carrying out DNA sequencing.

Once it has been determined that the species has not previously been named, the scientist must select a name and write a description. The name must follow certain Latin grammatical rules (though these allow room for creativity) and can be simple, descriptive, geographic, commemorative (i.e. named after a person), nonsensical, or some combination.

The description includes a thorough listing of physical characteristics, including variation in these that the worker has observed in the population. These proceed from general to specific (e g., first describing overall shape and size, then proceeding to describe each body part in more detail) and pay special attention to those characteristics that can be used to distinguish it from other species. The description should be objective and vivid and must include the distinguishing features by which the new species can be distinguished from others. The various provisions of International Code of Zoological Nomenclature provide vocabularies and specialized terminologies that are used while describing new species. Species are considered scientifically described when they have been given a two-part Latin name and have had a description published in a peer-reviewed scientific journal. The scientific name and description serve as the universal, formal reference for future identification.

DERIVATION OF NAME

The name of the genus and species may be derived from any source whatsoever. At the time of Linnaeus (1707-1778), Latin was used in Western Europe as the language of science and scientific names were in Latin or Greek, so Linnaeus also followed this pattern. The name of the genus must be unique in each kingdom. Species names are commonly reused and are usually an adjectival modifier to the genus name, which is a noun. Family names are often derived from a common genus within the family. The binomial nomenclature used for animals and plants is generally derived from Latin and Greek. Often, they are Latin words but they may also be derived from ancient Greek, from a place, from a person (preferably a naturalist) or it may be a name from local language etc.

Termination of Names

Taxa above the genus level are often given names based on the type of genus, with a standard termination. The termination used in forming these names depends on the kingdom and sometimes the phylum and class as given in the Table 4.1.

Spellings

The International Code of Zoological Nomenclature provides regulation for choice of names as well as their spellings. The name should be Latin or Latinized (i.e., spelled using Latin letters and treated in accordance with the rules of Latin grammar). In terms of etymology any derivation is allowed (Latin, Greek, other languages, even random combination of letters). All genus and family names begin with capital letter, while species group names are always written in lower case. Diacritic marks, apostrophe and diaeresis initially used in some name should be omitted in zoological publications. Hyphens, punctuation marks or digits in compound names are also omitted in the current zoological nomenclature.

Thus, each name is spelled as a single word (e g., *decimpunctata* instead of 10-punctata, *sanctijohannis* instead of *s-johannis* or *st. johannis* with the only exception of names including a Latin letter referring a physical character (such names are spelled with a hyphen, e g., *C-album*). All species names are binary thus consists of generic name followed by a specific name written separately without punctuation in between (e g., *Dysdercus koenigii* or *Periplaneta americana*).

The name of an animal subspecies is a combination of three names, i.e., a binomen followed by a sub-specific epithet also without punctuation marks (e g., *Heptagenia sulphurea dalecarlica*). If there are subgenera, the sub-generic name may be inserted in parenthesis between genera and specific names (e g., *Heptagenia (Heptagenia) sulphurea*). No other word can be inserted between generic and specific name.

Gender Agreement

In the species group gender agreement applies. The name of a species, in two parts, a binomen, say, *Loxodonta africana*, and of a subspecies, in three parts, a trinomen, say *Canis lupus albus*, is in the form of a Latin phrase, and must be grammatically correct Latin. If the second part, the specific name (or the third part, the subspecific name) is adjectival in nature, its ending must agree in gender with the name of the genus. If it is a noun, or an arbitrary combination of letters, this does not apply.

- For instance, the generic name *Equus* is masculine; in the name *Equus africanus* the specific name *africanus* is an adjective, and its ending follows the gender of the generic name.

- In *Equus zebra* the specific name *zebra* is a noun, it may not be 'corrected' to '*Equus zebrus*'.

- In *Equus quagga burchellii* the subspecific name *burchellii* is a noun in the genitive case ('of Burchell').

Table 4.1 Some derivatives of Latin/Greek words used in Zoological nomenclature

Word Latin/ Greek	Language (Greek: G, Latin: L)	English	Example
Archaeos, Archaeo-	G	Ancient	*Archaeopteryx*
brevis	L	Short	*Ceratogymna brevis*
cristatus	L	Crested	*Pavo cristatus*
decem	L	Ten	*Leptinotarsa decemlineata*
di	G	Two	Diptera
erectus	L	Upright	*Homo erectus*
familiaris	L	Common	*Canis familiaris*
gaster	G	Belly	*Drosophila melanogaster*
hexa	G	Six	Hexapoda
indicus	L	Indian	*Tapirus indicus*
lateralis	L	Side	*Pterogale lateralis*
morpho	G	Shape	Lagomorpha
octo	G	Eight	*Octopus*
orientalis	L	Eastern	*Blatta orientalis*
punctatus	L	Spotted	*Channa punctatus*
rufus	L	Red	*Canis rufus*
striatus	L	Striped	*Channa striatus*
uniflora	L	One	*Monotropa uniflora*

If a species is moved, therefore, the spelling of an ending may need to change. If *Gryllus migratorius* is moved to the genus *Locusta*, it becomes *Locusta migratoria*

According to this concept, an adjectival epithet must with extreme exceptions agree in gender with the name of the genus in which it is placed. The genus name *Xiphosoma* is neuter in gender and therefore the original spelling of the species should have been *madagascariense,* which is neuter form. Epithets that are noun are not changed. Changes in placement or confusion over proper Latin grammar lead to many incorrectly formed names appearing in print.

PRINCIPLES OF TYPIFICATION

For each taxonomic name a type is designated, which is the 'name bearer' of the species. For example, when a new species of any group is described, the taxonomist designates one individual as the **holotype** or the name bearer of that species which is deposited in the nearest museum and preserved. Museums are the main repositories of type specimens, with a key role of providing the archives of types that keep biological information anchored to organisms.

The description also includes information about the type material, which is an actual preserved individual of the species. This specimen is usually the one on which the description is based and is stored in a museum or collection, to be accessed if necessary to serve as a real-life reference. Sometimes the type material is an illustration or photograph, if the organism in question is very rare or threatened. Since type material also typically contains detailed information about the location and date that the specimen was collected and the name of the collector, it can serve as a historical artifact as well. For example, sampling tissue from museum specimens of sea birds can provide evidence that environmental levels of mercury have risen over time.

To do their work properly, taxonomists need to refer to the pre-existing type specimens and to the oldest published literature when they revise the taxa in a group of organisms. This requires access to information on the types, as well as access to all the previous taxonomic publications on their animals. Good taxonomic work requires reference to museum collections and libraries. The type specimen is chosen arbitrarily but once such specimen is designated it cannot be changed. A type specimen is preserved carefully, in case if the specimen is damaged or lost another specimen cannot replace a type specimen. The typification sub serves the following issues:

* Often the published description of the species concerned is insufficient to distinguish it from other species, a taxonomist studying the group at a later stage can examine the holotype and observe the details not mentioned in the original published description.

- At the time of observation, the scientist who described the new species was looking at several specimens appeared to belong to a single species. Subsequently, more extensive study revealed that the specimens actually belong to different species. Under such vexed situations, holotype helps to determine the name bearer of the species.

- It is relevant to mention here that as of 1993, the paleontologist Edward Drinker Cope is the type specimen for *Homo sapiens* Linnaeus, 1758. Robert Bakker formally described his skull and it was approved by the ICZN.

Kinds of Type Specimens

Each family group name has a type genus, each genus group name has a type species and each species group name has a type specimen. There are different names available to identify the types in taxonomic collections.

- Type genus refers to the 'name bearer' for family group taxa (taxa with a rank higher than genus up to super family).

- Type species is the 'name bearer' for genus group taxa (genus or subgenus); when a new genus is described, it is based on a particular species which becomes the type species of that genus. Thus, *A. ajax* is the type species of *Apatosaurus*. There are different type specimens discussed herein as under:

Primary Type It is the 'name bearer' for species names.

Holotype A single specimen designated by the author (s) for describing the species or subspecies at the time of publication of the original description.

Syntype (also known as cotype) It refers to group of specimens thought to represent a species, as designated or indicated by the author (s) of the original description.

Lectotype One of the syntype designated by the original or a subsequent author to function as the name bearer.

Neotype If the original primary type has been lost or destroyed, another specimen may be designated by a subsequent author as the name bearer of the species.

Secondary Type These are not recognized as the 'name bearer' of the species.

Allotype A specimen of the opposite sex from the holotype and designated as allotype, when the species was first described.

Paratype It refers to additional specimens that were examined when the species was first described but not designated as holotype or allotype. These specimens are valuable as reference materials that are deposited at museums or institutes. Paratype does not serve for nomenclature. They may however, serve as useful materials to select a neotype, if required.

Paralectotype A specimen that was once a syntype but is not the specimen later designated as lectotype. In other words, among syntype one of them can be designated (by same author or another author) as lectotypes, after which all the rest syntype become paralectotypes.

Topotype A specimen collected from the same locality from where the holotype has been collected, perhaps at the same time.

Monotype The genius has only one species included in it at the time the type species is designated.

Indication An original indication of a type is one that the author of the name indicates via illustration or other means and is referred as typus or typicus.

The type species of a generic name should be designated in the original description of the genus or for older names, where this had not been done originally in a later publication. Each name in the family group (superfamily, family, subfamily, tribe, subtribe etc.) has its type genus and is derived from the later name by change of ending. The types are referred as types of species, genera, families etc. A name may have no more than one type but a taxon includes as many types as many available names to it.

Significance of Type Specimens

The types are necessary to prevent transferring names from one taxon to another. For example, when a group once regarded as a single genus is divided into smaller genera. The type specimens are used to describe with species group names.

Voucher Specimens These are similar to type specimens from which the identity of organisms studied in the field or in laboratory can be verified. Authors of **voucher specimens** should cite the findings in suitable journal and deposit such specimens in natural history museum of the country where the specimen has been collected.

STATUS OF NAMES

These are used for different taxa and can have different status. In some cases, ICZN formally defines the names for the status. There may be different categories of names:

Valid Names

These refer to the correct names in scientific nomenclature. It is a name which is accepted as correct for the taxon and is not **synonym** or **homonym**. In some cases, it has certain deliberate change in spelling because the International Code specifies certain changes to fit the required format i e., it does not allow any digits, hyphens, diacritical marks. An example of invalid name is *Nessiteras rhombopteryx* (Loch Ness Monster); this proposed name is not a valid scientific name because it has no specimen attached with it. Further, gender of species name should be in conformity with the genus name.

Invalid Names

These are incorrect scientific names for taxa. These are homonyms and synonyms described herein as under:

Principles of Homonymy

It exists within the provisions of International Code of Zoological Nomenclature. In zoological nomenclature no species group (whether specific or sub specific) may be repeated within genus i.e., combined with the same generic name. Genus group name (generic and sub generic) should be unique within the animal kingdom (i.e., within zoological nomenclature). However, if a sub generic name is used, it does not affect the homonymy of the species group names. It refers that the name of each taxon must be unique. Consequently, a name that is junior homonym of another name must not be used as a valid name. These are the names that are the same and were proposed in the same genus-group taxon. It means that any one animal name, in one particular spelling, may be used only once within its group. This is usually the first published name; any later name with the same spelling (a homonym) is prevented from being used. If the rejected homonym has one or more available names (synonyms), the oldest is adopted as a valid name. The Principles of Priority and the First Reviser is applicable here. Genera are homonyms only if exactly the same - a one letter difference is enough to distinguish them. Some of the examples are as under:

- *Argus Scopoli*, 1763 (Lepidoptera : Lycaenidae:Polyommatinae)
- *Argus Scopoli*, 1777 (Lepidoptera : Nymphalidae:Satyrinae)
- *Argus Lamarck*, 1817 (Lepidoptera : Hespiridae)

Senior homonyms It refers to the available name on the basis of priority. Thus, of two homonyms: the first established, or in the case of simultaneous establishment the one given precedence under Article 24.

Junior homonyms According to the provisions of the Code, any name that is a junior homonym of available name must be rejected permanently and replaced by a new name. The junior names are those that are considered invalid on the basis of priority or because of a choice of the first reviser, or by a governing body of nomenclature in other words A preoccupied name (not in use) on the basis of priority or by a ruling by a nomenclatural body. Consider the case of *Rana tigrina* Linnaeus, 1758 and *Rana tigrina* Fabricius, 1759; in this case, the later name is junior homonym of the former. Names unavailable under the code are not considered homonyms. The homonymy may be classified as under:

Primary Homonym These are identical names given to different taxa (with different holotype) belonging to same genus. In a species group (species, subspecies) these are the names that are the same and were proposed in the same genus-group taxon. In other words, it refers to name with the same genus and the same species in their original combination. This occurs due to an error done by the author without knowing that the species name had already been occupied when he used it. Once the error is known, only the first published or senior homonym can be retained. A primary homonym is never eligible to become valid even if the taxa are separated to different genera.

Secondary Homonym The secondary homonym arises by transfer of one species to a new genus that contains a species with the same species name. These are species that are placed in the same genus subsequent to their publication and they have the same specific epithet. The senior secondary homonym is the older of the two names. This occurs as the later taxonomist decides that the two species belong to the same genus. The worker who has discovered secondary homonym should reject the junior name. A secondary synonym is only a temporary state; it is only effective in this classification. If the specimen under study is classified again, the secondary homonymy may not be produced and the involved name may be used again (Art.59.1). A rejected primary or secondary homonym must be replaced by the oldest available name or if no previously published name is available then by a new name. A name does not become unavailable or unusable if it was once in the course of history placed in such a genus where it produced a secondary homonymy with another name. This is one of the rare cases where a zoological species does not have a stable specific name and a unique species-author-year combination; it can have two names at the same time.

A secondary homonym can become valid, if the taxa are later separated in different genera. When original combinations of different species have different generic names with identical epithets but after being classified, such species end up in the same genus. According to International Code of Zoological Nomenclature, renaming junior secondary homonyms is reversible but renaming junior primary homonym is not possible. For example, Nunneley (1837) established *Limax maculatus* (Gastropoda), Wiktor (2001) classified it as a junior synonym of *Limax (Limax) maximus* Linnæus, 1758 from Southern and Western Europe. Kaleniczenko (1851) established *Krynickillus maculatus* for a different species from Ukraine. Wiktor (2001) classified *Limax maximus* Linnæus, 1758 and *Krynickillus maculatus* Kaleniczenko, 1851 in the genus *Limax*. This meant that *L. maculatus* Nunneley, 1837 and *K. maculatus* Kaleniczenko, 1851 were classified in the same genus, so both names were secondary homonyms in the genus *Limax*, and the younger name (from 1851) could not be used for the Ukrainian species. This made it necessary to look for the next younger available name that could be used for the Ukrainian species. This was *Limax ecarinatus* Boettger, 1881, a junior synonym of *K. maculatus* Kaleniczenko, 1851.

Junior Secondary Homonym It refers to species group names only. For example, Santschi in 1909 described the ant species *Tetramorium silvestrii*. Emery (1915) described the ant species *Triglyphothrix silvestrii*. Initially, it appeared to be correct, however, Bolton (1985) synonimised *Triglyphothrix* with *Tetramorium*, because there was no appreciable difference between these two genera. However, there several secondary homonyms appeared including *Tetramorium silvestrii* and *Triglyphothrix silvestrii*, which are different species with different types but having the same name. Bolton (1985) therefore replaced junior secondary homonym *Triglyphothrix silvestrii* by the new name *Tetramorium surrogatum*.

Hemi-homonym It refers to the identical names assigned to different taxa subject to different Codes (i.e., where only one of these name is available under each Code). However, in order to avoid confusion due to use of synonyms and homonyms, every taxon is given a single name, which is chosen among all the available names of each taxon and among all identical names suggested for different taxa. The naming of taxon is done under the provisions of International Code of Zoological Nomenclature.

Double Homonymy (genus and species) It is no homonymy: if the genera are homonyms and belong to different animal groups, the same specific names can be used in both groups. For example,

- Linnaeus gave the name *Noctua* Linnæus, 1758 to a lepidopteran subgenus.
- In 1764 Linnaeus established a genus *Noctua* Linné, 1764 for birds, overlooking that he had already used this name a few years ago in Lepidoptera. Thus, *Noctua* Linné, 1764 (Aves) is a junior homonym of *Noctua* Linnæus, 1758 (Lepidoptera).

- Garsault (1764) used *Noctua* for a bird and established a name *Noctua caprimulgus* Garsault, 1764 (Aves).

- Fabricius (1775) used a name *Noctua caprimulgus* Fabricius, 1775 (Lepidoptera), thus creating a double homonym. Thus, double homonymy is no homonymy, both names are available.

- Simialar is the case with *Noctua variegata* Jung, 1792 (Lepidoptera) and *Noctua variegata* Quoy & Gaimard, 1830 (Aves).

Synonyms

Different available names given to the taxon are referred as **synonyms**. The concept of synonymy applies only to names subject to the same Code, because the names unavailable under that Code are not considered synonyms. The synonyms are different scientific names that relate to the same taxon, for example two names for the same species. The rule of zoological nomenclature is that the first name to be published is the senior synonym; any others are junior synonyms and should not be used. The synonyms may be classified as under:

Subjective Synonyms *(Synonyma subjectiva)* Subjective synonyms are different names that have been applied to taxon as determined by the taxonomist or systematist. The concept of subjective synonymy is fundamentally different in rank based and circumscription based nomenclature, while in hierarchy based nomenclature, it simply does not exist. An example would include two species originally described as distinct but were later determined by a professional in the field that they are the same species, e g., *Campostoma nasutum* Giard, 1856.

Subjective Invalid Synonyms It includes names that have been made invalid because of a decision that is a matter of opinion.

Junior Subjective Synonyms These are different types thought to belong to the same taxon. A junior subjective synonym is the one which was described at a later date. For example, a taxonomist might compare the types for two different names. The two specimens look slightly different from each other but the taxonomist decides that these two are actually the same species; he therefore makes the more recent name a junior subjective synonym. This is a subjective decision because the other taxonomist may have a different opinion and he decides that the types are of different species and hence both names are valid. For example, *Kerria lacca lacca* (Kerr, 1782) is the accepted name for lac insect. Its junior synonyms are *Laccifer lacca, Carteria lacca* and *Tacchardia lacca*. However, the oldest name *Kerria lacca* is back in use.

Nominotypical subspecies and subspecies Autonyms

In zoological nomenclature when a species is split into subspecies, the originally described population is retained as the 'nominotypical subspecies' (Article 47) or 'nominate subspecies', which repeats the same name as the species. For example, *Motacilla alba alba* (often abbreviated *Motacilla a. alba*) is the nominotypical subspecies of the White Wagtail (*Motacilla alba*).The repetition of the species name is referred to in botanical nomenclature as the subspecies 'autonym', and the subspecies as the 'autonymous subspecies'.

Emendation
It refers to an available name whose spelling has been intentionally changed.

Justified
The mandatory correction of an incorrect original spelling containing diacritical marks, hyphens etc. A justified name is compatible with a name status of valid, temporary or synonym.

Unjustified
Intentional change in the original spelling of a name that is not justified is not approved officially.

Unavailable
A name that is not recognized under the rules of nomenclature. It may fall under following categories:

Misspelling An accidentally misspelled version of previous name, it is compatible only with a name status of synonym.

Incorrectly formed Name A name that does not conform to the ICZN standard, according to its taxonomic rank. In other words a name is unavailable because it fails to the conformthe rules set out in the ICZN and thus has the name status of temporary, *nomen nudum*, *nomen dubium*, or synonym.

Unnecessary replacement A name used as *nomen novum* for a valid taxon name.

Suppressed by ruling A name that has been specifically ruled by the International Code of Zoological Nomenclature to be unavailable is said to be suppressed. Depending upon the ruling the name may be partially, totally or conditionally suppressed.

Doubtful cases When biologists disagree over whether a certain population is a subspecies or a full species, the species name may be written in parentheses. Thus *Larus (argentatus) smithsonianus* means the American Herring Gull; the notation with parentheses means that some consider it a subspecies of a larger Herring Gull species and therefore call it *Larus argentatus smithsonianus*, while others consider it a full species and therefore call it *Larus smithsonianus* (and the user of the notation is not taking a position).

Other Types of Names

There are certain other types of name, discussed herein as under:

Misapplied Name

A name used erroneously because the specimens studied did not actually belong to the taxon, which the author erroneously considered that they were in.

Unnamed

A taxon that has not yet been given a name and description in accepted publication. An unnamed name is compatible only with a name status of temporary name.

Temporary Name

It is used until a valid name is provided.

Nomen nudum

A new name mentioned without description or indication or figure is a *nomen nudum*. It has no authorship and date and it is not an available name. If it is desired or necessary to cite the author of such an unavailable name, the nomenclatural status of the name should be made clear. It refers to a name that was not properly associated with actual specimens. These names are listed in plain Roman type. It is a name which has appeared in prints but has not yet been formally published by the standards of the International Code of Zoological Nomenclature.

Nomen dubium

It is used without sufficient information so that later authors are unable to determine the taxon. **Nomen dubium** is a dubious name. It is an available name but can not be assigned to a definite taxon due to shortcomings in the original description/diagnosis.

Nomen novum

It refers to the new name that is used to replace a valid taxon's original name in the event of homonymy. It is compatible only with a name status of valid, temporary, or synonym.

Nomen protectum

It is a name given precedence over its unused senior synonym or senior homonym which has become a nomen oblitum. A nomen protectum is compatible with name status of valid name or synonym.

Nomen oblitum

It refers to the name that has priority for a taxon but is not used due to continued usage of a widely used and accepted name. It is compatible with a name status of synonym. It is also known as forgotten name (after 1960 if not used for 50 years) [Law of proscription].

Nomen hybridum

These are name of hybrids, which are generally individuals not populations. Such names have no status in nomenclature.

Nomen triviale

In the late 1740s Linnaeus began to use a parallel system of naming species with *nomina trivialia*. *Nomen triviale*, a trivial name was a single- or two-word epithet placed on the margin of the page next to the many-worded 'scientific name'. The only rules Linnaeus applied to them was that the trivial names should be short, unique within a given genus, and that they should not be changed. Linnaeus consistently applied *nomina trivialia* to the species of plants in *Species Plantarum* (1st edn. 1753) and to the species of animals in the 10th edition of *Systema Naturae* (1758).

Vernacular name

It refers to general purpose name that are not proposed for zoological nomenclature. An individual can have any number of **vernacular names** in different languages.

Parenthesis

In taxonomy, it is conventional to mention the name of the person as a suffix, who described the species for the first time.

- The name of the author describing the particular species is written in parenthesis, when the species is transferred from one genus to another, it retains its original author and date.

- When a species is placed in a genus different to the one originally named, then the name of the author who had described it is placed in brackets even if, it is the same author who has described the both. Thus, *Brontosaurus excelsus,* Marsh, 1879 becomes *Apatosaurus excelsus* (Marsh, 1879).

Brackets

The date given for a taxonomic study is the date of publication. If the date printed in original work is incorrect, then the correct date is given in square brackets e g., [1900] as per recommendations 22 A (5) of the Code, but the brackets are omitted in the text. Square brackets are also used to include statements of misidentification in citing synonym. Date in round brackets e g., (1869) indicate the correct date of publication appearing in the work other than on the title page.

Asterisk

An asterisk {*} associated with a date implies that some research has been made and the date given is the nearest known.

Abbreviation

The generic name when used in the text for the first time is written in full for example, *Rana tigrina* which can be abbreviated to single capital letter e g., *R. tigrina* in subsequent usage in the text.

Punctuation Marks

The International Code of Zoological Nomenclature prohibits the use of any punctuation marks between the name of the species and the author. Comma (') is always used between the author's name and the year, for example, *Cranopygia bhalli* Kapoor, 1966. In this case, Kapoor is the original author of this species. In cases, when the authors in other literature make use of full stop, e g., *Phagocarpus immsi.* Bezzi, 1913 and *Phagocarpus immsi.* Munro, 1935. This implies that Munro in 1935 made use of the name validity published for the first time by Bezzi in 1913. No diacritic mark, apostrophe, ligature, dieresis or hyphen should be used in the scientific name [Article 32.5.2].

Use of Latin Propositions in and ex

The Latin proposition 'in' is used to connect the names of two person the second of which is the editor (or the main author) or a work in which the first author actually gave the name. For example, *Dacus indicus* Agrawal in Kapoor denotes that Agrawal was responsible for publication of the name *D. indicus* in a work edited by Kapoor or otherwise written.

Numerical in Compound Names

Zoological nomenclature does not favour the use of numerical and if a number of numerical adjective or adverb forming a part of a compound name is discovered, it is to be written as full word united with the remainder of the name, e g., *decemlineata* and not 10-*lineata and 7-punctata* is written as *septempunctata.*

The Latin proposition 'ex.' means 'from' or 'according to'. It is used to connect the names of two persons, the first author of which validly published a name, which was actually proposed (but not published) by the second author. For example, *Forcipula indica* Brindle ex Burr means that Brindle was responsible for the valid publication of the name, *F. indica,* originally proposed by Burr, which he himself never published.

Sensu stricto (s.s)

It refers to the use of taxon 'in the strict sense'. It is abbreviated as *s.s.* of the type of its name. It is added after a taxon to mean the taxon is being used in the sense of the original author, or without taxa which may otherwise be associated with it. For example, *Homo erectus sensu stricto* is used for the Asian *H. erectus.*

Non

The latin word '*non*' or nec means not and are used to denote the actual authority in nomenclature.

Sensu lato (S.l)

The Latin word *'sensu lato'* (s. l., sens. Lat.) refers to wide-ranging sense or broad sense. It is abbreviated as s.l. For example, *Homo erectus sensu lato* for the larger species comprising both the early African populations (*H. ergaster*) and the Asian populations. Similarly, there is *Xiphinema americanum sensu lato.*

Partim or pro parte (*pp*)

It is a Latin word which means that only a part of a taxon as circumscribed by previous author is being referred to by the writer. For example, *F. indica* new species Brindle for *F. pugnax* Burr (nec Kirby, 1891), proparte or in part, means that part of the *pugnax* species of Burr (i.e., some specimens only) belongs to the new species *indica* described by Brindle. The authority of taxon does not change, if it's distinguishing features or circumscription is changed. In case the author for bringing about such alteration in the taxon is to be mentioned, he should be mentioned after the original author in the manner with the use of the Latin word partim. For example, *Taenia solium* Linnaeus, partim Goeze, means that Linnaeus is the original author while Goeze made alterations circumscription.

Use of Certain Suffix

Use of certain suffix is governed under the provisions of Article 58 of the Code. These are used for species group names formed from modern personal names. For example,

- se of ae, oe or e (e.g. *caeruleus, coeruleus, ceruleus);*.
- use of ei, i or y *(e.g. cheiropus, chiropus, chyropus);*
- use of i or j for the same Latin letter *(e.g. iavanus, javanus; maior, major);*
- use of -i or -ii, -ae or -iae, -orum or -iorum, -arum or -iarum as the ending in a genitive based on the name of a person or persons or a place, host or other entity associated with the taxon, or between the elements of a compound species group name for example, smithi, smithii; patchae, patchiae, fasciventris, fasciiventris).

Use of suffix 'ensis' or 'iensis'

These are used for species group names based on a geographical name, e g., *gorakhpurensis* for specimens collected from Gorakhpur; likewise *bengalensis* for Bengal and *bangalorensis* for Bangalore, e g., *Copelatus bangalorensis* is a species of diving beetle.

HISTORICAL BACKGROUND OF ZOOLOGICAL NOMENCLATURE

First attempts to establish rules of nomenclature date back to the late 1700s. In 1842 Strickland's 'Rules for Zoological Nomenclature' were published, in 1877 - the Dall's code, in 1905 - 'Regles Internationales de la Nomenclature Zoologique'. The first edition of the ICZN was published in 1961, the second (some what revised) in 1964, the third (amended) in 1985; currently the forth (amended) edition is in force, published in English and French in 1999.

Since late 1830s written nomenclatural rules were compiled in various countries (Allen, 1897) and Strickland (1878). It was only in first and second International Zoological Congress (Paris, 1889, Moscow, 1892), the need was felt to have commonly accepted international rules for all disciplines and countries to replace conventions and unwritten rules that varied across disciplines, countries and languages.

The International rules on Zoological Nomenclature was first proposed in 1895 in Leiden during 3rd International Congress for Zoology and it was officially published in 1905 in three languages viz., English, German and French. Subsequent development resulted in the publication of first edition of **International Code of Zoological Nomenclature** in 1961, second in 1963, third in 1972. The present edition is 4th and is effective since 2000. The ICZN Commission takes its power from a general biological congress (IUBS-International Union of Biological Sciences).

The system of rules and recommendations on the formal naming of animals published in the form a book originally adopted by the International Congress of Zoology and since 1973, by the International Union of Biological Sciences (IUBS).

The starting point of zoological nomenclature is January 1, 1758, considered to be publication date of the 10th edition of *Systema Naturae* (Linnaeus, 1758). The reason why the tenth edition has been chosen is due to the fact that in this edition Linnaeus for the first time used binary nomenclature in a consistent fashion. All names published before 1758 are automatically considered to be unavailable (including names used by Linnaeus himself in the previous editions of *Systema Naturae*). Accordingly, all names used in 10th edition of *Systema Naturae* are considered published for the first time in 1758, and Linnaeus is regarded as their author (though many of these names had been in use well before Linnaeus).

PREAMBLE TO THE INTERNATIONAL CODE OF ZOOLOGICAL NOMENCLATURE

The International Code of Zoological Nomenclature is a widely accepted convention in zoology that rules the formal scientific naming of organisms treated as animals. The rules principally regulate the following:

- How names are correctly established in the frame of binomial nomenclature.
- Which name must be used in case of name dispute?
- How scientific literature must cite names?

The object of the Code is to promote stability and universality in the scientific names of animals and to ensure that the names of each taxon are unique and distinct. All its provisions and recommendations are binding and none restricts the freedom of taxonomic thought and actions. Priority of publication is the basic principle of zoological nomenclature; however, under conditions prescribed in the Code, its application may be modified to conserve a long accepted name in its accustomed meaning. When stability of nomenclature is threatened in an individual case, the strict application of the Code may under specified conditions be suspended by the International Commission of Zoological Nomenclature. The rules of the ICZN are considered and supposed to be among the most rigorous for scientific publication.

The task of keeping correct information on animal names is immense and critical, as almost all information on the living world is linked through names. The Code envisages the precision and consistency in the use of terms. Both the Preamble and the glossary are the integral parts of the Code's provisions. The International Commission of Zoological Nomenclature is the author of the Code.

Features of Zoological Nomenclature

The Zoological Nomenclature must possess following features:

Uniqueness
The name given to a particular animal should not be given to any other individual because this would create unnecessary confusion. However, if different names are given to the same genera, the valid name should be accepted following provisions of Law of Priority.

Universality
As per provisions of zoological nomenclature, every individual is given a single name under one language, since the use of vernacular names or any other non-conventional name would result in much ambiguity.

Stability
The nomenclature should also be stable because any change in the name of an animal is likely to create quite confusion and difficulty to locate a particular organism. Thus in order to keep the stability of nomenclature, even the provisions of Zoological Nomenclature are waived, i. e., not strictly adhered to. It has been mentioned in the preamble of the International Code of Zoological Nomenclature that the object of the Code is to promote stability of nomenclature. However, the stability may be disturbed due to the following:

- Several species (e g., sibling species) may be referred to by a single name prior to a more detail analysis.
- Two or more authors use the same name for different genera often in unrelated groups.
- Wrongly placed species should be shifted to another genus.
- Same taxon is named in a different way by various authors working in different countries.

Explanatory Note to the Code

The International Code of Zoological Nomenclature adopted by the International Commission on Zoological Nomenclature (ICZN) and is ratified by the Executive

Committee of the International Union of Biological Sciences (IUBS) acting on behalf of the Union's General Assembly. The Commission may authorize official texts in any language and all such texts are equivalent in force and meaning (Article 87).

• The Code proper comprises the Preamble, 90 Articles (grouped into 18 chapters) and the Glossary.

• Each Article consists of one or more mandatory provisions, which are sometimes accompanied by the recommendations and/or illustrative examples.

• In interpreting the Code, the meaning of a word or expression is to be taken as that given in the Glossary (see Article 89).

• The provisions of the Code can be waived or modified in their application to a particular case when strict adherence would cause confusion but this can only be done by the Commission acting on behalf of all the zoologists and using its plenary power (Article 78 and 81) and never by an individual. The commission takes such action in response to proposals submitted to it. It is relevant to mention that the acronym 'ICZN' refers to Commission and not to the Code.

• There are three Appendices, the first two of these have the status of recommendations and the third is the Constitution of the Commission.

Range and Authority of Code

The International Code of Zoological Nomenclature applies to both living and extinct animals (Article1). There are however, separate Codes for botanical and bacteriological nomenclature and Article-2 of the Code spells out the relation between the Codes and status of name transferred between kingdoms.

Freedom of Taxonomic Thought

It is argued that rules of nomenclature do not restrict the 'freedom of taxonomic thought or action'. This basic principle is applied to specific situation throughout the International Code of Zoological Nomenclature.

• Article 10g permits a zoologist to continue using a name considered by another author to be junior homonym.

• Article 11e provides that subjective action of an author in placing a name in synonymy should not determine the nomenclature to be used by others.

• Article 64, authorizes a zoologist to choose (as the type of a new family) that genus which he considers to be the most typical or characteristic genus of that family but 'not necessarily having the oldest name'.

It is relevant to mention that any mistake committed by the taxonomist should not affect the nomenclature except when this is expressly authorized by the Code. For example, the erroneous assignment of a species to a genus does not constitute the proposal of a new species name in that genus (Article 49), the naming of an individual variants that do not form a taxon has no validity in nomenclature (Article 45e) and the incorrect placing of a name in synonymy does not invalidate it (Article 10g).

CATALOGUES OF ZOOLOGICAL NAMES

The principles of Zoological Nomenclature cannot be consistently applied unless catalogues of published names are available. The Zoological Record published since 1864 by the Zoological Society of London, initially independently; but in partnership with Biosis- a U.S. Company, since 1981 is of great help to the taxonomists. It contains the details of published names of described genera and each volume contains literature published during the year. In addition to it, following is the list of other catalogues:

• Agassiz, (1848) gave an early catalogue listing names of classes, orders, genera and species, which however, was not complete.

• Index Animalium, 1902 (contain early generic and species name published during 1758 -1850).

• Nomenclator Zoologicus (1939-1966) and Nomenclator animalium generum et. subgenerum, 1926-1940 (genus group names published since 1758).

A new animal name published contrary to the provisions of the Code may deemed simply 'unavailable', if it fails to meet certain criteria or fall entirely out of the providence of science (e g., the 'scientific name' for Loch Ness Monster is not a valid name). The rules of the Code determine which names are potentially valid for any taxon including the ranks in the family group, genus group and species group.

First edition of International Code of Zoological Nomenclature was published in 1961, followed by second edition in 1964, the third in 1985 and the current 4th edition was published in 1999. It is also relevant to mention that the Code does not recognize any case law. If there is any dispute regarding nomenclature, it is decided first by applying the Code directly and not by reference to the precedent.

INTERNATIONAL COMMISSION OF ZOOLOGICAL NOMENCLATURE

The International Commission on Zoological Nomenclature (ICZN) is an international organization of experts on scientific nomenclature dedicated for 'achieving stability and sense in the scientific naming of animals'. Founded in 1895, it currently comprises 28 members from 20 countries, mainly practicing zoological taxonomists. The office of the commission is situated in the Natural History Museum, London. Zoologists attending General Assemblies of the

International Union of Biological Sciences (IUBS) or other International Congresses elect members. The work of the Commission is supported by a small Secretariat based at the Natural History Museum in London and is funded by the International Trust for Zoological Nomenclature (ITZN), a charitable organization. The Commission assists the zoological community 'through generation and dissemination of information on the correct use of the scientific names of animals'. The Commission ensures names of animals are stable and universal by applying the rules of the *International Code of Zoological Nomenclature*. The preamble of the International Code of Zoological Nomenclature says. "The objects of the Code are to promote stability and universality in the scientific names of animals and to ensure that the names of each taxon are unique and distinct. All its provisions and recommendations are subservient to those ends and none restricts the freedom of taxonomic thoughts or action". The commission serves following important functions:

• The ICZN provides and regulates a uniform system of zoological nomenclature ensuring that every animal has a unique and universally accepted scientific name. Thus, the ICZN maintains an international standard in animal nomenclature.

• The ICZN publishes the International Code of Zoological Nomenclature having the formal rules 'universally accepted as governing the application of scientific names to all organisms which are treated as animals'.

• The Commission also provides rulings on individual nomenclature problems brought to its attention, in order to achieve internationally acceptable solutions and stability published in the Bulletin of Zoological Nomenclature.

• The International Commission on Zoological Nomenclature (ICZN) acts as adviser and arbiter for the zoological community by generating and disseminating information on the correct use of the scientific names of animals. In addition to it, there is a **ZooBank** - a proposed register of the scientific names of every animal species (For details, please see chapter 21).

It is also relevant to mention, that several million species of animals are recognized and more than 2000 generic names and 15000 new species names are added to the zoological literature every year. With so many names, problem is bound to occur, which are well addressed by the ICZN. The Commission operates through its quarterly journal-the *Bulletin of Zoological Nomenclature* (BZN) in which problems needing a formal decision by the Commission is published for discussion by the zoological community. The BZN provides a platform for presentation of Cases, public input and final results of Commission consideration that is transparent, accessible and archived.

Universal Rules and Powers of Commission

Nomenclature rules are framed in a manner so as to promote name stability. However, no single Code can reflect established criteria for every species. Some of the salient features of the Commission are outlined as under:

• Under ICZN rules, the valid name is always the oldest one among available names of a taxon. In some cases, a well known taxon may have generally accepted and long used name other than its oldest available name.

• The widely circulated name is to be replaced with the technically correct one. However, this refinement generally creates difficulty in renaming the name, which is familiar and then the ICZN issues a special ruling to establish such generally accepted name as the correct one surpassing the older unused name or else declaring it unavailable. The rules in the Code determine which names are potentially valid for any taxon including the ranks of subspecies and superfamilies.

The Commission is mandated by its scientific membership of the International Union of Biological Sciences (IUBS), and members are elected by zoologists attending General Assemblies of IUBS or other international congresses. Casual vacancies may also be filled in between. Nominations for membership may be sent to the ICZN secretariat at any time. The Commission's work directly affects biodiversity studies, farming and horticulture, medical & veterinary research, paleontology and biostratigraphy, entomology and conservation. It may be added that ICZN is an associate participant to the Global Biodiversity Information Facility (GBIF) and a scientific member of the International Union of Biological Sciences (IUBS).

ICZN AMENDMENT ON ELECTRONIC PUBLICATION

According to Minelli (2013) creation and use of the scientific names of animals are ruled by the International Code of Zoological Nomenclature. Until recently, publication of new names in a work produced with ink on paper was required for their availability. It was on 4th September, 2012 the ICZN announced an amendment to the International Code of Zoological Nomenclature allowing for electronic publication of the scientific names of animals. According to Zhang (2012) the amendment to the International Code of Zoological Nomenclature (ICZN 2012: 1-7) allows publication in "widely accessible electronic copies with fixed content and layout" of new names and nomenclature acts in zoology after 2011 under the following conditions (in addition to article 8.1.1-8.1.3 of ICZN 1999):

1. the work must be registered in ZooBank before it is published,
2. the work itself states the date of publication with evidence that registration has occurred, and
3. the ZooBank registration states both the name of an electronic archives intended to preserve the work and the ISSN or ISBN associated with the work.

With this amendment, the rule of zoological nomenclature have been aligned with the opportunities (and needs) of modern digital era. However, a possible cause of nomenclature instability remains as it is. These could be completely removed if the Code compliant publication of new names will be identified with their online registration, under suitable technological and formal (legal) conditions. Future developments of the ZooBank may provide the tool required to make this definitive leap ahead in zoological nomenclature.

It is speculated that the new rules will open the door to electronic publication and facilitate a truly web based taxonomy. With this amendment, the rules of zoological nomenclature have been aligned with the opportunities and needs of digital era.

INTERNATIONAL TRUST FOR ZOOLOGICAL NOMENCLATURE

The International Trust for Zoological Nomenclature (ITZN or 'the Trust') established in 1947, registered in England as a non profit company limited by guarantee. It acts on behalf of the Commission to raise and administer funds in accordance with the policy of the Commission and the International Union of Biological Sciences (IUBS).

NOMENCLATURE vis-a-vis TAXONOMY

It is to be made clear that taxonomy and nomenclature are two different aspects. Taxonomy in fact is the science, hence cannot be dictated by any set rules of an organization. Unlike taxonomy, nomenclature is not a science as the names are artificially produced. Thus, the nomenclature can be subjected to rules adopted by a competent body and is binding for any worker in the field. Such rules, however, are framed by the applicability of International Codes. The Code is meant to guide only the nomenclature of animals, while it is for the zoologists to classify the new taxon.

TAXONOMIC INFLATION

Taxonomic inflation is the rapid accumulation of scientific names due to processes other than new discoveries of taxa. The main processes are splitting "elevation of taxa to a higher level (creating inflation at the higher level) or taxonomic error of some sort. According to Issac et al., (2004) species numbers are increasing rapidly. This is due to taxonomic inflation - a phenomenon where known subspecies are raised to the status of species as a result of change in species concept, rather to new discoveries. Macro-ecologists and conservation biologists depend heavily on species lists treating them as accurate and stable measures of biodiversity. Further, taxonomic uncertainty is due to the evolutionary nature of species. The elevation of subspecies to species caused by the increasing use of phylogenetic species concepts (PSCs) is responsible for most growth in the number of species. This phenomenon has been termed as 'taxonomic inflation' (hereafter taxonomic inflation s. s.). The effect of the phylogenetic species concept (PSC) on taxonomic inflation and its implications for biodiversity studies, conservation and macroecology is enormous.

MODERN DEVELOPMENTS

Linnaeus had established a system of binomial nomenclature and a new methodology in classification referred as cladistics taxonomy by which taxa are named in an evolutionary manner. This also reflects the Darwinian principle of common descent of different groups of animals.

The botanical code (now renamed the International Code of Nomenclature for algae, fungi and plants), changed its rule to allow e-publication without any additional requirements for archiving or registration. Its current version is specifically designed to regulate the naming of clades, leaving the governance of species names up to the rank-based codes. In 1990, three domain systems were given, in which **Archaea** and **Bacteria** were separated, which were initially kept together in to a single kingdom Bacteria (Monera).

Several issues now confronting the zoological community make desirable the development of a 5th edition of the International Code of Zoological Nomenclature. The International Commission on Zoological Nomenclature would like to take following into consideration for developing 5th edition of the Code.

• To make the process more transparent, by allowing broader participation of members of the zoological community.

• To preserve a record of the deliberations leading to the text ultimately adopted.

• To develop more examples and explanatory sections, making the Code easier to use and understand.

Animals comprise the vast majority of the world's recognized living species and currently stands at around 1.9 million, growing at a rate of about 20,000/year. For each of these groups there are as many as 2 - 10 legitimately published names due to past debate and poor information exchange. Estimates of the total of living animal biodiversity are 4–20 times this number (8–50 million species), with fossils adding many more. When the task of cataloguing biodiversity approaches completion, this vast amount of information will be linked through names.

Molecular systematics has provided new insights about the relationships between different groups of individuals. Advances made in the field of molecular biology including the PCR techniques (polymerase chain reaction) which allows the analysis and expansion of DNA molecule along with DNA barcode will go a long way in correct identification of the species (For details, please see Chapter 21).

A great strength of the Code is that most of its Rules are automatic in use, thus allowing any two taxonomists working in different continents to establish the same valid nomen for the same taxon without recourse to any committee, board or court. However, several important changes in the Code are needed in the years to come. Most of them are made urgent by the pressure exerted by the projects of alternative nomenclatural systems. These projects often start from real questions and problems, but propose inadequate solutions.

Biology is now facing a new paradigm, which results from the confrontation of the taxonomic impediment with the biodiversity crisis (Dubois, 2005). The 'grand biological challenge of our age' (Wheeler et al., 2004) is to speed up considerably the collection, inventory and description of the living species of our planet before they get extinct.

Chapter 5

MICROTAXONOMY AND MACROTAXONOMY

Taxonomy deals with the description, identification, nomenclature and classification of living beings on the basis of similarities and differences, and phylogenetic relationship. The taxonomist brings order into the perplexing diversity of nature in two steps. The first is the discrimination of species, an endeavor referred to as microtaxonomy. The second is the classification of species into related groups, an activity referred as macrotaxonomy. Consequently taxonomy, the combination of the two, was defined by Simpson (1961) as "theory and practice of delimiting kinds of organisms and of classifying them". A species is widely recognized to be a fundamental unit in a biological hierarchy, composed of organisms. Further, the component organisms of a species are genetically similar to one another, and are virtually similar to one another.

According to Godfray and Marks (1991) species is a fundamental unit of biology. It marks the boundary between macroevolutionary and microevolutionary processes, the virtual cessation of genetic contact among populations. A species also occupies a particular niche or has a unique way of life in natural world. Finally the species is an enduring entity through time, whose longevity transcends that of any of its component organisms. These properties of a species have all been emphasized to a greater or lesser extent by different schools of biological philosophy. Moreover, taxonomy at the species level is very different from taxonomy at the level of higher taxa. The two aspects of taxonomy *viz.,* microtaxonomy and macrotaxonomy are being discussed herein as under:

MICROTAXONOMY

Microtaxonomy refers to the taxonomic status of individuals at the level of species. A vast array of organisms with amazing diversity of individuals makes it difficult to classify them. Thus, the task before the systematist is to identify this enormous assemblage of animals and arrange them in an orderly sequence. The diverse forms of life is due to the variation which is observed at two stage viz., initially between the individuals and then in the reproductively isolated populations. This fact must be taken into consideration for the purpose of classification. The species problem is tackled in microtaxonomy through delimitation of species with the analysis of genetical and ecological barriers between populations and with the investigation of the ranges of morphological and genetic variability of populations in relation to their environments. In addition to delimiting taxa of specific rank, microtaxonomy faces the problem of classifying groupings of populations at the intraspecific level. The experimental taxonomy focuses on patterns of variation at the infraspecific level. Microtaxonomy, therefore, embraces a wide range of views and this is in sharp contrast to macrotaxonomy, which deals specifically with the grouping of species into higher taxa.

The principle of microtaxonomy is fraught with problems at the level of species. It is quite difficult to distinguish different species. Moreover there are certain species, which do not show complete features to be designated as species. This issue is further aggravated by the confusions between **phenon, taxon** and **category**. A brief discussion of certain terms associated with the use of micro-taxonomy is given herein as under:

Phenon

Camp and Gilley (1943) used this term to describe the phenotypically homogenous samples at the level of species. Initially when species was described merely on the basis of similarity, biologist used to define species solely on the basis of differences. Linnaeus had described members of one sex as one species and that of other as second species. However when the biological status of species was ascertained, the members of opposite sex were placed under single biological species.

In fact the population of most species of animals and plants contain several different **phena** as a result of sexual dimorphism, age, variation, polymorphism and other causes. In cases where animals show sexual dimorphism, males and females belong to two different phena one having totally males while the females constitutes the other phena. This leads to understand that, these two phena belong to one species. It is absolutely necessary to have maximum possible information for the phena in question. This helps a lot in classifying the animal species correctly. This is quite essential because species show their own ecological features and life cycle, specific to their own group. Further, the phenotype of the animal population often varies due to locality, seasonal changes or habitat which they occupy. Differences between phena may be due to species differences or intraspecific variations. There are many phena, which are not correctly identified as true biological species. For example, male and female in sexually dimorphic groups of insects, worker and sexual castes in social insects and life cycle stages in the parasitic forms etc.

Most species contain several phena. Often a phenon of one species resemble a corresponding phenon of another species much closely than to any other phenon of same species. For example, females of many species of insects and birds resemble to females of closely related species than to male members of their own species. Thus correct interpretations of morphological features help in the accurate assignment of phena.

Taxon

In phylogenetic or cladistics taxonomy, taxon refers to monophyletic lineage at any level. According to Simpson (1961) " A taxon is a group of individuals recognized as a formal unit at any level of a hierarchical classification". Taxon is singular (e g., a species or family) while taxa are plural (e g., multiple species, genera, families etc.). A taxon has several attributes like its circumscription, diagnosis, rank and position in a scheme of classification. It is difficult to name a taxon, which may show all such attributes at one time because any change in the classification includes changes in attributes. For example, in different classification, taxa of the same rank may have different circumscription and so on. It is a named taxonomic group of any rank considered sufficiently distinct by taxonomists to be formally recognized and assigned to a definite category. Thus species is not taxon but a category for example; common cockroach *Periplaneta americana* is a taxon.

It is necessary to identify the taxon for proper description. Taxa are also identified within any large genus by grouping of species. Similarly, when demes and geographically isolated populations within species are recognized as subspecies, they attain the status of taxa. The higher taxa show quite large variations and represent group of animals, e g., insects and vertebrates. On the other hand, smaller taxa include different individuals like grasshoppers, beetles, birds and rodents.

Category

A category represents a rank or level in a hierarchal classification. According to Mayr (1961), it is a class whose members are all taxa that are assigned a given rank. For example, the species category is class, whose members are the species taxa. Terms like species, genera, family and order designate categories, (For detail, please see Chapter 15).

Species and Classification

It is well understood that every phyletic line and every higher taxon is the product of speciation. The species being the keystone of evolution possess high evolutionary significance. Leaving case of hybridization, it is evident that speciation has not affected the process of classifying the individuals. In the new concept of systematics, emphasis has been laid on the variation due to geographical isolation, origin of polytypic species, species and sub-species concept, taxonomic classification status of incipient species and the role of non-morphological characters in separating new species.

MACROTAXONOMY

It is the branch of taxonomy that deals with the classification (or grouping) of organisms above species level. Fortunately, most species seem to fall into natural, easily recognized groups, such as mammals and birds or butterflies and beetle. It involves recognition, description and delimitation of species. Under this scheme, the basic principles of classification, theories of biological classification, taxonomic procedures, taxonomic features, **cladistics**, **phenetics** and evolutionary classification are considered.

In macrotaxonomy, emphasis has been laid as how classification can reflect both similarity and descent. According Scott-Ram (1990) the results of macrotaxonomy appear to be of little relevance to microtaxonomy, indicating that the disputes in macrotaxonomy are too divorced from an experimental basis. Mayr and Aschlock (1991) have made the conceptual shift with regard to species; they continue to view higher taxa explicitly as classes and have considered the entire efforts of macrotaxonomy as a problem how best to classify organisms according to their similarities rather than working on common ancestry of species. Thus, the macrotaxonomy is a two step process. In the first step, species are sorted into relatively homogeneous group that reflect their overall similarity. Once the taxa have been formed according to this non evolutionary criterion, they are tested for monophyly.

Chapter 6

PHYLOGENETIC SYSTEMATICS

Phylogenetics (Greek: *phylon* -tribe, race and *genetikos*-relative to birth, from genesis-birth) is the study of evolutionary similarity among individuals and deals with the relationship between the organisms, the origin of such relationship and the reasons behind this pattern. Classifying the diversity of life based on phylogeny is referred as phylogenetic systematics. **Phylogenetic systematics** is a method of classification based on the study of evolutionary relationships between groups of organisms, and the integration of proper names of groups of organisms into a hierarchical system which reflects their phylogeny. It is an attempt to understand the evolutionary interrelationships of living things, which helps to unravel the way in which life has diversified and changed over a period of time.

There is an amazing diversity of life, both living and extinct. The classification of such immense number of individuals demands that it should be meaningful and based on evolutionary relationship. It has been clearly observed that the life on earth is integrated by evolutionary history. Relationships between taxa are defined in terms of common ancestral species. Two species are more closely related to one another than to a third species if they share a more recent common ancestor that either does with the third. These two taxa are considered as sister groups. In Phylogenetic taxonomy groups of individuals are classified according to their degree of evolutionary relatedness. In fact history of individuals is some thing, which has happened once and has left behind certain clues of the grand evolutionary march. The systematists use these clues in reconstructing evolutionary history. Thus, it is the science of the reconstruction of phylogeny.

In phylogenetic systematics, the pattern of events that have led to the distribution and diversity of life is reconstructed. An individual species terminates either by extinction (extinction of all populations), or by disintegration by speciation (splitting into two or more new descendant species), in which the stem-species does not die in a literal sense but it looses its own individuality and hands it equally over to all its descendant species, like a cell that is dividing into two new cells. A stem-species is thus defined as portion of a phylogenetic tree between two successive splitting events. Species that are not stem-species became either extinct before they could produce descendant species or they are still existing in the recent time horizon as potential stem-species which are neither yet extinct nor disintegrated by speciation yet (incomplete species). According to this view so-called 'surviving stem species ' (species that maintain their individual identity after budding of descendant species) are excluded by definition. Since species are recognized as individual spatio-temporal entities of nature, their ontological status as such is absolutely independent of any characters that might be necessary to recognize and distinguish them. Uniting a stem-species with only one of their descendant species, because of mere phenetic similarity, would be totally arbitrary and would furthermore automatically lead to the formation of paraphyletic groups that are dismissed by phylogenetic systematics for good reasons. One-egged twins are also not regarded as the same individual only because they look alike. Like a cell that is dividing into two new cells, a stem-species lives on in all of their descendants, irrespective of their appearance and similarity. The split of a stem-species into descendant species, caused by spatial separation and subsequent divergent evolution, is of course a continuous process (like the division of a cell) and consequently there is a fuzzy zone in which it will always be rather arbitrary to decide if two populations still belong to the same biospecies as subspecies or already constitute separate biospecies. This certain arbitrariness is by no means a weak point of the biospecies concept or even a proof for the artificial status of species, but a direct corollary of all continuous natural processes.

RECAPITULATION THEORY

It is also known as the **biogenetic law** or **embryological parallelism** - and often expressed as "ontogeny recapitulates phylogeny" - is a hypothesis that in developing from embryo to adult, animals go through stages resembling or representing successive stages in the evolution of their remote ancestors. In other words, ontogenesis is the mechanical cause of phylogenesis. The term **ontogeny** refers to the embryonic development of individuals and phylogeny is the general process of evolutionary change. It has been observed, that during the course of embryonic development, an individual refers back to the racial history of the past. The concept originated in the 1790s among the German Natural philosophers and, as proposed by Étienne Serres (1824-26) became known as the 'Meckel-Serres Law'. Ernst Haeckel (1866) proposed that the embryonic development of an individual organism (its ontogeny) followed the same path as the evolutionary history of its species (its phylogeny). Haeckel used the Lamarckian picture to describe the ontogenic and phylogenic history of the individual species, but agreed with Darwin about the branching nature of all species from one, or a few, original ancestors.

CLADISTIC ANALYSIS

The word **cladistics** is derived from the Greek word 'clade' meaning branch. Cladistics is a particular method of hypothesizing relationships among organisms. Like other methods, it has its own set of assumptions, procedures, and limitations. Cladistics is now accepted as the best method available for phylogenetic analysis, for it provides an explicit and testable hypothesis of organism relationships. It is a system of biological taxonomy based on the quantitative analysis of comparative data and used to reconstruct cladograms summarizing the (assumed) phylogenetic relations and evolutionary history of groups of organisms. A **clade** in cladogram is a group of organisms that include all the descendants of a common ancestor. For example, birds, dinosaurs, crocodiles and their extinct relatives form a clade. Before describing cladistics analysis, an idea about the different terms encountered in its interpretation seems desirable.

Apomorphy

It refers to the character which is present in ancestral species and its descendants. It is a derived state of characters. In other words, apomorphies refers to a condition where related group of organisms are recognized because they share a set of unique features i.e., shared derived (apomorphic) traits. Further, an apomorphic trait that is restricted to a single species is termed as **autapomorphic**. In other words, attributes shared in common are taken to indicate a shared evolutionary relationship. Thus, presence of feathers is unique to birds.

Plesiomorphy

It refers to ancestral or primitive features shared more widely than in a group of interest are referred as **plesiomorphic** trait and the phenomenon is known as plesiomorphy. An evolutionary trait that is homologous within a particular group of organisms but is not unique to the members of that group and therefore cannot be used as a diagnostic or defining character for the group. For example, vertebral column is present in zebra, cheetah and orangutans but the common ancestor in whom this trait first appeared is so distant that many other animals also share this trait. Therefore, possession of vertebrae sheds no light on the phylogenetic relations of these three species.

Symplesiomorphy

The possession of a character state that is primitive (plisiomorphic) and shared between two or more taxa is referred as **symplesiomorphic** characters. However, shared possession of symplesiomorphic features does not indicate the relationship between the taxa.

Synapomorphy

It refers to features which are shared between two or more taxa (i.e., the features are shared, derived). The possession of apomorphic features by two or more taxa in common is known as synapomorphy. If the two groups share a character state that is not primitive one, it is believed that they are related evolutionary. Further, only **synapomorphic** character states can be used as evidence to show that the taxa are related. It should be made clear that apomorphy and plesiomorphy are relative concepts. Their status depends on their position in the phylogeny. A character is an apomorphy at one branch of the tree but is plesiomorphic to all the branches after that. For example, hair is a mammalian feature (an apomorphy of mammals) but is primitive in squirrels (a plesiomorphic trait).

In a cladogram all organisms lie at the endpoints and each split is ideally binary (two ways). Each branch is called a clade. The entire individuals contained in one clade share a unique ancestor for that clade and do not share features with any other animal. Each clade possesses a series of characteristics features present in its member but not in other forms it diverged from. These identifying features are called synapomorphies (shared derived characters), e g., hardened forewings are a synapomorphy of beetles.

Hypothesis of Cladistic Analysis

The cladistic analysis is based on the characters of individuals. The traits may be anatomical, physiological, behavioral or genetic. Accordingly, only shared derived characters could possibly provide some clues about phylogeny. Hennig (1966) gave a new concept of taxonomy showing relationship between individuals based on development of hypotheses. Hennig's theory revolutionized the concept of taxonomy. He proposed that while classifying individuals and working out their relationship, certain hypotheses can be deduced. This can be exemplified by taking groups of three species viz., coelacanth, lungfish and salamander.

First Hypothesis
The coelacanth and the lungfish are more closely related to each other, than either to the salamander.

Second Hypothesis

The salamander and the lungfish are more closely related to each other than either is to coelacanth.

Third Hypothesis

The coelacanth and the salamander are more closely related to each other than either is to fish.

It is relevant to mention here, that once a new individual is found, the taxonomist considers it with the help of certain characters as mentioned above. Based on these assumptions, the hypotheses are worked out which help in deducing the relationship among different individuals.

Basics of Cladistic Analysis

As discussed above, cladistists use shared derived characters to establish evolutionary relationships. A derived character is a feature that apparently evolved only within group under consideration. Cladistic analyses are based on following three basic facts:

Organisms within a group are descended from common ancestors

In cladistics, groupings of individuals depend on their evolutionary relationship. For example, consider the case of jellyfish, sea star and human beings. Jellyfish and sea star are marine, radially symmetrical and without backbone. Phenetists might place jellyfish and sea star together in a group. However, this does not reflect any evolutionary relationships as sea stars are more closely related to humans than they are to jellyfish being kept under deuterostomia (animals where the mouth arises as a second opening opposite the blastopore. In this group anus is formed at a place where the blastopore is closed. The deuterostome phyla include Echinodermata, Hemichordata and Chordata.

Thus, in cladistic analysis, the emphasis is placed on the presence of shared derived traits. This implies that all living organisms are related to each other and share a common ancestor. With the help of available information, a meaningful pattern of relationships between different groups of organisms can be studied.

Bifurcating Pattern of Cladogenesis

This assumption is perhaps the most controversial i.e., that new kinds of organisms may arise when existing species or populations divide into exactly two groups. There are many biologists who hold that multiple new lineages can arise from a single originating population at the same time or near enough in time to be indistinguishable from such an event. While this model could conceivably occur, it is not currently known how often this has actually happened. The other objection raised against this assumption is the possibility of interbreeding between distinct groups. This, however is a general problem of reconstructing evolutionary history and although it cannot currently be handled well by cladistic methods, no other system has yet been devised which accounts for it.

Changes in Characteristics

One of the most important implications of cladistic analysis is that, characteristics of organisms change over time. It is only when the traits change; different groups can be worked out. It is relevant to mention that the 'original' state of characteristics is known as **plesiomorphic** and the changed state is referred as **apomorphic.** There is bifurcating or branching pattern of lineage splitting. According to this concept, when a lineage splits, it divides into exactly two groups. However, sometimes objections are raised over this view:

- There are situations contrary to this supposition. For example, multiple new lineages have arisen from a single originating population at the same time or near in sufficient time to be indistinguishable from such an event.

- Sometimes, there is possibility of interbreeding between distinct groups. While such exceptions may exist for many groups they are relatively rare and so this assumption often holds true.

METHODOLOGY OF CLADISTIC ANALYSIS

The phylogenetic systematics involves study of organisms which share a common evolutionary history and are more 'closely related' to members of same group than to other groups of individuals. Cladistic analysis is a particular method of hypothesizing relationship among organisms. It has its own set of assumptions, procedures and limitations. The ancestors in the cladistics analysis generally give rise to only two descendant species. The formation of two descendant species is referred as **splitting event**. After this splitting event, the ancestor is supposed to 'die'. This relationship expressed with the help of tree diagram known as **cladogram**. The cladogram represents the pattern of ancestry and descent. To bring about a real cladistics analysis, the cladogram is constructed.

Construction of Cladogram

Following steps are crucial for the construction of a cladogram:

Selection of Taxa

The taxa whose relationship is to be worked out is chosen. The taxa must be a clade.

Determination of Characters

Once the taxa are selected then the characters are worked out to find which character states are primitive and which are derived. The most common method involved is the out group comparison method. In this method, if a taxon which is not a member of the group of animals being classified, has a character state similar to some of the organisms of the group then that character state can be considered to be plesiomorphic. The outside taxon is referred as **out-group** and the organisms which are classified are known as the **in-group**. For each member of the group, certain observable traits (characters) are examined and it is also observed that whether each taxon has a particular character or not.

Polarity of Characters is Determined

Characters are shared similarities of features in different organisms. They are recognized, established and supposed to be inheritable and homologous. Differences between homologous characters in different organisms are called as character states. Once the characters are marked, it is worked out whether these characters are primitive i.e., present in ancestral forms (plesiomorphic) or derived in each taxon i.e., apomorphic. It is suggested that the traits shared with more distantly related organisms are likely to be 'ancient' or ' plesiomorphic'.

Grouping of Taxa by Synapomorphies

The taxa are grouped based on shared derived or 'changed' characters state shared by two taxa.

Parsimony

If there is any doubt it is sorted out with the help of **parsimony**. In this method choice of a phylogenetic tree is made that minimizes the number of evolutionary changes necessary to explain divergence. It may be added that the similar structures may evolve independently in separate lineages facing similar selective pressures (convergent evolution); this may be a rare event. Thus, in case of any doubt that pathway is selected which minimizes the number of times a feature may be postulated to have arisen (or lost) separately. It must be made clear that a cladogram is not an evolutionary tree. While constructing cladogram, following points must be given due emphasis:

- All taxa go on the endpoints of the cladogram and never at the nodes. All the nodes must have a list of shared derived characteristics (synapomorphies) which are common to all taxa above the node (unless the character is modified).

- All synapomorphies appear on the tree only once, unless the character state derived separately by parallel evolution.

Out-group Comparison

This process involves placing of new characters into well-supported phylogenetic trees (character optimization on a given cladogram). Being based on the principle of parsimony the outgroup comparison represents an indirect method to polarize a new character. The existence of evidence (synapomorphies) for the monophyly of the group in study is essential, and the knowledge of some of their closer phylogenetic relatives. If a character has two or more states within the (monophyletic) group in study, the state that is also occurring in the close phylogenetic relatives, which are not members of the monophyletic group in study, is regarded as plesiomorphic state. For example, the monophyly of mammals is documented by a number of derived characters (e g., hairs, mammary glands, heterodont dentition, secondary jaw articulation etc.). It is also acknowledged that amniote vertebrates (lizards, snakes, turtles, crocodiles and birds) are more closely related to mammals that all other recent organisms. Within mammals there are two different modes of reproduction: oviparity (egg laying) in monotremes (Prototheria) and viviparity in marsupials (Metatheria) and placental mammals (Eutheria). Since ovipary is also the mode of reproduction in the other amniotes and even in most other animals, the oviparity of monotremes has to be regarded as a symplesiomorphy. However, it cannot demonstrate the monophyly of Prototheria, while viviparity is a shared derived character (a potential synapomorphy), that could demonstrate the monophyly of the taxon Theria (Metatheria and Eutheria).

This method of phylogenetic out-group comparison is something fundamentally different from the so-called out group method in computer-cladism in which an unrooted parsimonious tree is calculated and a posteriorly rooted by choice of one of the analysed taxa as an out-group (often a hypothetical ancestor with all character states coded as '0').

Interpretation of Cladogram

The cladogram is shown graphically as a binary tree, where every fork has exactly two branches and the fork represents each **clade**. A clade in cladistics analysis means a group of organisms that include an ancestor and all the descendants of that ancestor. In a cladogram, there will always be one less clade than the number of groups. It is constructed in such manner that the number of changes from one character state to the next is minimized. This can be illustrated with the help of different cladogram. In Fig. 6.1, small dots (A, B, C) represent the nodes of the tree. The stems (clades) of the tree end with taxa under consideration. At each node a splitting event has occurred. The nodes represent the end of ancestral taxon, the stems or clades represent the species that split from the ancestor. The two taxa that split from the nodes are known as **sister taxa**, because they share a close common ancestor. In this cladogram, the node C shows that humans and gorilla are sister taxa and are more closely related to each other than either to baboon or chimpanzee, which is an out-group.

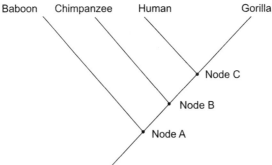

Fig. 6.1 Dots represent nodes of cladogram showing chimpanzee, human and gorilla more closely related to each other than to baboon

The second cladogram (Fig. 6.2) shows that humans and gorillas split from the chimpanzees. Therefore, the sister taxon of chimpanzee is human/gorilla ancestor. A sister taxon can be an ancestor and all its descendents. In Fig. 6.3, the cladogram shows that chimpanzees and gorillas are sister taxa. It has been argued that only shared derived characters could possibly provide us information about phylogeny. Organisms that share derived characters can be grouped through phylogenetic analysis or to be more precise by cladistics. Taxa that share many derived characters are grouped more closely together than those that do not.

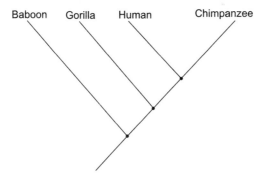

Fig. 6.2 Cladogram showing chimpanzee, human and gorilla more closely related to each other than to baboon which is an out group in this cladogram. Humans and chimpanzees are sister taxa

Some important characters of members of different vertebrate groups have been considered and represented in Table 6.1. A careful perusal of the characters would reveal that, frog shares all major traits with the out group (i.e., they show mostly ancestral or plesiomorphic characters, except that they have legs and slightly enlarged brains. The presence of legs and brain are the two apomorphic traits widespread in vertebrate lineage. It is held that amphibians have branched from the main vertebrate lineage relatively early in the evolutionary history. If the traits of turtle are analyzed it will be observed that they posses hard shell i.e., **carapace** and an increased brain size which suggests that their lineage branched next from ancestral lineage.

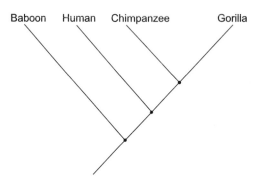

Fig. 6.3 Cladogram showing chimpanzee
and gorilla as sister taxa

Further in kangaroo, mouse and human beings, eggs develop inside the mother. As such they share a common ancestor not shared by frogs and turtle. Regarding distribution and extent of hair, mice and kangaroo are similar and the bipedal gait is the characteristic feature of kangaroo and human beings. Thus, humans and mice may be linked more closely and kangaroo more distantly based on development of placenta and presence of hair, even in lesser quantity.

Based on different features given in Table 6.1, a cladogram is constructed (Fig. 6.4).

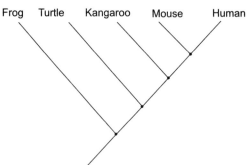

Fig. 6.4 Cladogram showing relationship
between different mammalian groups

It is observed that during the course of evolutionary history of mammals attainment of bipedal posture has evolved twice: once during the marsupial lineage (in kangaroo) and the other during the course of human evolution. Thus, humans and kangaroos are supposed to have a common bipedal ancestor not shared by mice. Further the tendency towards development of placenta along with other anatomical changes (i.e., absence of marsupium) is supposed to have evolved twice, once during mouse pedigree and secondly in the human lineage. Different chordate characters drawn from the study of fossils and out-group is summarized in Table 6.2 which shows some of the characters shared by different vertebrate groups. It is observed that **amniotic egg** and **postorbital fenestra** in the skull was present in primates, rodents, rabbits, crocodiles and relatives of dinosaurs and birds (Fig. 6.5 and Fig. 6.6).

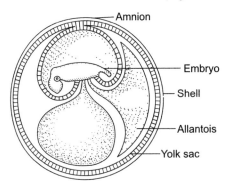

Fig. 6.5 Amniotic Egg

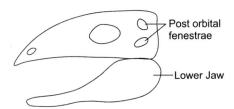

Fig. 6.6 Reptilian skull showing post orbital fenestrae

Characteristic Features of Cladistics

It is believed by cladistists that new species arise by bifurcation of the original lineage and the ancestral species no longer exists after this bifurcation, so each branching event results in two species. The cladistics analysis must possess following characteristics:

- All species in grouping must share a common ancestor.
- All species derived from a common ancestor must be included in taxon. The grouping in cladistics analysis can be made as under:

Table 6.1 Presence of certain traits in different vertebrate groups

Character	Out group	Frog	Turtle	Kangaroo	Mouse	Human
Dorsal Nerve cord	Present	Present	Present	Present	Present	Present
Legs	Absent	Present	Present	Present	Present	Present
Nature of Egg	Laid in water	Laid in water	Covered by shell	Develops inside mother	Develops inside mother	Develops inside mother
Nature of Development	Occurs in egg	Occurs in egg	Occurs in egg	Occurs in marsupium	Placenta supports the development	Placenta supports the development
Hair	Absent	Absent	Absent	Present	Present	Present to some extent
Presence of Pouch	Absent	Absent	Absent	Present	Absent	Present
Bipedal posture	Absent	Absent	Absent	Present	Absent	Present

Table 6.2 Certain distinguishing features of chordates traced back by studying fossils and out-group closely related to the vertebrate clade

Group	Vertebral column	Bony skeleton	Limb condition?	Amniotic egg?	Hair?	Two post orbital fenestrae
Sharks & Relatives	Present	Absent	Absent	Absent	Absent	Absent
Ray finned fishes	Present	Present	Absent	Absent	Absent	Absent
Amphibians	Present	Present	Present	Present	Absent	Absent
Primates	Present	Present	Present	Present	Present	Absent
Rodents & Rabbits	Present	Present	Present	Present	Present	Absent
Crocodiles & relatives	Present	Present	Present	Present	Present	Present
Dinosaurs & birds	Present	Present	Present	Present	Present	Present

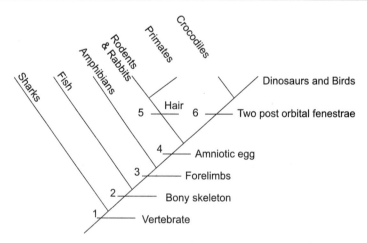

Fig. 6.7 Cladogram showing relationship between different vertebrate groups

A cladogram can be constructed by keeping presence of amniotic egg into consideration (Fig. 6.7).

Monophyletic Grouping

In a hierarchical system of descent the ancestor (stem species) and all its descendants (descendant species) together constitute a closed community of descent referred as monophyletic group. For example, all birds and reptiles are derived from a single common ancestor, so this grouping is known as **monophyletic** (Fig. 6.8 a).

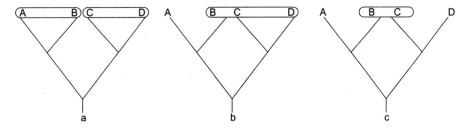

Fig. 6.8 Cladogram showing relationships of taxa (a) monophyletic (b) paraphyletic (c) polyphyletic. Term A, B, C and D represent different taxon

Paraphyletic Grouping

This type of grouping is based on the foundation of shared primitive characters e g., Protozoa, Pisces and Mamamlia etc. The members possess a common stem species and may even include all parts of the phylogenetic tree between this stem species and the recent representative but they do not include all the descendant species of this stem species. **Paraphyletic** groups are created when organisms are grouped on the basis of symplesiomorphic characters. Classical examples of paraphyletic grouping include fishes, whose descendants include tetrapods (amphibians, reptiles, birds and mammals) and reptilia (turtles, lizards and crocodiles) whose descendants include birds. However, neither tetrapods nor birds are included in the groups named Pisces and Reptilia respectively.

All major ancestral taxa of mammals viz., Syanapsida, Therapsida, Theriodontia and Cynodontia are considered praphyletic. Modern reptile is a group that contains a common ancestor but not all the descendants of that ancestor because birds are excluded, so it is referred as paraphyletic (Fig. 6.8 b) and neither tetrapods nor birds are included in the groups named Pisces and Reptilia.

Polyphyletic Grouping

A taxon is said to be **polyphyletic** that does not contain the most recent common ancestor of its member. Species that do not share an immediate common ancestor are 'lumped' together while excluding other members that would link them. In other words, a non-monophyletic group is defined on the basis of shared but not homologous character states (convergence). For examples, pterosaurs (flying reptiles), bats and birds are grouped as 'flying vertebrates'; the old taxon Pachydermata includes hippo, rhino and elephants and homeotherms include birds and mammals (Fig. 6.8 c).

When polyphyly is caused by parallel acquisition of some apomorphies in several independent lineages derived from the same ancestral species, it is referred as **parallelophyly.** In this case, parallel but independent evolutionary change may occur resulting in the derived taxon. Under such circumstances, all the members of the derived taxon are descended from the same nearest ancestral taxon in which the tendency originated. Gardiner (1982) referred it to be convergent polyphyly, which is applicable when a 'phenetic' taxon actually consists of two or more groups of species that evolved independently by converging with each group having derived from a different nearest ancestral taxon.

Consider the case, where coelenterates and echinoderms have been placed in **radiata** due to **radial symmetry.** Radiata is polyphyletic group because coelenterates and echinoderms are unrelated. It is relevant to mention here that the class reptilia are traditionally regarded as paraphyletic group which excludes birds (Aves) which are descendants of reptiles. Paraphyletic group are often erected from (sym-pleisiomorphies) i.e., on the basis of ancestral similarities instead of synapomorphies i.e., derived similarities (Fig. 6.9).

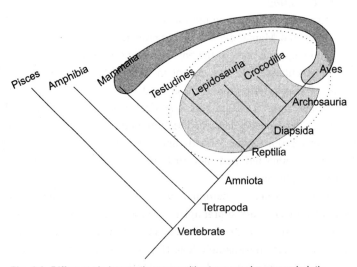

Fig. 6.9. Different phylogenetic groups (the taxa can be monophyletic, disphyletic or polyphyletic group).

Monophyletic   Paraphyletic   Polyphyletic

Holophyletic Grouping

It refers to a group that consists of all the descendants of its most common ancestor.

It is relevant to mention that every holophyletic taxon is also monophyletic but many classical monophyletic taxa are not holophyletic as they do not include some of the ex-groups which they have given rise. A holophyletic taxon is recognized by cladistics, exclusively based on shared possession among its species and one or more synapomorphies characters. A holophyletic taxon which corresponds to a traditional monophyletic taxon such as birds is reasonably homogenous.

In many cases, holophyletic taxa are heterogeneous group due to long phyletic lineage and based on some relatively undifferentiated stem groups giving rise to highly divergent ex-groups. Holophyletic taxa are delimited by combining relatively undifferentiated stem groups with their highly derived ex-groups which are usually very heterogeneous.

With the development of advance techniques in molecular biology, it has now become easier to gather large amounts of DNA or amino acid sequences to estimate phylogenies. Large data can be gathered by increasing the number of taxa in the character matrix because greater the number of taxa, more forceful is the resulting phylogeny. This favors incorporation of data from fossils into phylogenies as far as possible. Zwickl and Hills (2002) has shown that increasing taxon sampling in phylogenetic inferences provides greater accuracy and help in the interpretation of data easily.

EVOLUTIONARY TREE, PHYLOGENETIC TREE AND CLADOGRAM

The phylogeny represents a true evolutionary history of organisms. Phylogenesis is a general process of evolutionary change (anagenesis) together with the general process of speciation (cladognesis), while the term phylogeny refers to the resulting history of organisms on earth as singular historical facts. By constructing evolutionary trees we can trace the relationship of the group of organisms. The phylogenetic tree is based on combination of the fossil record, morphology, embryological patterns of development, chromosome number and structure of DNA. A cladogram only represents the relative degree of phylogenetic relationship of the analyzed taxa as well as their monophyly. The cladogram represents a hypothesis about the actual evolutionary history of a group. A Linnaean system of classification might place the birds and the non-avian dinosaurs into two separate groups. However, the phylogeny of these organisms reveals that the bird lineage actually branched off from the dinosaur lineage and so in phylogenetic classification, the birds should be considered a part of the group Dinosauria.

PATTERN CLADISM

It was given by Nelson and Platnick (1981) that a hierarchical order of organisms can be discovered from the pattern of their characters alone, without any recourse to the theory of evolution. This separation of 'pattern and process', which are regarded as two opposite aspects of nature which can not both be considered in the biological system. The discovery procedure of pattern-cladistics is a mere computer-aided parsimony-analysis of the character pattern, using a large set of taxa and equally weighted and unpolarised characters. Only that cladogram is accepted, that requires the smallest number of character transformations or steps (most parsimonious tree-MPT). The computer is primarily calculating an unrooted tree, that is a posteriorly rooted by choice of one of the analysed taxa as outgroup, and by designating the root between this outgroup and the remaining part of the tree. By this procedure of outgroup-rooting (not to be confused with an a priori character polarisation by a true outgroup comparison) and subsequent most parsimonious optimisation of the characters on the resulting cladogram, the characters finally become polarised and homologised and are then interpreted in terms of symplesiomorphies and synapomorphies.

BIOSYNTHETIC PHYLOGENY

It refers to the study of phylogeny on the basis of advanced sequencing techniques in molecular biology. By employing these techniques, it has become possible to gather large amounts of data regarding sequences of DNA or amino acids, to deduce certain relationship between groups of organisms. The biosynthetic mechanism involved in the synthesis of different amino acids and enzymes used in certain physiological processes going within the living organisms. This is important due to the advancement made in this field and also over the belief that classification should reflect genotypic relationships as evidenced from the genome i.e., the total gene content of an organism (For details, please see Chapter 8).

SIGNIFICANCE OF PHYLOGENETIC SYSTEMATICS

It was described by Hennig (1966) for the reconstruction of phylogenetic trees and the discovery of monophyletic groups on the basis of shared (homologous) derived character states (synapomorphies). Phylogenetics is mostly understood as the reconstruction of the phylogenetic relationships between such species. An important consequence of the theory of phylogenetic systematics is that new species can only originate by speciation (splitting of stem-species into two or more new descendant species) but never by a successive phenotypic change of a single evolutionary line, which would mean a gradual transformation of one species into another by a gradual shift of the gene-frequencies in an undivided gene-pool. The dismissal of a speciation by gradual transformation of species without multiplication of species by splitting can be easily explained with the example; a caterpillar that is transforming into a butterfly also remains the same individual during its whole life cycle, although its appearance drastically changes. New individuals of butterflies only originate by reproduction of this individual.

It is relevant to mention that cladistics is explicitly evolutionary. Perhaps the most important feature of cladistic is its use in testing long-standing hypotheses about adaptation. Phylogenetic classification has an upper hand over Linnaean system on the following grounds:

• The phylogenetic classification gives an idea about the organisms and their evolutionary history.

• The phylogenetic classification does not attempt to rank organisms. On the other hand, Linnaean classification ranks groups of organisms into kingdoms, phyla, orders etc. It is practically difficult to work out the evolutionary relationship of groups whose fossil records are unavailable. The cladistics line of approach relies upon the fact that animal groups have evolved by descent with modification. However, shared primitive characters are not used as basis of classification, since these may be lost or modified during the course of evolutionary process. Further, the cladistic classification is not based on the degree of evolution after the branching of evolutionary line. For example, chimpanzees and orangutan are more similar to one another than either to humans; however, cladistics taxonomy groups human and chimpanzee together, since they share a more recent common ancestor than chimpanzee and orangutan. The significance of cladistics may be summarized as under:

• Cladistics analysis is useful for creating a system of classification as it helps to explain the evolutionary theory in classification.

• It also provides hypothesis about the relationship of individuals and their features.

• Cladistics help to elucidate the way in which characters change within group over time and the direction of the change as well as the relative frequency with which the characters change. It also helps to compare the descendants of a single ancestor to work out the patterns of origin and extension in these groups or to look at relative size and diversity of the groups.

• Cladistics also helps to observe and analyze the patterns of many traits including parasitism, geographical distribution of animals and pollination by insects and animals.

Chapter 7

PHYLOGENETIC TREE

Systematics describes the pattern of relationships among taxa and helps to understand the evolutionary history of life. However, history is something we cannot see; it has happened once and has left only clues as to the actual events. Scientists use these clues to build hypotheses or models of life's history. According to modern evolutionary theory, all organisms on earth have descended from a common ancestor which means that any set of species extant or extinct is related.

Phylogenetics is the study of evolutionary relationship among living organisms using morphological, physiological and molecular characteristics. In phylogenetic studies, the most convenient way of visually presenting evolutionary relationships among a group of organisms is through illustrations called phylogenetic trees. A taxonomic concept would not be clear without knowing the evolutionary progress of different animal groups. This can be achieved with the construction of phylogeny. The species evolve by **divergence** due to varying evolutionary rates. The phylogeny represents the quantum of evolution, direction of change of characters, character complexes, whole organisms, populations and the sequence of different directional changes. This relationship can be represented with the help of tree diagram, referred as phylogenetic tree also known as evolutionary tree, cladogram or dendrogram which graphically represent the evolutionary history related to the species of interest. Further, the evolutionary history of any group is not complete without the study of genealogical relationship.

Biologists estimate that there are about 5 to 100 million species of organisms living on earth today. Evidence from morphological, biochemical and gene sequence data suggests that all organisms on earth are genetically related and a vast evolutionary tree, the Tree of Life, can represent the genealogical relationships of living things. It represents the **phylogeny** of organisms, i.e., the history of animal lineage as they change through time. It implies that different species arise from previous forms via descent and that all organisms from the smallest microbe to the largest plants and vertebrates are connected .

by the transfer of genes along the branches of the phylogenetic tree. Biologists aim to describe a single **tree of life** that reflects the evolutionary relationships of living things. However, evolutionary relationships are a matter of ongoing discovery and there are different opinions about how living things should be grouped and named.

UNDERSTANDING PHYLOGENETIC TREES

Before exploring and estimating phylogenetic tree, it is important to have a basic familiarity of the terms and elements encountered in the study of phylogeny. These are being discussed herein as under:

Node

It represents a taxonomic unit. This can be either an existing species or an ancestor. In other words node represents speciation event during the course of evolution.

Branch

It defines the relationship between the taxa in terms of descent and ancestry.

Topology

A tree topology represents the evolutionary relationships but it does not represent time or genetic distance. The topology is the graph and the leaf labels. In other words, it represents the branching pattern of the tree.

Branch length

It represents the number of changes that have occurred in the branch.

Root

The root of a tree is the node that represents the common ancestor of all taxa in the tree.

Distance scale

It is the scale that represents the number of differences between organisms or sequences.

Clade

It is a group of two or more taxa or DNA sequences that includes both their common ancestor and all of their descendents.

Operational Taxonomic Unit (OTU)

Terminal node represents the data under comparison i.e., **operational taxonomic units** (OTUs). The taxonomic level of sampling selected by the user to be used in a study, such as individuals, populations, species, genera, or bacterial strains is taken into account.

Hypothetical Taxonomic Units Internal nodes are the inferred ancestral units.

Out Group

A taxon outside the group of interest is referred as out group. All the members of the group of interest are more closely related to each other than they are to the out-group. Hence, the out-group stems from the base of the tree.

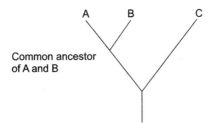

Fig. 7.1 Taxon A and B are sister groups
taxon C is an out group to A and B

In Fig. 7.1 Taxon A and B represent sister groups, i.e., they are closest relatives of each other in comparison to taxon C which is the out-group to A and B. The tree represents relationship of organisms on the basis of certain attributes such as genes, proteins or organs which are derived from a common ancestor. Species are therefore arranged based on ancestral relationship by constructing phylogenetic tree, where root represents ancestral types while branches indicate derived types. In construction of a simple phylogenetic tree, a number of groups that are different species or populations are considered. Phylogenetic studies help to work out evolutionary relationship among different groups of organisms which can be shown with the help of tree diagram referred as phylogenetic or evolutionary tree. If a molecular evolutionary tree has to be considered then the differences between species are measured in terms of genetic data such as changes in gene frequencies, amino acids sequences, nucleotide differences or **restriction map** data for **mitochondrial DNA** are considered. A matrix of such values is obtained for making **dendrogram** or micro-evolutionary tree.

Features of Phylogenetic Tree

The phylogenetic tree shows a definite evolutionary sequence of events or relative position of divergent lineages. By visualizing a phylogenetic tree an evolutionary thought is developed about the origin and affinity of group shown in the evolutionary tree, which provides an idea about the direction of evolutionary change. Phylogenetic trees are usually based on a combination of following evidences:

- Fossil records
- Morphology
- Embryological pattern of development
- Chromosome and DNA

It is relevant to mention that the evolution of one population from another involves change in some character more often a complex of characters, which result from changes in gene frequency. The phyletic line that connects the ancestral and derived populations characterized by phenotypic change and the quantum of evolution which they have undergone can be assessed by the differences in between them. Thus the degree of evolutionary relationship is calculated as the sum of the characters state differences of all the characters used to describe the two populations. Phyletic lines even short ones diverge from other phyletic lines. If this divergent aspect of evolution is considered, phenotype of any two populations can be compared and the extent of evolutionary progress can be estimated.

It is relevant to mention here that the fossil record may provide some information about the absolute time; however, it is not necessary for the reconstruction of phylogeny. Total length of tree is measured by simply calculating the sum of the character state differences of the populations involved.

Further phylogeny of the group of populations is their arrangement in the form of the dendrogram that has the greatest chance of being true based on data employed. It is suggested that the phyletic tree with the shortest total length produces the most likely set of relationships of the populations compared; regardless of the time that has elapsed during the course of evolution.

KINDS OF PHYLOGENETIC TREE

Phylogenetic trees may also be either rooted or unrooted. In rooted trees, there is a particular node, called the **root**, which is the oldest point in the tree and the common ancestor of all taxa in the analysis. An unrooted tree only specifies the relationship among species, without identifying a common ancestor or evolutionary path (Fig.7.2 and 7.3).

The nodes in trees represent taxonomic units (**species, subspecies** and **order**) and external nodes represent units directly compared (e g., extant species). The branches of the tree define the relationship among OTUs and branch length reflects the differences among OTUs in terms of time since divergence and percentage difference. A clade is a group of all OTUs having a common ancestor. In rooted tree, there exists a particular node known as root, which is the common ancestor of all OTUs and the tree represents branching order for different species showing respective lineages. The branching pattern of tree is called as **topology**.

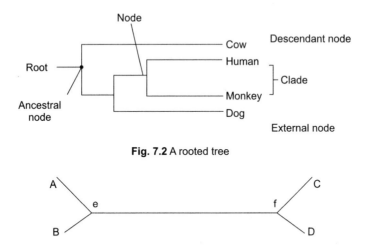

Fig. 7.2 A rooted tree

Fig. 7.3 An unrooted tree

Relationships among four homologous genes: A, B, C and D have been shown in unrooted tree (Fig. 7.4 a, b, and c). The topology of this tree consists of four external nodes (A, B, C and D), each representing one of the four genes. The clade length indicates the degree of evolutionary difference between the genes. This tree shows the relationship between genes A, B, C and D but does not indicate anything about the series of events that led to the diversification of these genes or about their common ancestor. The topology of this tree consists of four external nodes (A, B, C and D), each representing one of the four genes.

The clade length indicates the degree of evolutionary difference between the genes. This tree shows the relationship between genes A, B, C and D but does not indicate anything about the series of events that led to the diversification of these genes or about their common ancestor. In un-rooted tree it is difficult to find out the root of the tree i.e., the common ancestor representing different groups cannot be inferred from such tree without establishing ancestry.

Further, rooted tree can be developed from un-rooted tree by including an out group in the input data or introducing additional assumptions about the relative rate of evolution on each branch, such as, an application of molecular clock hypothesis. This can be exemplified by representing four genes from human, chimpanzee, gorilla and orangutan. In this case, the out-group could be represented by baboon, which is known to have branched away from the four species much before the common ancestor of the species (Fig. 7.5). Here the nodes are connected to each other nodes and the tips go as straight line. This gives V-shaped appearance to the cladogram. Both the rooted and unrooted phylogenetic tree can be of following types:

- Bifurcating tree has a maximum of the two descendants arising from each interior node.
- Multifurcating tree may have more than two branches.
- Labeled tree has specific values assigned to its leaves.

An unlabeled tree only defines topology and the leaves do not represent any particular group. Different types of phylogenetic trees are discussed herein as under:

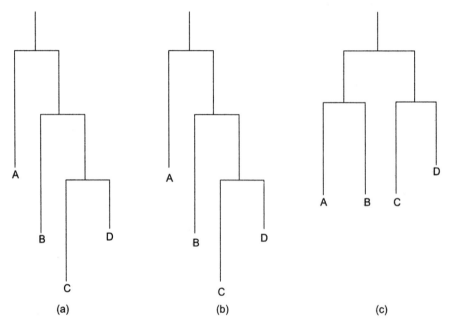

Fig. 7.4 Development of rooted tree, where A, B, C and D represent different taxon

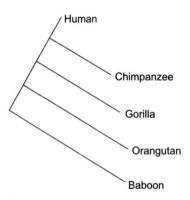

Fig. 7.5 Cladogram showing development of rooted tree

Cladogram

The cladogram represents phylogenetic relationship between different groups of animals. All the groups shown in cladogram through different clades share a common ancestor. The paths from root to the nodes correspond to the course of evolutionary events. For construction of a cladogram, as many as possible characters are examined, and then the similar features are analyzed. The common characters are taken together to produce the shortest tree with least degree of **convergence** or character reversal and the maximum amount of resemblance of characters in these branches. Thus the degree of relationship between populations is assessed.

Phenogram

These are phylogenetic trees where all the animals on it are related descendants but it is very difficult to specify the common ancestor (root). It shows the taxonomic relationships among organisms based on overall similarity of many characteristics without regard to evolutionary history or assumed significance of specific characters usually generated by computer. A phenogram is an unrooted tree. The paths between nodes do not specify an evolutionary time. The number of tree topologies of a rooted tree is much higher than that of unrooted tree for the same number of OTUs. The phenogram shows that nodes are connected to other nodes and to other tips by a horizontal then a vertical line (Fig. 7.6).

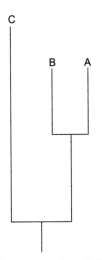

Fig. 7.6 Phenogram, where C, B and A
represent different taxon

Phylogram

This type of tree indicates phylogenetic relationship between the taxa. It provides temporal changes and also shows the degree of anagenetic divergence between the taxa. It represents number of character changes through its branch length. Here the historical relationship among individuals both by branching order and by distance measure is presented. It differs from cladogram as the branches are drawn proportional to the extent of inferred character change. The phylogram indicates both cladistic relationship and the relative extent of progressive **anagnesis** which has taken place between the internodes (Fig. 7.7).

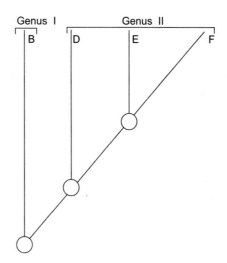

Fig. 7.7 Phyllogram (After Mayr, 1991) where, B, D, E and F represent different genus

Dendrogram

Phylogeny represents the quantum of evolution, character complexes, whole organisms, populations, the direction of change and its sequence. These parameters can graphically be represented in the form of phylogenetic tree which is also referred as **dendrogram.** The dendrogram represents the attributes of taxa and do not carry loop. It is a special kind of cladogram in which tips of trees are equidistant from the root representing the results of hierarchical cluster analysis. The dendrogram shows evolutionary changes from ancestral to descendant forms based on shared characteristics.

This is tree like plot where each step of hierarchical clustering represented as fusion of the two branches of the tree into single, thus indicating the phylogeny. In this case, the terminal dots represent contemporary species. Further, the branches do not represent their time dimension. The lines below are not composed of species but the relationship among the species, represented by dots. The numbered stems group the species in to phylogenetically valid higher taxa. The dendrogram does not represent true phylogenetic relationship (Fig. 7.8).

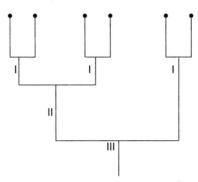

Fig. 7.8 A dendrogram with on true dimension. The terminal dots represent contemporaneous species, also without representation of their time dimensions. (Redrawn from Principles of Animal Toxonomy-simpson, 1961)

Curvogram

In this case nodes are connected to other nodes and to the tips by a curve which is some what ellipse like starting horizontally and then curving upwards to become vertical (Fig. 7.9).

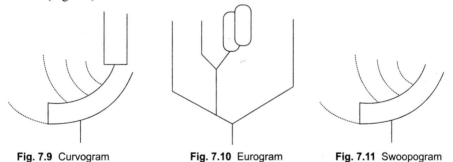

Fig. 7.9 Curvogram   **Fig. 7.10** Eurogram   **Fig. 7.11** Swoopogram

Eurogram

In this type of graphic representation, the nodes are connected to other nodes and to the tips by diagonal lines that goes outwards and give at most one-third of the way up to the next node then turns sharply straight upward and is vertical (Fig. 7.10).

Swoopogram

This type of phylogenetic tree connects two nodes or a node and a tip using two curves that are actually each one-quarter of an ellipse. The first part starts out vertical and then bends over to become horizontal. The second part, which is at least two-thirds of the total, starts out horizontal and then bends up to become vertical. Thus the two lineage split apart gradually then more rapidly and finally both turning upwards (Fig. 7.11).

Chronogram

It is a phylogenetic tree that explicitly represents evolutionary time through its branch length. It is also known as ultra metric tree.

UNDERSTANDING PHYLOGENY

Phylogeny can be defined as the evolutionary history of a group or lineage, the origin and evolution of higher taxa, or the natural process or repeated irreversible splitting of populations (Lincoln et al., 1998; Wägele, 2005). Reconstruction of **phylogeny** is very important aspect of systematics. All the available information using the modern scientific methods is used to verify the relationship between organisms. Data obtained from comparative morphology and biochemistry is used to study the divergence and convergence which are the two major pattern of evolutionary change. In divergence, the organisms that share a common ancestor became more different with the passage of time and the degree of difference can be used to workout the relationship. For example, structure of vertebrate forelimb and stout fleshy fins of lobed fin fishes, wings of birds and bats and the human hand are built on **pentadactyl** plan and such structures are referred as **homologous** organs. On the other hand, in **convergence** organisms that do not share a common ancestor become superficially more similar to each other as they share a common habitat. Similar body shapes and fins of sharks (fish), porpoises (mammals) and penguins (birds) illustrate the case of analogy and such organs are known as **analogous** organs. Thus homology is a stamp of heredity, whereas analogy merely signifies coincidence of function. It must be made clear that the analogous features must be avoided when trying to reconstruct phylogenetic relationship. Now computer based application software known as **Phylodendron** is available to construct phylogeny. DNA sequence analysis from different groups of animals reveal following interesting results:

- Pandas and bear are related.
- Whales and porpoises are not related, yet the DNA analysis shows that they are most closely related to even-toed ungulates like pigs, camels and antelopes (Artiodactyls).
- New world vultures are not closely related to old world vultures but rather have evolved independently from storks.

CONSTRUCTION OF PHYLOGENETIC TREES

Phylogenetic trees, a convenient way of representing evolutionary relationships among a group of organisms, can be drawn in various ways. Branches on phylogenetic trees may be scaled (*top panel*) representing the amount of evolutionary change, time, or both, when there is a molecular clock or they may be unscaled (*middle panel*) and have no direct correspondence with either time or amount of evolutionary change. Phylogenetic trees may be rooted (*top* and *middle panels*) or unrooted (*bottom panels*). In the case of unrooted trees, branching relationships between taxa are specified by the way they are connected to each other, but the position of the common ancestor is not. For example, on an unrooted tree with five species, there are five branches (four external, one internal) on which the tree can be rooted. Rooting on each of the five branches has different implications for evolutionary relationships. There are following basic types of phylogenetic reconstruction methods.

Distance Based Methods

These methods are guaranteed to reconstruct the true tree of their estimates of pair wise distances sufficiently close to the number of evolutionary events between the pair of taxa. For many models of bio-molecular sequence evolution, estimation of sufficiently accurate pair wise distances is possible. This method is based on following:

- Estimation of pair wise distances
- Edge weighted tree using distances are computed

Maximum Parsimony

The term **parsimony** refers to non-parametric statistical methods for estimating phylogenies. Under this method the preferred phylogenetic tree requires the least number of evolutionary changes. The data used in maximum parsimony analysis is in the form of 'characters' for a range of taxa. A character is an attribute on the basis of which taxa are observed. These attribute can be physical, morphological, molecular, genetic, physiological and behavioral etc. The variation in the character must be heritable. Each character can have only two or more states and is divided into discrete character states into which the variations observed are classified. In this case, matrix of discrete phylogenetic character is considered to infer one or more phylogenetic trees for a set of taxa, commonly a set of species or reproductively isolated populations of a single species.

However, it is submitted that coding of character is not an exact method for deducing phylogenetic analysis. Taxa similar to each other with reference to a particular character are marked. Sometimes the characters do not represent the possible variation. The character coding is done on similarity. For example, if **antenna** is **serrated** or **pectinate**, then these two different traits might be lumped together as single character, viz., shape of the antenna; this character may be subdivided to study the nature of antenna. However, it is relevant to mention that the character analysis often creates confusion, dispute and error in phylogenetic analysis. For example, the 'shape of the antenna' is one such character which would be applicable only when it is present and fully developed.

Characters may be ordered or unordered. For a multistate character, un-ordered characters can have an equal evolutionary events changing from one state to another. These events do not require an intermediate state. Ordered characters have a particular sequence in which the states must occur through evolution. It requires passing through an intermediate state. However marking character often becomes a controversial issue. One school of thought groups character on evolutionary basis while the other consider characters which show a clear transition in the phylogenetic tree.

Sometimes values are assigned in considering a particular character. Further, some character may reflect a true evolutionary relationship among individuals and may be given priority in comparison to other character, which may not be important in establishing any evolutionary relationship. Moreover, some workers consider all the characters equally. Changes in character state can also be considered individually for example, sequence of bases in nucleotide viz., A-T, T-A, G-C, C-G.

Maximum Likelyhood

The process of finding a phylogenetic tree using maximum likelihood involves finding the topology and branch lengths of the tree that will give the greatest probability of observing the DNA sequences in the data. After each step, the likelihood of each tree is examined. The tree that gives us the largest likelihood is then chosen to be examined in the next step.

The maximum likelihood method requires a possible model of evolution for estimating nucleotide substitution. The likelihood of a hypothesis is defined as the probability of the data given in that hypothesis. Under this model, potentially large number of parameters can be considered which provide a basis to estimate the differences in the probability of particular states, the probabilities of particular change and the differences in the change among character.

Bayesian Phylogenetic Interference

It uses the likelihood function and is based on priori belief about the expected results of prior probability test and gives a revised estimate of probabilities based on the result of posterior probabilities test. The interpretation of Bayesian posterior probabilities, the automatic production of a confidence set of trees and the relative computational case of the Markov-Chain Monte Carlo approach (broadly comparable in computational time to a single ML analysis) are rapidly bringing Bayesian analysis into mainstream.

Distance Methods

Non-parametric distance methods were originally applied to phenetic data using a matrix of pair-wise distances. The distances are then worked out to produce a tree (a phylogram with informative branch lengths). The distance matrix can be obtained from different sources including measured distance or morphometric analyses. Further, various pair wise distance formulae applied to discrete morphological characters or genetic distance from sequence, restriction fragment or allozyme data are also used for this purpose. For phylogenetic character data, raw distance values are calculated by simply counting the number of pair wise differences in character state (**Manhattan Distance**).

Limitations in Construction of Phylogenetic Tree

These are based on sequenced genes or genomic data in different species and can provide an evolutionary insight, yet they do have certain limitations. Phylogenetic trees do not necessarily (and likely do not) represent actual evolutionary history. The data on which they are based is noisy due to **horizontal gene transfer** and **hybridization** between species that were not nearest neighbours on trees before hybridization and convergent evolution. While making phylogenetic tree, there are certain limitations which are discussed herein as under:

- Deducing the animal history is particularly difficult because all the animal phyla have evolved simultaneously within a short time before and during **Cambrian** and since have diverged on separate lines. This implies that all the clades on the phylogenetic tree are long and branched so closely at their bases that it is difficult to determine their relationship.
- Back mutations mask the changes that preceded them and make branches look shorter.
- Complete gene sequence of many bacteria has been developed which reveal that they have descended from a common ancestor.
- The analysis based on a single gene or protein taken from a group of species can be problematic because trees constructed on the basis of an unrelated gene or protein sequence often differ from the first.
- It is relevant to mention that **lateral gene transfer** and **recombination** result in different cases; hence, the phylogenetic tree fails to represent true relationship among different groups of organisms.

- Since extinct species are not direct ancestors of any present species, they should always be represented at terminal nodes, if included in phylogenetic trees.

- In convergent evolution, two species from different genealogies come to resemble each other and the structures that resemble each other superficially (and may serve the same function) are called analogous. Further convergent evolution occurs at molecular level, for example:

  - Cows and langur monkey both synthesize a **lysozyme** that shares the same activity but comparison of their amino acid sequence indicates that each has evolved from a different ancestral molecule.

  - Cows and the bacterium *Yersinia* both synthesize **tyrosine phosphatase** with similar three-dimensional structures around their active site and similar activity. However, each has evolved from a very different ancestral molecule.

  - The bacterium *Bacillus subtilis* synthesizes a enzyme **serine protease** that acts just like the one synthesized by the mammals. However, it has an entirely different primary three-dimensional structure as well. There is considerable varying rate for evolution. Thus a branch based on molecules that have evolved rapidly would seem longer than otherwise.

STATISTICAL TESTS

In order to evaluate the significance of any particular branching pattern, statistical tests-analysis or re-sampling methods are applied. In phylogenetic analysis, **bootstrap** is the most popular method. Its value should always be provided along with topology of the tree in the form of numbers at the internal nodes, which defines the statistical support for any given cluster. The bootstrap support value should of course be equal to or higher than 95% for a very well supported cluster.

PHYLOGENETIC ANALYSIS TOOL

There are several good online tools and databases that can be used for phylogenetic analysis. These include PANTHER, P-Pod, PFam, TreeFam, and the PhyloFacts structural phylogenomic encyclopedia. Each of these databases uses different algorithms and draws on different sources for sequence information, and therefore the trees estimated by PANTHER, for example, may differ significantly from those generated by P-Pod or PFam. As with all bioinformatics tools of this type, it is important to test different methods, compare the results, then determine which database works best (according to consensus results, not researcher bias) for studies involving different types of datasets.

COMMON GENES AND NEW FAMILY TREE

Morphology - the form and structure of animals has provided the basis for the taxonomists to determine their place on the family tree. The new genetic evidences suggest that in animal kingdom, there are three primary lines of descent that first diverged from a common ancestor at least 540 million years ago and that gave rise to most animals (with the exception of jelly fish and sponges) living today.

It is also possible to determine which critical genes were 'switched on' hundreds million years ago, by analyzing the genes of the living descendants of animals found in the fossil record. Studies of genetic organization of different animals from divergent groups of animal kingdom reveal that they all possess **Hox genes** with slight but significant variations. It is inferred that ancient common ancestor that existed around 600 million years ago conferred these genes on animals that subsequently evolved in different directions.

Hox genes organize cells into different body parts and determine the number and placement of legs, wings and other appendages in the body. Hox genes are part of genetic toolbox that is responsible for pattern formation and development. It has been found that there are same critical organizations of Hox genes in an unsegemented marine worm related to insects, marine **lampshell** and in a segmented worm related to earthworm and leeches. Thus the presence of Hox genes in unrelated groups of animals suggests that these critical body organizing genes were present in common ancestor. In other words, it can be said that the genetic machinery was quite complicated in earliest animals and now it has been expressed in different ways.

Sometimes back, most biologists agreed that all organisms evolved from a single ancestral cell that lived 3.5 billion or more years ago. More recent results, however, indicate that the family tree of life is far more complicated than it was believed and may not have had a single root at all. It is suggested that eukaryotes have genes, besides those involved in cellular respiration and photosynthesis, which could come from the **alphaprotobacterial symbionts** that gave rise to the mitochondria and chloroplasts which clearly developed from bacterial lineage. In the beginning the phylogenetic trees were based on following:

- Anatomical structures
- Pattern of development

With the advancement of molecular biology, the phylogenetic trees are being constructed based on protein strucutre, **DNA-DNA hybridization, chromosome**

painting and comparing DNA sequence. Thus the evolutionary history of the molecules can be reconstructed and accordingly of the animal species to which they belong. For this the **genetic code** is utilized to determine the minimum number of nucleotide substitutions in the DNA of the gene needed to derive one protein from the other and a powerful computer program to search for the shortest path linking the molecule together. Over the past few years, a new tree has been proposed on the basis of similarities found in genes.

However, there are some proteins like **fibrinogen**, which has evolved more rapidly than **cytochrome** *c* and these can be used to decipher the recent evolutionary event. Clotting of blood occurs due to the formation of **fibrin** with the help of fibrinogen. Once removed, fibrin peptides have no further function and are free from the rigors of **natural selection** and as such have diverged rapidly during the course of evolution.

GENE TREE VERSUS SPECIES TREE

Speciation is a process rather than a short event and there is no definite demarcation when two groups considered as **sub-populations, ecotypes, subspecies** or **species** actually diverged. On the other hand, gene divergence is a long lasting event where mutations continuously increase **polymorphism**, while natural selection and random **genetic drift** eliminate some of them.

The phylogeny, i.e., the evolutionary history of a set of organisms, has become an indispensable tool in the post-genomic era. Emerging techniques for handling essential biological tasks (e g., gene finding, comparative genomics and haplotype inference) are usually guided by an underlying phylogeny. The performance of these techniques, therefore, depends heavily on the quality of the phylogeny. Almost all phylogenetic methods, however, assume that evolution is a process of strict divergence that can be modeled by a phylogenetic tree. While the tree model gives a satisfactory first-order approximation for many families of organisms, other families exhibit evolutionary events that can not be represented by a tree.

Phylogenetic tree based on **biosequence** comparisons represent time and order of divergence of the sequence. Sequence divergence does not necessarily coincide with species divergence and would normally precede it in regions under selection for polymorphism. In some genes, **divergence** starts after **speciation.** With the discovery of lateral gene transfer, it is possible to get an ideal model for evolutionary tree. Biologists aim to describe a single tree of life that reflects the evolutionary relationships of living things. However, evolutionary relationships are a matter of ongoing discovery and there are different opinions about how living things should be grouped and named. Thus, the evolutionary tree as it stands out today is far from perfect and the phylogenetic events that gave rise to myriad of complexity and diversity in animal organization are extremely complex to be described.

SIGNIFICANCE OF PHYLOGENETIC TREES

Phylogenetic analysis has become a fundamental element of 21st century biology, impacting all disciplines from molecular biology to ecosystem structure, from population genetics to the Tree of Life. A phylogenetic tree or evolutionary tree is a branching diagram or tree showing the inferred evolutionary relationships among various biological species or other entities based upon similarities and differences in their physical and/or genetic characteristics. According to Baun (2008) one of the most profound discoveries of evolutionary biology is the fact that all living species are connected through descent from a common ancestor. Thus, there is an underlying unity to life. At the same time, however, there is tremendous diversity in the living world, which is the result of the accumulation of different traits in different organisms. Interestingly, the tree metaphor not only offers a way to keep track of the features of different organisms, but it also provides guidance in how to conceptualize the broad sweep of biological diversity. Because evolutionary trees depict common ancestry, they also contain information on the degree of relatedness of the terminal nodes. Thus, the phylogenetic trees provide a systematic approach towards understanding phylogeny and have a profound impact on our understanding of the emergence of life and the characterization of the evolutionary process in its most general form.

Chapter 8

MOLECULAR PHYLOGENY

Initially morphology was used to determine the relationship between the organisms and systematics helped in classifying the animals into categories based on overall similarities. Phylogenetic analysis resulted in unraveling the branching pattern of evolution and helped in the construction of **family trees**. It was observed that classification strictly based on cladistics is somewhat complex and inconvenient too. With the advancement made in the field of molecular biology, it has become quite easy to work out the phylogenetic relationship with great precision.

Molecular **phylogeny** helps to build the 'relationship' which enables to draw similarities between organisms and work out their probable course of evolution. It employs **nucleotide sequences** (or some characters depending on sequences) from several organisms to compute the tree requiring the fewest of steps; which represent the individual **mutations** (substitutions, insertions, deletions etc.) required to transform the sequence of one species into another. The reliability of the trees can be tested by comparing the tree structure drawn from different sources. If the trees are identical, the process is reliable and this requires appropriate data sets and procedures. Every living organism contains **DNA, RNA** and **proteins**. Closely related organisms have a high degree of similarity in the molecular structure of these substances, whereas, the molecules of organisms distantly related usually show a pattern of dissimilarity.

With the unraveling of DNA sequences, homologous molecules were discovered in different organisms and soon it became evident that the basic molecular framework of all living individuals is the same. Molecular **sequencing** amply clarifies the evolutionary history of the molecules and the life form to which they belong. By comparing homologous molecules from different animals, it is possible to establish their degree of similarity, thereby revealing a hierarchy of relationship. It helps to work out the relationship on the basis of molecular sequences between the individuals. Certain relationships of individuals which otherwise can not be inferred from morphology can easily be deduced by comparison of macromolecular data. This includes sequencing of nucleotides and amino-acids, **DNA-DNA hybridization**, immunological analysis and protein **electrophoresis**. The molecular phylogenetic techniques attempt to determine the rates and patterns of change occurring in DNA and proteins and to reconstruct the evolutionary history of genes and organisms. It involves the study of DNA using different organisms to analyze their evolution as well as the evolution of DNA.

ESSENTIALS OF MOLECULAR PHYLOGENETIC ANALYSIS

The molecular phylogenetic analysis involves finding out the nucleotide sequence (or some character which depends upon the sequence of nucleotides) from different organisms to compute the phylogenetic tree. The molecular systematics is based on following concepts:

- It is an essentially cladistics approach, hence the classification must correspond to phylogenetic descent and that all taxa must be monophyletic.
- It often utilizes the **molecular clock** assumptions that quantitative similarity of genotype is sufficient measure of genetic divergence that occurred at a time immediately before the present. However, in relation to speciation, this assumption could be wrong, if either:

i. Some relatively small genotypic modification acted to prevent interbreeding between two groups of organisms.

ii. In different subgroups of the animals being considered, genetic modifications proceeded at different rates.

- In animals, it is often convenient to use mitochondrial DNA for molecular systematic analysis. However mitochondria are inherited only from mother and this is not fully adequate because inheritance in the paternal line may not be detected. The different molecular techniques in the construction of phylogenetic tree are discussed herein as under:

Database for Phylogenetic Analysis

Phylogenetic trees present evolutionary model of particular group. These models are based on following:

- The sequence is correct and originates from the specified source.
- The sampling of taxa is adequate to resolve the problem under study.
- Each position in a sequence alignment is homologous with every other in that alignment.

- The sequences are homologous, all descended in some way from shared ancestral sequence.
- The sequence variability in the sample contains phylogenetic signal adequate to resolve the problem under study.

In brief, phylogenetic analysis involves the following steps:

- Alignment-building the data models and extracting a dataset.
- Determining the sub-situation model to consider sequence variation.
- Tree building
- Tree evaluation

Tree Building

For DNA based phylogenetic tree building, the **homologs** which are the sequences that have common origins but may or may not have common activity are compared. The sequences that share an arbitrary level of similarity determined by alignment of matching bases are homologous. These sequences are inherited from a common ancestor that possessed similar structure. It is difficult to trace the ancestor because it has been modified through descent. The homologs can be categorized as under:

Orthologs

These are homologs produced by speciation and represent genes derived from a common ancestor that diverged because of the divergence of the animals and tend to have similar function. The phylogenetic classification of proteins can be visualized with the help of data base of clusters of orthologous groups of proteins (COGs). Each COG includes proteins that are thought to be orthologs or connected through vertical evolutionary descent. The COGs may be used to detect similarities and differences between species for identifying protein families, predicting new proteins functions and to point the potential drug targets in disease causing species. Consider gene 'A' in the ancestral species. Following duplication and modification, A_1 and A_2 of gene A was fixed in ancestor. The ancestor species diverged into species X and Z. The two variant species X and Z independently evolved into lineages into A^1X, A^2X and A^1Z and A^2Z species (Fig. 8.1).

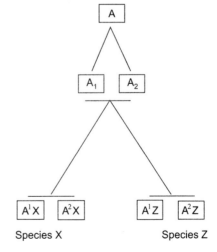

Fig. 8.1 Comparison of orthologous groups sequences

Paralogs

These are homologs derived from a common ancestral gene that duplicated within an organism then diverged and tend to have different functions. Paralogs are created by gene duplication. Once a gene has been duplicated, all subsequent species in the phylogeny will inherit both copies of the gene.

Xenologs

The homologs produced as result of **horizontal gene transfer** of genes between organisms. In this case, individuals inherit genes from unrelated organisms like prokaryotes, whereas in lateral or vertical gene transfer individuals inherit genes from parent. The function of xenologs can be variable depending on how significant the change in context was for the horizontally moving gene. In prokaryotes, e g., in bacteria, gene exchange is an important evolutionary process. In eukaryotes, horizontal gene transfer is of rare occurrence as such tree structure is not affected.

HomoloGene

It is a database of both curated and calculated orthologs and homologs. Computed orthologs and homologs are identified from **BLAST** (Basic Local Alignment Search Tool) nucleotide sequence comparisons between all UM gene clusters for each pair of organisms. HomoloGene also contains a sort of triplet clusters in which orthologous clusters in two organisms are both orthologous to the same clusters in a third organism. HomoloGene can be searched via **Entrez retrieval system**.

Entrejz Genome

Entrez is an integrated database retrieval system for DNA and protein sequences derived from several sources. The whole genomes of over 120 individuals can be found in Entrez Genomes. The genomes represent completely sequenced organisms in different domains of life viz., bacteria, **archaea** and **eukaryotes, viruses, viriods, plasmids** and **eukaryotic** organelles. Data can be accessed hierarchically starting from either an alphabet listing or a phylogenetic tree for complete genomes in each of the six principal taxonomic groups. The hierarchical relationship can be worked out and multiple views of the data, pre-computed summaries and links to analysis appropriate for that level is worked out. In addition, any gene product (proteins) that is member of a COG is linked to the COGs data base.

MEASUREMENT OF IMMUNOLOGICAL DISTANCE

The intensity of response of antigen to the antibodies of related species is measured. Greater response shows that the two groups have close evolutionary relationship. The antigen responses between species are often less marked in one direction than in other. Further, as the **antigen-antibody** reactions are complex and involve the interplay of a number of genes; as such, the intensity of response may not truly reflect the time of evolutionary divergence. In these circumstances, it is suggested that after testing several antibodies the result obtained should be pooled. The protein phylogenetic analysis is being discussed herein as under:

Protein Sequences

Evolutionary relatedness or divergence can be measured by comparing the amino acid sequence of proteins common to various organisms. Studies of amino acid sequences of homologous proteins from different species reveal similarity in protein structure and thus show that the species are evolutionary related. The degree of similarity between amino acid sequences of homologous proteins from different species is correlated with the evolutionary relationship of the species. In many cases, positions in the amino acid sequence in all the species are similar and they are referred as **invariant residues**. Further, in other positions, there may be considerable variation in the amino acid from one species to another, this is known as **variable residues**. With the help of information on number of amino acid residues differences between homologous proteins of different species, **evolutionary maps** can be constructed. It is suggested that the number of residues that differ in homologous proteins from any two species is in proportion to the phylogenetic differences between the species. Thus amino acid sequences provide a tool for establishing homologies from which genealogies can be constructed and phylogenetic trees drawn. Following two molecules help to draw a phylogenetic basis of interrelationship between different groups of organism:

Hemoglobin Hemoglobin (Hb) is well known respiratory pigment, having two pairs of polypeptide chains viz., alpha (α) and beta (β), which serves to transport oxygen to different parts of the body. The amino acid sequence difference between **myoglobin** and hemoglobin are sufficient enough to suggest that the genes coding for these proteins diverged long ago in the history of vertebrate evolution.

From the degree of amino acid sequence similarity among the various **globin** genes and the evolutionary history of the organisms in which globin genes are found, the timings of duplication events have been worked out. This hypothesis is supported by the fact that the hemoglobin molecule of primitive chordates consists of only one polypeptide chain. The major differences between the alpha (α) and beta (β) chains are probably because of the early divergence (about 400 million years) of these chains from the ancestral hemoglobin molecule. Further, little differences between the beta and gamma chains (γ) are found only in mammals. This suggests that the gamma gene diverged from the beta gene more recently (100 million years) during the grand evolutionary march of mammals. Around 40 million years ago, the beta

lineage underwent gene duplication again during the evolution of primates yielding
the delta (δ) polypeptides (Fig. 8.2).

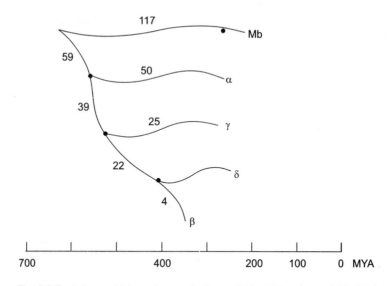

Fig. 8.2 Evolutionary history of genes for hemoglobin (Hb) and myoglobin (Mp)
Where alpha (α) beta (β) delta (δ) and gamma (γ) represent component of Hb
molecule. The dots represent the stage at which ancestral genes were duplicated
giving rise to new gene line. The number given at each branch indicates the
minimum number nucleotide substitutions required to account for the amino acid
differences between the proteins for each branch. The time estimates are based in
paleontological and morphological studies of the vertebrates (Redrawn after
Prakash 2000).

The hemoglobin molecules of different animals which differ in amino acid
contents have been shown in the Table 8.1. The information gathered from the amino
acid difference between homologous proteins of different species allow the
construction of evolutionary map, which clearly indicate the origin and sequence of
development of different animals and plants during the course of evolution.

Cytochrome It is iron containing mitochondrial proteins involved in transfer of
electrons during biological oxidations and is found in **mitochondria** of every
aerobic eukaryotic animal, plant and **protista**. It is a part of **respiratory chain**
through which electrons are passed to oxygen during cellular respiration. The amino
acid sequence of many of these has been determined and on making comparison, it is
found that they are related. It is relevant to mention that cytochrome is an ancient
molecule that has evolved very slowly. Even after more than 2 billion years, one-
third of its amino acids are unchanged. This conservation is of great help in working
out the evolutionary relationships between distantly related creatures like fish and
humans. Human cytochrome *c* contain 104 amino acids and 37 of these have been
found in equivalent positions in every cytochrome *c* that been sequenced. It is

assumed that each of these molecules has descended from a precursor cytochrome in a primitive microbe that existed over 2 billion years. In other words, these molecules are homologous. For example, 48 amino acid residues differ in the cytochrome *c* molecule of horse and yeast which is very widely separated species, whereas duck and chicken which are closely related differ in two amino acid residues only.

Table 8.1 Different groups of individuals showing variation in the amino acid contents of Beta chain

S.No.	Animal	Amino acid number
1.	Human beta chain	0
2.	*Gorilla*	1
3.	Gibbon	2
4.	Rhesus monkey	8
5.	Horse, cow	25
6.	Mouse	27
7.	Gray kangaroo	38
8.	Chicken	45
9.	Frog	67
10.	Lamprey	125
11.	Sea slug	127
12.	Soyabean (leghemoglobin)	124

GENETIC COMPLEMENT

Animal cells possess DNA in the nucleus and also in their organelles-like mitochondria and chloroplasts. Knowledge on molecular biology tells us that the amount of DNA per chromosome set is constant for each species. Since DNA is the basic genetic material, it is believed that if the DNA composition of all species is known, their evolutionary history can also be traced. The DNA consists of four nucleotides (**deoxyribonucleotides**) viz., **adenine** (A), **guanine** (G), **cytosine** (C) and **thymine** (T). The double helix consists of pairs of nucleotide in chains which run anti-parallel to each other. In this chain, A always pairs with T and G always pairs with C. The information from DNA is transcribed to **messenger RNA** (m RNA) which in turn serves to translate a particular protein by sequencing **transfer RNA** (t-RNA), each bringing a particular amino acid over the surface of ribosome. Thus the sequence of DNA bases has a direct control over the synthesis of proteins and it in turn allows all organisms to develop, grow and adapt themselves for the perpetuation of their race. It is relevant to mention that sequencing alone is not the end result of analysis. With the help of **polymerase chain reaction** (PCR) it is possible to trace the ancestry as well as the relationship between different groups of organisms. The result of molecular phylogenetic analysis is expressed with the help of phylogenetic tree (For details, please see Chapter 7).

Nucleic Acid Phylogeny

The DNA molecule has been the document of evolutionary history. Nucleotide sequencing is much faster and less expensive than peptide sequencing. Even very small quantity of DNA in **fossils** over 100 million years old, can be successfully multiplied a billion fold by PCR techniques; this provides enough material for DNA sequencing. Comparisons of DNA sequences of various genes between different individuals tell us a lot about the relationships of different groups of animals that can not be deduced by morphology alone. As the **genome** evolves by gradual accumulation of mutations, the extent of nucleotide sequence differences between a pair of genomes from different organisms ought to indicate how recently these two genomes shared a common ancestor. Nucleic acid phylogeny can be obtained by the measurement of deoxyribonucleic acid, the relative amount of repetitive and non-repetitive DNA, DNA-DNA hybridization and **chromosome painting**.

DNA Hybridization Technique

Genetic relationship among species can be assessed on the basis of 'hybridization' between single stranded DNA taken from different species. This technique helps to measure the degree of similarity between the genomes of different species and therefore it is possible to workout genealogy of species by comparing the genetic resemblance between lineages that diverged millions of years ago.

The principle of the technique is simple and utilizes the basic physical and structural property of DNA. When the DNA is heated almost to boiling the hydrogen bonds binding the two strands of DNA together in a double helix break, the DNA is denatured and gets separated into two **complimentary strands**. Then it is allowed to cool and incubated between 50 and 60 C°. The single strand associates with the complimentary partners and reforms the double helix. Given enough time, all single strands will find partners and the DNA will reorganize virtually perfectly as the strands are exact matches for each other. The double helix rebuilds when the hydrogen bonds between complimentary base pairs are reformed. The techniques can be described in brief, herein as under:

- The total DNA is extracted from the cells of each species and purified
- The DNA is heated so that it becomes denatured into single strands (**ssDNA**)
- The temperature is lowered just enough to allow the multiple short sequences of repetitive DNA to rehybridize back into double stranded (**ds DNA**).
- Mixture of ssDNA (representing single DNA) is passed over a column packed with **hydroxyapatite** (HA); ssDNA does not flow right through.

- The ss DNA of species A is made radioactive.
- The radioactive ssDNA is then allowed to hybridize with non-radioactive ssDNA of the same species (A) as well in separate tube with the ssDNA of species B.
- After hybridization is complete, the mixture (A/A) and (B/B) are individually heated in small (2°–3°C) increments. At each higher temperature, an aliquot is passed over hydroxyapatite. The radioactive strand (A) that has separated from DNA duplexes pass through the column and the amount is measured from their radioactivity.
- A graph showing the percentage of ss DNA at each temperature is drawn.
- The temperature at which 50% of the DNA duplexes (dsDNA) get denatured ($t^{50}H$) is determined.
- Thus, DNA-DNA hybridization provides genetic comparisons integrated over the entire genome.

DNA hybridization technique enables us to find out the similarities between the species based on DNA molecule. From the above experiment, it follows that if the DNA strands of two species do not bond together, it is due to the genetic differences between the species. If the two strands differ, there would be imperfect match between the strands and thus the bonds between them would be weaker and can easily be broken by little heating, while closer matches require more heat to separate the strands, as similar DNA molecules 'melt' at higher temperature.

Comparison of DNA Sequences

DNA sequencing provides a direct record of the genealogy of present species. Furthermore, DNA also has significant advantages over proteins for molecular systematics due to several reasons:

- The genotype is studied and not the phenotype.
- One or more sequences can be selected to specifically explain the evolutionary question (s) being asked.
- The techniques are applicable to all types of DNA.
- The DNA can be prepared from relatively small amounts of tissue and is comparatively stable.
- Proteins are the expression of genes, so it is quite advantageous to compare actual gene sequences.
- DNA is much easier to sequence than proteins. DNA sequences may be compared directly or for those regions, that code for a known protein sequences.
- Introduction of **BLAST** helps to rapidly scan high database in matter of seconds to study homologies or sequence similarity and to statistically evaluate the resulting matches. Thus BLAST works by comparing unknown sequence against the database of all known sequences to determine possible matches. After comparing the

difference between sequences, BLAST assigns a 'score' based on sequence similarity. The scores in BLAST have well defined statistical interpretations. BLAST uses a special algorithm or mathematical formula and thus easily detects relationships among sequences that shows only isolated regions of similarity. The comparison of DNA has several advantages over proteins due to the following reasons:

- Sequencing of DNA is much easier than proteins.

- Sites in the genes are prone to changes during evolution than protein sequences because of the following reasons:

- Nucleotides that produce synonymous **codons** are important. For example, even if the amino acid at position 20 in two proteins is the same, the codon for that amino acid might be different in the two species.

- **Introns** and the flanking sequences are relatively free to vary without affecting the final protein product. In other words, these regions of the genome are under much less pressure of natural selection.

- In nature, DNA is much more stable than proteins. This has also helped the DNA sequencing of fossils. For example, Egyptian mummies over 2000 years old and human remains in Florida that are at least 7000 years old, yielded samples of DNA that were successfully cloned and sequenced. In 2006, DNA sequence from 80,000 years old fossil elephant was sequenced.

Chromosome Painting

The number, size, morphology and banding pattern of chromosomes can be studied with the help of staining techniques. The entire **genome** can also be compared by a technique known as chromosome painting. It is of great help in establishing phylogenetic relationship among different groups of organism. Phylogenetic relationship among different families of insect order Trichoptera has been worked out by Kiauta (1967). Dipteran flies provide an excellent material for karyological study. It has been carried out detailed herein as under:

- A **fluorescent label** is attached to the DNA of individual chromosome of one species (e g., human).

- The chromosome of another species is exposed to it.

- It will be observed that the homologous regions of gene will hybridize taking up the fluorescent label and the 'painted' chromosome can be examined under the microscope.

This method is a modification of fluorescence *in-situ* hybridization (FISH) and is also referred as **Zoo-FISH**. With the help of this technique it has been possible to locate those large sections of human chromosome 6, which includes hundreds of genes in the **major histocompatibility complex** (MHC) having their homologous genes in the following:

- Chromosome 5 of the chimpanzee.
- Chromosome B_2 of the domestic cat
- Chromosome 7 of the pig.
- Chromosome 23 of the cow.

MOLECULAR CLOCK

It is an important tool in molecular systematics and is used to find out the time when two species actually diverged. It is sometimes referred as a **gene clock**. The molecular clock hypothesis postulates that DNA sequence evolution is roughly constant over time in all evolutionary lineages. A detail analysis of the divergence of sequences in human mitochondrial DNA has been carried out. The results support the view that modern humans originated in Asia.

The concept of molecular clock is based on the fact that a given biological molecule shows a relatively constant rate of change over time irrespective of taxonomic lineage within which it evolves (Zuckerkandl and Pauling, 1962). For example, cytochrome c appears to have evolved at similar rates in groups like vertebrates, fungi and plants. These organisms are phylogenetically quite diverse, their genes coding for cytochrome exhibits similar rate of evolution. If the rate of evolution of a particular molecule is nearly constant over time the degree of divergence between homologous genes in different taxa can be used to estimate the time at which their evolutionary lineages had separated. This can be made clear by taking the following example:

Once a homologous gene 'A' has been sequenced, for example in two species (1 and 2) and the rate of evolution of this gene is known through prior calibrations (say 2% per million years- MY), then the percent difference in the DNA sequence of gene 'A' between the two species allows the calculation of age of their last common ancestor. For example, in this case, if species 1 and 2 differed by 10% in their DNA sequence of gene A, then the common ancestor of these would be expected to have lived around 2.5 MY. These two lineages would have taken this long time to diverge at a rate of 2% per million years to accumulate 10% difference in gene 'A' (Mayr, 2001). It is relevant to mention that molecular clock method must be applied with caution as sometimes the clock is not constant. Often mosaic evolution also occur where evolutionary changes in a taxon occur at different rates for different structures; organs or other components of the phenotype. Different models have been proposed using statistical methods, maximum likelihood techniques and **Bayesian modeling**. Some of these models take into account rate variation across lineages to get better estimates of divergence times. These models have been referred as 'relaxed molecular clocks' (Drummond et al., 2006) and have been developed from viral phylogenetic and ancient DNA studies.

IMPORTANCE OF MOLECULAR PHYLOGENY

Molecular phylogenetics includes a combination of molecular and statistical techniques to infer evolutionary relationships among organisms or genes. The similarity of biological functions and molecular mechanisms in living organisms strongly suggests that species are descended from a common ancestor. It entails the study of structure and function of molecules and how they change over time to infer these evolutionary relationships. The primary objective of molecular phylogenetic studies is to recover the order of evolutionary events and represent them in evolutionary trees that graphically depict relationships among species or genes over time. This is an extremely complex process, further complicated by the fact that there is no single correct method to approach all phylogenetic problems.

Molecular phylogenetic trees are generated from character datasets that provides evolutionary content and context. Character data may consist of biomolecular sequence alignments of DNA, RNA or amino acids, molecular markers, such as single nucleotide polymorphisms (SNPs) or restriction fragment length polymorphisms (RFLPs), morphology data or information on gene order and content. Evolution is modeled as a process that changes the state of a character, such as the type of nucleotide (AGTC) at a molecular phylogenetics specific location in a DNA sequence; each character is a function that maps a set of taxa to distinct states.

It is evident now that traditional taxonomic practices are indeed inadequate for species identification and that molecular techniques can unravel a host of hidden taxa which cannot be deciphered by alpha taxonomy. However, before molecular taxonomic tools are applied, initial identification by traditional methods needs to be carried out by a taxonomist. What has all too often happened is that molecular phylogenetics have been studied without any of the biological aspects of the said taxa and this has led to many 'false draws and blind alleys' (Godfrey and Knapp 2003) in the analysis of data, as has happened in -Strepsiptera. For instance, if two morphologically identical specimens are sequenced and show different molecular characterization, questions such as: whether they come from the same habitat, are sibling/incipient/ hybridizing species, and (if a parasite) do they parasitize the same species of host need to be addressed.

These are crucial biological questions vital for ecologists, conservationists and policy makers alike. Otherwise, the molecular data are meaningless and molecular phylogenetics divorced from traditional taxonomy is indeed an improvised study. Taxonomy is therefore not a dead-end, but the beginning of a larger study of biological diversity which has an endless number of users and end-users. Combining morphological and molecular data is thus vital and ecological geneticists and evolutionary biologists have to work in close collaboration with taxonomists, as their work compliments each others. Without the historical background - knowledge of interesting anatomical structure and behaviour, the study of molecular phylogenetics will be of no interest to biologists (Wheeler, 2002) and unless deep studies in comparative morphology coupled with ethology and ecology is not done, there will soon be little to explain with molecular phylogenetics.

The molecular phylogeny is an important tool to test the validity of theory of evolution. It is also suggested that all part of the genome should evolve in parallel and exhibit the same taxonomic pattern. Molecular studies, especially sequencing of mitochondrial genes have expanded enormously during the last decade and have helped to characterize biodiversity and biogeographically patterns of different groups of animals. Sometimes, morphological evidence alone has not been able to establish phylogenetic relationship, in such cases an assessment of molecular and morphological evidence has often proved helpful. Evolution required molecular phylogeny to be consistent with classical phylogeny. The molecular studies help to identify the actual clades to which different species and higher taxa belong. The determination of phylogeny enables to understand the evolution of various molecular, cellular and development characters shared by any of the three kingdoms. It will result in the development of more accurate taxonomic procedures and would finally lead to a unified classification for all ramifying patterns of life.

Chapter 9

EVOLUTIONARY TAXONOMY

Taxonomy refers to the study of classification, principles, procedures and rules thereof. It is a part of scientific practice known as systematics which entails the study of evolutionary relationships between different groups of animals and the way by which they are grouped together. It may be mentioned that traditional taxonomy involves groupings of individuals on the basis of similarities and it is also held that phylogeny is the basis of biological classification, explaining the fact that present day diverse forms of life arose from a common ancestor. As such in taxonomy evolutionary relationships are used to classify organisms. There are different methods of classification and a taxon or group of organisms can be classified in several different ways.

Darwin (1859) in *Origin of Species* observed that arrangement of the groups within each class in due sub-ordination and relation to other groups must be strictly genealogical in order to be natural. Since then, different taxonomists have given their scheme of classification based on the Darwinian principle of evolution.

FEATURES OF EVOLUTIONARY TAXONOMY

Evolutionary taxonomy can be considered to be a blend of phenetics and cladistics, which classifies organisms partly according to their evolutionary branching and somewhat according to their overall similarity. To work out evolutionary relationships, different characters are analyzed. An out group analysis is conducted in which a group of three species which are supposed to be related are examined. Following inference can be drawn:

- If a character is present in two of the three species, then the character is said to be **synapomorphic** character.
- If a character is present in all of the species, then it is said to be **pleisomorphic** character.

The evolutionary classification utilizes certain fundamental principles for delimitation and ranking, detailed herein as under:

Distinctness

This feature enables to recognize two clusters of species. Greater the gap between two clusters of species, better would be the justification for recognizing both as separate taxa. The size of the gap is measured in terms of biological significance of the difference. The gaps between the taxa are produced as result of **evolution**, **speciation**, extinction, **adaptive radiation**, unequal rates of evolution and other evolutionary phenomenon. It is also argued that the gap among higher taxa is due to extinction and divergent evolution.

Degree of Difference

This refers to the extent the two clusters of species differ from each other. If more characters are considered the more reliable would be the process of delimiting the taxa.

Ecology and Evolution

Variation of species is due to the process of evolution. Further in order to grow and reproduce, each species has its own adaptive features. Whenever a species encounters new ecological conditions, it tries to adapt itself in that ecological niche.

Grade Characteristics

According to Huxley (1939, 1940) groups of animals, similar in general organization as distinct from groups of common genetic variation are referred as grades. Grades refer to shared morphological level and are used to characterize the morphology of a group of species that occupies some adaptive zone. The phenomenon of grade characteristics can best be illustrated by origin of life.

Prokaryotes include diverse forms of life like **archaebacteria, eubacteria** etc. and are characterized by the absence of nucleus and other cellular organelles. The prokaryotes represent definite evolutionary grade and are referred as **Monera**. It is relevant to mention that after the origin of eukaryotes some 1.5 billion years ago, vast majority of single celled organisms originated; which gave rise to different groups of animals including fungi and plants. In spite of diverse lineages, all these organisms are unicellular.

Equivalence

While ranking of a taxon, it is recommended to have equivalence in related taxa at that level and for the categories immediately above and below it. The taxa, which diversified rather early in the course of evolutionary history show quite large differences in their molecular sequences.

Stability A classification should be stable and help in the retrieval of information. It should provide sufficient information to all the taxonomists under some meaningful classification scheme which are accepted by and large.

SIZE OF TAXON

Taxon is characterized by the number of species it constitutes. Branching and **divergence** result in speciation. Continuous speciation i.e., branching without visible divergence result in production of large taxa, while divergence without branching leads to evolution of **monotypic** or series of superimposed monotypic higher taxa. Some systematists are of the view that the different taxa should be of equal size. On the contrary, if the taxa are of large size, it would be rather difficult to get any useful information from the classification. Hence, it is recommended that the size of the gap which justifies separation of higher taxon should be inversely correlated to the size of the **taxon**:

$$\text{Size of gap between taxa} \quad = \quad 1/\text{ size of taxon}$$

It has been observed that higher taxa of organisms contain some 20 to 30% monotypic genera. For example, certain animals like giant *Panda* and *Ostrich* represent an entire family or order. On the other hand, certain well-defined natural genera contain over 1000 or even 20000 species e g., genus *Drosophila* has more than 150 species.

Splitters and Lumpers

It is argued that for want of any reliable method it is quite difficult to measure the size of the taxon. There are two groups of systematists viz., **splitters** and **lumpers**. Splitters try to express each shade of difference and every degree of relationship through formal recognition of separate taxa and their elaborate categorical ranking. On the other hand, lumpers try to produce a classification in which the emphasis is placed on relationship.

Sister Groups

Two monophyla or two species which together constitute a monophyletic group originated from the same speciation event have been referred as sister groups by Hennig (1966). The discovery of **sister group** relationships is one of the primary objectives of phylogenetic systematics. As the fossil record of most phyla is incomplete, the term sister groups stands true if it is applied to recent taxa. The fossil relatives are referred to as **stem group** representatives. A stem group is usually a paraphyletic group, since it includes all fossil relatives that existed between the split of a recent monophylum from its sister group and its division into recent subgroups.

During the course of evolutionary history, the characters that evolve in only one of the two sister groups are referred as **autapomorphic** character. For example, all the distinguishing features of birds that evolved after they diverged from their archosaurian sister groups are referred as autapomorphic characters or autapomorphies. On the other hand characters by which a species or group differs from its sister species or group, are known as synapomorphic characters. This can be explained by taking the example of birds with their reptilian sister groups.

Mosaic Evolution

In taxonomy, characters are weighed into derived and ancestral ones and the later seem to be more important in evolutionary scheme of classification. The ancestral characters provided much insight about the total **genotype** of individuals under the classification scheme. Different portions of genotype may evolve at very different rates; this is referred as mosaic evolution. Naturally therefore, a taxon may acquire different characters, as a result of this, due caution must be given to the selection of characters. Mosaic evolution is responsible for the fact that different classifications are generated whenever new sets of characters are used.

Concept of Evolutionary Taxonomy

Evolutionary taxonomy attempts to classify organisms using combinations of phylogenetic relationship and overall similarities. It is somewhat different from cladistics, where all the taxa in a scheme of classification should descend from a common ancestor. In other words, all taxa in a cladistic scheme of classification should always include all the descendants of a single ancestral node. Evolutionary taxonomy has also been referred as gradistic taxonomy or classical taxonomy. It is based on combination of branching and divergence. Evolutionary systematic school has also been referred as Simpson-Mayr school. Traditional taxonomy has admitted numerous taxa which are not accepted on the basis of cladistic analysis as they exclude some of the descendant of the ancestors. For example, traditional taxa such as Pisces and Reptiles suffer from cladistic point of view because tetrapods have descended from fish, and birds arose from reptiles.

The difference between cladistics and evolutionary taxonomy can be illustrated by taking the example of birds and crocodilians. According to traditional evolutionary taxonomy, birds (class Aves) and crocodiles (class Reptilia) are kept in separate groups based on overall morphology. However, it must be added that the birds and reptiles have a common **thecodont** reptile ancestor (Fig. 9.1). It is relevant to mention that thecodonts were primitive reptiles which gave rise to dinosaurs and **pterosauria** (flying reptiles), in addition to crocodilians and birds.

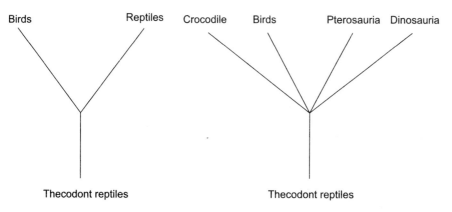

Fig. 9.1 Origin of birds and reptiles from thecodont reptiles

Fig. 9.2 Origin of different vertebrate groups from thecodont reptiles

According to cladistics taxonomy, all the above four derived taxa are placed in the same higher taxonomic category, since they share a common ancestor. On the other hand, evolutionary taxonomy do take into account the phylogenetic affinity of birds with other reptiles, but place birds in a separate class because they have reached a new structural grade by developing feathers. Like birds and unlike pterosaurs and dinosaurs, crocodilians possess four-chambered heart and thus appear more close to birds than to thecodont reptiles. However, evolutionary taxonomists discard the similarity while placing crocodilians in class reptilia. Based on strict cladistics analysis, crocodilians should be included in the same class with birds.

As discussed in the preceding paragraphs, traditional taxonomy i.e., evolutionary taxonomy is not based on genealogical relationships. In fact, it was based on similarity i.e., similar animals were placed in the same taxonomic groups or at least in groups that were close to each other. The animals so grouped may or may not have any genealogical relationship. Origin of different vertebrate groups from a common ancestor has been shown in Fig. 9.2. This can be further explained by taking the examples of birds. A careful perusal of this evolutionary tree would reveal that birds arose from archosaurian reptiles. Further this stock of ancestral reptiles also gave rise to pentadactyls, dinosaurs and crocodilians. The birds and crocodilians share a number of derived similarities that originated in the archosaurian lineage, after it had branched off from other reptilian lines. Moreover, crocodilians are quite similar to other reptiles in having several ancestral characters. Birds on the other hand, developed many specialized features like the development of feathers, possession of strong hind legs, ability to maintain body temperature, high rate of metabolism, nesting and song behavior and finally having developed the remarkable ability of migration, which enabled them to become the dominant denizens of the sky.

In can be concluded that whenever a clade i.e., a phyletic lineage enters a new adaptive zone, it leads to drastic reorganization of its members, as such due emphasis must be given to the acquisition of new autapomorphies than to sister groups. Genealogy (family history) may not provide a true classification because the extent of differences between different taxa or groups may be due to the degree of modifications which they may have undergone.

Thus traditional taxonomy takes into consideration the evolutionary pattern as well as overall similarity while classifying different groups of animals. The evolutionary taxonomy accepts the Hennigion-Cladistic methodology as adequate for reconstructing phylogenetic trees while retaining paraphyletic groups. It is based on homologies but do not distinguish between ancestral and derived homologies. Shared derived homologies cannot (if correctly identified) contradict one another; but different ancestral homologies can end up in distributed contradictory groupings. The evolutionary taxonomy encompasses following two sources of information:

- Branching Point
- Degree of Differences

However, the above actions cannot be applied simultaneously. As discussed previously, in practice, the taxa are recognized by the totality of their shared characters and then tested for **monophyly**. Further, in evolutionary taxonomy, due consideration is given to autapomorphic characters. The evolutionary basis of classification as given by Darwin (1859) based on his remarkable *Origin of Species* includes following two criteria:

- Common descent.
- Extent of evolutionary change.

According to Darwin, a classification must reflect the common ancestry of each member of taxon. However, Darwin cautioned the unequal degree of subsequent divergence. It has been the objectives of taxonomists to develop a classification having following features:

- The classification must be stable, yet flexible.
- It must reflect the evolutionary history of taxa that are included in the scheme of classification.
- It must be convenient for the storage and retrieval of information.
- It should help in comparative studies.
- It should be easy to use.

According to Mayr (1961) evolutionary taxonomy is a system that starts with observed similarities and differences among groups of organisms however, its prime objective is to workout the relationship of organisms based on their evolutionary history as under:

- Descent from a common ancestor.
- Gradual acquisition of new characteristics.
- Adaptations to new environment.

An evolutionary classification will therefore only be objective in so far as shared derived homologies are used to construct it. If it contains shared ancestral homologies, the classification will be less objective.

EVOLUTIONARY TAXONOMY versus CLADISTICS

According to evolutionary classification, taxa should be grouped according to the degree of morphological similarity (grade) even if this grouping includes species that do not share the same immediate common ancestor. Cladistic taxonomy supports that the taxa should be grouped strictly on phylogenetic affinity (clade) without regard for differences in morphology i.e., all taxa in the same category should have the same common ancestor. On the other hand, evolutionary taxonomy simply requires that the classification be consistent with phylogeny, even if does not depict it exactly.

There are minor differences between evolutionary taxonomy and cladistics taxonomy so far their applicability is concerned. For example, in evolutionary taxonomy, synapomorphies are used to identify and reject phylogenetic taxa and does not contribute for the creation of taxa. At the time of cladistics theory of taxonomy proposed by Hennig (1966), most major taxa of animals had already been described and reasonably well characterized. The evolutionary taxonomists used these taxa as a basis of cladistics analysis. Thus, the essence of evolutionary taxonomy lies in working out whether all the members of a taxon are descendant of common ancestor or not. Evolutionary classification differs from cladistics in following respects:

- The taxa are not delimited on the basis of holophyly but through an evaluation of similarities and differences, provided they are monophyletic.
- Plesiomorphic characters are given due consideration.
- Monophyly is not used as a method for delimiting taxa.
- Sister groups are ranked on the basis of degree of difference.
- Monophyletic taxa that have given rise to one or several ex-groups (for example, reptiles have given rise to birds and mammals) have been given due recognition in evolutionary classification.
- While constructing a scheme for evolutionary classification, evidences are also derived from biogeography, ecology and paleontology.

EVOLUTIONARY TAXONOMY versus *PHYLOGENETICS*

It is relevant to mention the differences between phylogenetic and evolutionary taxonomy which can be explained with the help of classification. According to phylogenetic taxonomy, crocodilians, dinosaurs and birds are kept in **Archosauria**. Further, dinosaurs are kept in a more restricted lineage **Dinosauria** and birds alone are grouped in Aves. On the other hand, if we follow the concept of traditional taxonomy crocodilians and dinosaurs are kept within class **Reptilia** and birds in class **Aves** (Fig. 9.3).

Under this scheme of classification, it is difficult to draw any genealogical relationship. It would not be out of place to mention that cladistics taxonomy considers monophyletic lineages which includes an ancestral form and all its descendants. Evolutionary classification differs from phenetics in following respects:

- It has been suggested that new species arise as a result of natural selection.
- The taxa are the products of evolution, which cannot be ignored in the delimitation and ranking of taxa.

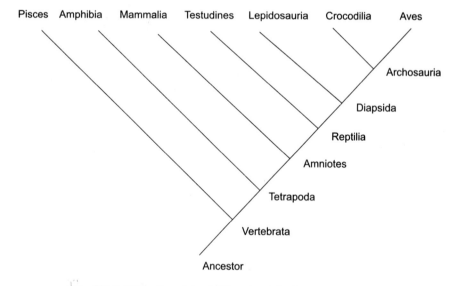

Fig. 9.3 Showing origin of different vertebrate groups

LIMITATIONS OF EVOLUTIONARY TAXONOMY

The major limitations of evolutionary taxonomy lie in the fact, that it requires an accurate judgment about how much information to use for overall similarity and branching pattern to use. The evolutionary taxonomy enables to work out relationship between different groups of animals. The evolutionary taxonomy suffers from a great deal of infirmity, being discusses herein as under:

• It is just a matter of speculation to consider overall similarity and branching pattern. This judgment is always highly subjective.

• Different criteria like **phylogeny**, divergence and **adaptation** level are used for outlining a classificatory scheme however, it is very difficult to assess as which criteria has actually been used for a particular taxon. Further, there are inconsistencies inherent in **paraphyletic** approach for example, separating class Aves (Birds) from Archosauria.

• Evolutionary classification is based on both phenetic and phylogenetic relationship. It is difficult to differentiate in a scheme of classification to work out the groups that are phenetic and phylogenetic.

• Evolutionary taxonomy suffers from ambiguity of phenetic taxonomy and its argument for excluding convergence and divergence.

• In an evolutionary scheme of classification, it is difficult to consider different criteria like differences, similarities and descent with modification. However, in cladistics taxonomy, a cladogram is initially constructed and then different taxa are assigned to different clades. The evolutionary taxonomy subserves following two important aspects:

• Study of monophyly of sorted taxa.

• Delimitation of taxa which are formed by the common principle of classification.

It follows from the above scheme that in evolutionary taxonomy, pattern of variation in individuals (phenetic taxonomy) and monophyly of the phenetically delimited provisional taxa is taken into account. Further, while making phylogenetic analysis, monophyly of taxa is examined (origin by common descent). Moreover, synapomorphies provide valid evidence for a taxon being natural, and it also enables to assess whether the characters are ancestral or derived.

SIGNIFICANCE OF EVOLUTIONARY TAXONOMY AND ITS DECLINE

Evolutionary taxonomy arose as a result of the influence of the theory of evolution on Linnaean taxonomy. The idea of translating Linnaean taxonomy into a sort of dendrogram of the animal and plant kingdoms was formulated toward the end of the 18th century, well before Darwin's book *Origin of Species* was published.

Evolutionary systematics remained the standard paradigm in paleontology and evolution until the 1980s, when it was supplanted by phenetics and especially cladistics. Both evolutionary systematics and cladistics use evolutionary trees, but differ radically the way the tree is drawn. Phylogeny in evolutionary taxonomy allows for groups to be excluded from their parent taxa (e g., dinosaurs are not considered to include birds but to have given rise to them). It assumes that ancestor-descendant relationships can be inferred from nodes on phylogenetic trees and considers paraphyletic groups to be natural and discoverable, and at times designated as ancestors (Mayr, 1942).

Evolutionary taxonomy is essentially traditional taxonomy with evolution taken into account. Evolutionary systematics also makes possible the organizing of organisms into groups (taxa) and hierarchies of such groups (classification systems) in contrast to cladistic which instead identifies clades and produces cladograms; so both systems can be correct by their own standards.

Evolutionary systematics, also known as gradistic taxonomy, is based on a combination of branching and divergence. This approach accepts the Hennigian cladistic methodology as adequate for reconstructing phylogenetic trees but retains paraphyletic groups (e g., Reptilia). Also, unlike cladistics, with it's reliance on a hypothetical most recent common ancestor that is never actually described or discovered (a missing link that is always missing), evolutionary systematics gives illustrations of the actual evolution of one species or higher taxon into another. Admittedly, evolutionary systematics suffers from a number of shortcomings. For example, the use of different criteria viz., phylogeny, divergence and adaptation level to define a particular taxa, as well as inconsistencies inherent in the paraphyletic approaches (e g., separating Aves (birds) from Archosauria (flying reptiles).

With the rise of phenetics and statistical methods, evolutionary systematics was criticized for being based on imprecise, subjective, and complicated sets of rules. It was argued that the resulting phylogenies became impossible to reproduce other than by the specialists themselves and there was a call for more repeatable and objective methods Since it was difficult to distinguish between homology and homoplasy, evolutionary systematics was replaced by cladistics and molecular phylogeny.

The purpose of evolutionary taxonomy has been to provide a meaningful classification of maximum utility through maximum use of evolutionary theory. However, it has not been able to accomplish its objective because of the fact that, the evolutionary systematists have relied upon descriptive rather on theoretical definition of terms. It seeks to classify organisms using a combination of phylogenetic relationship and overall similarity. This type of taxonomy considers taxa rather than single species, so that groups of species give rise to new groups.

Chapter 10

NUMERICAL TAXONOMY

The classification of taxonomic units into various groups by numerical methods is known as numerical taxonomy. Numerical taxonomy also known as taximetrics refers to the numerical affinity or similarity between taxonomic units and ordering of these units into taxa based on large number of quantitatively measured characters. It envisages the study of the relationship of taxa by the application of numerical similarity value to characters and rank them into categories based on the overall similarity. Sokal and Sneath (1963) referred numerical taxonomy as phenetics which stresses on the use of as many characters as possible to classify individuals based on overall similarity using morphology or other observable traits, regardless of their phylogeny or evolutionary relations using numerical measurements.

Classifications by numerical taxonomy are based on many numerically recorded characters. These may be represented numerically or may be coded in such a way that the differences between them are proportional to their dissimilarity. It is relevant to mention that similarity or resemblance refers to the attributes that an organism possesses, without reference to how these attributes arose. It is expressed as proportion of similarities and differences, for example, in existing attributes, and is called as phonetic relationship. Here the similarity refers to the resemblances both in phenotype and genotype. In this case, the classification is made on the basis of a multivariate analysis of observable differences and similarities between taxonomic groups. Classifications based on numerical taxonomy reflect degrees of similarity rather than evolutionary relationships.

Heywoods (1967) defined numerical taxonomy as the "numerical evaluation of the similarity between groups of organisms and ordering of these groups into high-ranking taxa based on these similarities". According to Sneath and Sokal (1973), numerical taxonomy aims to develop methods that are objective, explicit and repeatable both in the evaluation of taxonomic relationship and in raising taxa. It is interesting to add that phenetics include various forms of clustering and co-ordination. This helps to reduce the variation shown by the organism. Phenetic studies do not provide any idea about evolution. According to McNeill (1978) there is misconception to equate phenetics with numerical taxonomy since most phenetic taxonomy especially in case of plants is not numerically based.

FEATURES OF NUMERICAL TAXONOMY

In this scheme of classification, organisms are classified into taxonomic categories by comparing the extent of similarities they share between one another. Here the classification is made on the basis of a multivariate analysis of observable differences and similarities between taxonomic groups. Similarity between organisms is considered while classifying different groups of animals. In other words, it is the numerical evaluation of the affinities of units and ordering these units into **taxa** based on affinities and similarities between them. Here the information content of taxa is converted to numerical quantities and then studied. The term 'numerical' refers to the number, depicting number, consisting of number or expressed by number, i.e., anything expressed by number. With the help of numerical taxonomy the earth's biodiversity can be analyzed by using morphology and physiology of organisms.

It is submitted that numerical taxonomy is based on phenetic evidence as judged from individual's **phenotype** rather than from its supposed **phylogeny**. Thus phenetics is a process by which taxa are clustered together based on the number of their similarities or differences, depending upon the numerical coefficient employed. It does not take into account the plesiomorphic and apomorphic characters. It is relevant to mention that it is quite difficult, rather practically inconvenient to take into consideration all the characters at a time. Thus it is suggested to group the individual **taxa** into a number of manageable units whose characters are generally constant. As these characters are somewhat stable the grouping of such taxa is of significant value.

BASIS OF NUMERICAL TAXONOMY

Numerical taxonomy relies on the principle of natural phenetic classification. It provides methods that are objective, clear and repeated. It is based on the concept put forward by Adanson (1757) having information rich taxa having as many features as possible. In this case, every character is given equal weight and overall similarity between two taxa is a function of the similarity of many characters on which the comparison is based. Further, the taxa are constructed based on diverse characters correlations in the group studied. It entails mathematical methods to evaluate observable differences and similarities between taxonomic groups, also referred as **taximetrics**. It utilizes many equally weighted characters and employs clustering and similar algorithms to yield objective groupings. It can be extended to give phylogenetic or diagnostic systems and can be applied to many other fields.

Phenetics

It literally means, as observed in the phenotype and **phenetic** taxonomy refers to the classification observed by similarity as followed in traditional methods. Here the groupings are based on overall similarity and key characters. Phenetic taxonomy has a more restricted meaning being equated with numerical taxonomy, a method where animals with greatest number of common characters are put together. In phenetic taxonomy, similarities can be evaluated on an objective numerical basis. However phylogenetic relationship between different taxa is observed, if some or more characters used in numerical analysis have become similar due to **convergence**. This method fails when convergence is common or when only small numbers of independent characters are available.

The Adansonian Principle

Adanson (1763) gave the concept of numerical taxonomy which was later modified by Sneath (1957). Some of its basic concepts are summarized as under:

- Taxonomy is strictly an empirical science.
- Every character is of equal importance in creating natural taxa.
- Several characters in different groups of organisms are measured.
- These measurements are used to calculate the similarity between all pairs of organisms.
- The taxa should include maximum information and should be based on many possible characters.
- Overall similarity or relatedness between any two taxa is a function of the similarity of many characters in which they are being compared. Thus, affinity is estimated independently of phylogenetic considerations.
- The similarity coefficient is a number between zero (0) and one (1), where '1' is absolute identity and '0' indicates total dissimilarity.
- These similarity coefficients are then tabulated in matrix form with one coefficient for every pair of taxonomic entities.
- The individuals are then clustered in such a way that the most similar ones are grouped close together and the different ones are linked distantly. Thus distinct taxa can be constructed because of diverse character correlations in the groups under study.

• The measured traits are either converted into integers or input directly as numerical data. The data are then mathematically processed using an **algorithm** that generates a similarity or distance as the case may be. As quite large data is involved in this process, the calculations are subjected to specially developed computer programmes.

• The data matrix is graphically represented in the form of **phenogram** and principal **co-ordinate plot**. However it should be made clear that the phenogram that results from such analysis do not necessarily reflect genetic similarity or evolutionary relatedness, as the evolutionary history of several groups of animals is rather obscure.

PROCEDURES IN NUMERICAL TAXONOMY

Initially numerical taxonomy included phenetics and cladistics as well. However, later more emphasis was given to analyze the characters. In numerical taxonomy mathematical methods are used to group taxonomic units into taxa on the basis of their phenetic characteristics. A key assumption in numerical taxonomy is that all the different characters upon which the classification is based are equally important (Sneath and Sokal, 1973). Following methods are being discussed herein as under:

Distance Methods

In this method, the similarities or differences (distance) of taxa with respect to other taxa is estimated. It is based on working out with individual taxon. With the development of molecular methods like DNA-DNA hybridization; this method has become more realistic and is much in use.

Character Data Method

Taxa are determined based on shared derived characters. These methods include **parsimony**, compatibility and maximum-likelihood methods. Successive steps in numerical taxonomy (Sneath, 1962) are being discussed herein as under:

Operational Taxonomic Units (OTUs)

The basic unit of study in numerical taxonomy is OTU which represents the sample unit from which the data is collected. Thus, OTU can be single population, entire population or single species. A terminal node in phylogenetic analysis is an individual and is referred as OTU. OTU includes variety of entities like individual organisms, populations, species, genus or higher taxa classified by numerical methods. However, it is not possible to refer to each of several taxonomic groups to develop a scheme for classification. To overcome this technical difficulty, **species** as a unit is used in numerical classification.

Characters

The taxonomic characters of individuals may be studied on the basis of morphology, physiology, ecology, habitat, food, hosts, parasites, behaviour, population dynamics and geographical distribution. In numerical taxonomy these attributes may be binary, qualitative multistate or quantitative multistate. It is also suggested to consider all the characters which vary within group, instead of relying only on diagnostic characters.

Binary Characters

A binary character possesses two alternative states, e g., a particular feature in any animal may be present or may not be present at all e g., presence or absence of **respiratory siphon** in dipteran larvae. It is relevant to mention that the characters are given two codes, for example, zero (0) and one (1) or plus (+) and minus (-). The characters which are present are designated as 1 or with plus sign (+) and those traits which are absent, are represented by 0 or with minus sign (-). On the other hand, if a particular character is not observed in an individual because either the character is missing in that specimen or the specimen is damaged, as such that feature is written as not counted or not considered (NC). For example, information about one particular trait viz., presence or absence of respiratory siphon in two different species of mosquito larvae is given in Table 10.1.

Table 10.1 showing two states of characters

Characters	Taxa	
Respiratory Siphon	Present in *Culex* larva	Absent in *Anopheles* larva

Qualitative Multistate Characters

These characters may either be qualitative or quantitative. This feature can be analyzed in the following manner:

The coat color in dogs may be converted into series of binaries e g., coat color may be white, brown or black, which can be further worked out into possible alternatives, i. e., white or not white, the later leading into brown and not brown, this also includes black which is not brown.

Quantitative Multistate Characters

These characters cannot be arranged in some order; as such it is difficult to establish a reliable sequence. In these cases, the qualitative characters are converted into some other new characters. It is suggested to convert them into two state characters.

Measurement of Quantitative Characters

Numerical taxonomy considers the principal of natural phenetic classification. In this case there is no reference to evolution or to a possible common ancestor. Individuals are objectively arranged in **polythetic** groups taking into account all their features. According to Sneath and Sokal (1973) in a polythetic group, organisms are placed together that have the greatest number of shared character states and no single state is either essential to group membership or is sufficient to make an organism a member of that group.

Aschlock (1979) proposed two different terms to make the working of character analysis rather easy. Accordingly, he coined two different terms, **signifier** and **signifier state**. The word signifier refers to the character that varies, e g., eye colour and signifier state i.e., the actual colour of the eye, which may be blue or brown. Mayr (1969) used the term **character state** to avoid any confusion.

Blackith and Blackith (1968) considered around 92 characters including internal and external ones to workout the relationship between Australian and South American Orthopteroid insects. The study showed following results.

• The South American Proscopiidae seemed to split off at an early stage from the stem of **Caelifera** which is distinct from it.

• Distinction between **Blattidae** and **Mantidae** within **Dictyoptera** is less than suborder rank.

• Most Orthopteroid orders share almost half the 92 characters.

According to the scheme of classification used in numerical taxonomy, individuals are not considered on one or several uncertain characteristics but the distinguishing features of body are taken into account and the individuals are arranged into different groups based on such features.

For example, if we have to classify **bacteria**, due emphasis has to be given to study different physiological and structural features of bacteria. It is difficult to measure the **quantitative characters** in case of cell size of the colony etc. As such **qualitative characters** are commonly used in microbiology. The data obtained in such cases are represented in the form of distance matrix. The coefficient of this matrix is calculated by different methods. Further, hierarchical clustering methods like UPGMA (unweighted pair grouping method with average) are used for this purpose. It was one of the first clustering algorithms but now discarded due to its inability to group taxa that have evolved at unequal rates.

STRUCTURE OF TAXONOMIC SYSTEM

A taxonomic system can be constructed based on resemblance among operational taxonomic units (OTUs). The OTUs are grouped based on similarities. It may be recalled that the prime task of the systematist is to develop a scheme of classification which is simple and reflects the affinities of the individuals studied. Classification in numerical taxonomy is based on matrix of resemblances and it consists of various methods which are designed to disclose and summarize the structure of matrix.

Essentials of Taxonomic System

Before attempting to construct a taxonomic system, it would be meaningful to discuss the necessary steps summarized as herein as under:
- The taxonomic system should be natural
- The individuals should be classified based on correlation and **resemblances** (Q-technique) and the **correlation** among characters (R-technique).
- Since it is difficult to remember all the characteristic features of vast assemblage of individuals and higher taxa, it is convenient to group them into smaller taxa group which can be handled easily.
- Due to high constancy and mutual inter-correlation such groups enable in developing the scheme of classification more effectively. For example, if a bug is collected during a field trip, its important taxonomic features may be studied. For example, mouthparts (piercing and sucking type), **wings** (may be equal or unequal, forewings having **hemelytra**), structure of legs (walking) and segmentation of tarsi (3 or 4 segmented) and it may be parasitic or plant feeding. After further analysis of the characters, the bug may be assigned the correct systematic position. Thus, the principal aim of the taxonomic system is to develop practically convenient method involving economy of memory.
This is achieved by following:

Monothetic Methods

In this case, the features are employed to find out the gap between the taxa.

Natural Methods

Here the taxa are clustered according to their all attributes which are considered simultaneously. This method appears to be natural but it is quite difficult to estimate the matrix of affinity values.

Nested Hierarchy

In this method due emphasis is given to group large number of taxonomic units into fewer composite groups of higher rank and when these groupings are perfect, it gives the desired result. For example, if a given genus is assigned to one family then the family can be assigned to only one order and so on.

It is interesting to note that some characters are present in all the members of one group. For example, all the members of class mammalia possess dermal mammary glands and only mammals possess them. Similarly, all birds possess feathers and only birds possess them. It follows that it is easy to delimit and classify different taxa based on only one distinguishing character. The distribution pattern of organisms is depicted in Fig. 10.1 A- F). A random distribution of organisms in two-dimensional phenetic space is shown in Fig. 10.1 A and Fig. 10.1 B shows uniform distribution. In both the cases, the hierarchy can be developed with the help of arbitrary lines only. A clumped distribution that help in the development of hierarchy is observed in Fig.10.1C, here the dotted lines indicates a nested hierarchy. Hierarchical arrangements where two higher levels are clumped have been shown in Fig. 10.1 D. Further, Fig. 10.1 E shows a hierarchical intermediate level case, while; Fig. 10.1F represents parts of the distribution that could be arranged hierarchically.

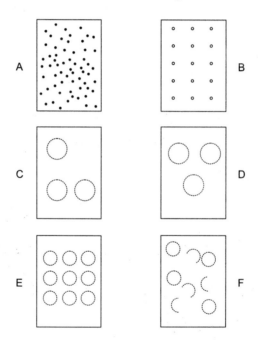

Fig. 10.1 Distribution pattern of organisms 1 (A-F)
where (A) random (B) uniform (C) nested (D) clumped
(E) hierarchical (F) parts of distribution of organisms

In random distribution of individuals, the **probability** of occurrence of an entity is independent of the occurrence of other entity in a sampling unit. It may also help in development of some hierarchical groups, since some parts are clustered by chance but there are too many intermediate types for a satisfactory hierarchy. It is relevant to mention that while making clusters of OTUs, a dividing line is drawn to emphasize the distinct gaps in the combination of characters observed. There is sufficient gap at the level of species. It is argued that the taxonomic system should be 'natural' in an empirical sense and ought to be of high predictive value. It can be arranged in the form of nested hierarchy and dividing lines are to be placed at gap in combination of characters observed. The taxa must possess the property of naturalness, as the taxonomic system should be natural.

Establishment of Resemblance and Description of Taxa

It is generally agreed that establishment of taxa based on adequate representation of characters result in natural classification. Numerical taxonomy results in the establishment of a scheme of classification based on matrix of resemblance. The structure of matrix can be designed and obtained with the help of different techniques. The coefficient of similarity between different taxa can be studied by different methods like **coefficient** of association, **correlation** and distance. It followed almost naturally that the taxa should be based on many characters, since it help in deducing overall phonetic similarity.

FREQUENCY DISTRIBUTION OF SIMILARITY COEFFICIENT

Phenetic relationship among OTUs can be observed by an assessment of frequency distribution of the similarity coefficients. It has been observed that a bimodal distribution supports the similarity values for two strains of bacteria; one peak represents the **interspecific** similarity values, while the other shows the **intraspecific** values. Further variation in similarity coefficients serves as a useful indicator of homogeneity of the OTUs.

The similarity matrix can be calculated by comparing character states of several taxa to determine the overall similarity or differences between each pair of taxa. Several characters of different taxa are analysed for differences, which are codified and assembled into data matrix. The coefficient of similarity i.e., the percentage of characters in which the taxon shows similarity or dissimilarity is calculated by comparing all character states of each taxon with those of every other taxon or OTUs. The coefficients are assembled into a similarity matrix and presented in tabular form.

Euclidean Distance

This is one type of taxonomic distance and the coefficient is used to measure the distance between two taxa in multidimensional space and is used in phenetic studies. In mathematics, the Euclidean distance or Euclidean metric is the 'ordinary' distance between two points that one would measure with a ruler, and is given by the

Pythagorean formula. By using this formula as distance, Euclidean space (or even any inner product space) becomes a metric space. The associated norm is called the Euclidean norm. The Euclidean distance between locations often represents their proximity, although this is only one possibility. It is used in field analysis for ethological studies and animal movement.

DIFFERENTIAL SHADING OF THE SIMILARITY MATRIX

In numerical taxonomy, establishment of resemblance is most important. Initially the information regarding different characters is collected which may require retrieval or it may have to be discovered entirely or partly again. Differential shading of matrix is the most common method, where the similarity coefficients are grouped into five to ten evenly spaced classes. Each of the class is represented by a degree of shading in the squares of half matrix. The highest value is generally shown darkest and the lowest values lightest. The half matrix can be seen as a pattern of different shades By careful rearrangements of squares these clusters can be more sharply defined as shown in Fig. 10.2 to Fig.10.4.

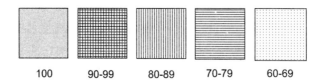

| 100 | 90-99 | 80-89 | 70-79 | 60-69 |

Fig. 10.2 Distribution of similarity coefficient in five classes
represented by different degree of shading in the square of matrix

It is relevant to mention, that the groups of related OTUs based on high similarity coefficients can be analyzed by different numerical techniques, such as elementary **cluster analysis**, clustering by single lineage (Sneath, 1957), clustering by complete lineage (Sørensen,1948), clustering by average linkage group method (Sokal and Michener,1958) central or nodal clustering (Rogers and Tanimoto, 1960). The different cluster method can then be compared and evaluated by means of **co-phenetic** correlation-coefficient (Sokal and Rohlf, 1962).

A correlation is computed between the similarity coefficients which help to draw a **dendrogram**, the so-called co-phenetic values, which are a matrix of coded similarity values extract from dendrogram. Equal weight given for every character helps to establish the taxonomic value of different traits under study. It must be made clear that the taxonomic character is a property or feature that varies from one kind of organism to another or anything that can be considered as a variable, independent of any other thing considered at the same time.

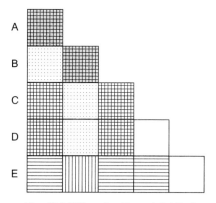

Fig. 10.3 Different pattern of distribution

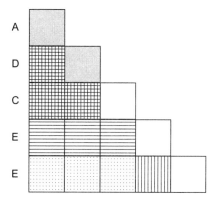

Fig. 10.4 Different pattern of distribution

If bacterial taxonomy is to be considered by employing the numerical taxonomy methods, a variety of characteristic features would have to be considered, e g., gram stain, shape of the cell, motility, size, chemistry of the cell wall, aerobic and anaerobic capacity, nutritional capacity, nutritional capabilities, **immunological** characteristics etc. The relationship is then established on similarity coefficient. If 10 features are considered then assigning a scale of 10, the different strains of bacteria are matched against this scale, e g., if the two species A and B share 8 characters out of 10, then similarity coefficient would be given by formula:

$$S_{AB} \; 8/10 \; = 0.8$$

Where S is similarity coefficient and A & B represent two species of bacteria

DATA ANALYSIS

To analyze the data for numerical taxonomy, character with states of each character are coded i.e., assigned a number ranging from 0 to 1 or with minus (-) or plus (+) sign, meaning thereby a particular character may be present or absent. For example, in insects the antennae may be **pectinate, serrated, monoliform** or **filiform**. In this case, quantitative multistate characters may be represented by several states for example, 1, 2, 3, 4 and so on. After the different characters of several taxa are worked out, the data matrix is presented in a tabular form which represent data matrix of different taxa under study.

Table 10.2 Data matrix of different taxa under study

Character (n)	Taxa (t)			
	A	B	C	D
1	+	-	+	NC
2	+	+	-	+
3	+	-	NC	+
4	NC	+	+	-
5	-	NC	+	C
6	+	+	-	+
7	NC	+	NC	-
8	+	-	-	+

Where 'n' refers to different characters and t represents taxa under study. Plus sign (+) denotes presence of a particular character, while it's absence is indicated by minus sign (–), NC=character could not be compared (Naik, 1984).

Here an arbitrary level is selected on the scale of similarity coefficients. All coefficients above this level are written down and lines or links connecting the OTU's (Operational taxonomic Units) which are represented as points indicates the relationships expressed by these coefficients. Selection of a very high coefficient of similarly as criterion for clustering would yield only a few small clusters.

DISTANCE ANALYSIS

Distance between different taxa can be analyzed by measuring the distance at the molecular level. Further, the character data can also be transformed into distance data. These data are then gathered into matrix of distances. This is further worked out by cluster analysis. It can be explained by, Wagner distance method and Other distance method.

CLUSTER ANALYSIS

A taxonomic system can be constructed based on resemblances among OTUs (Operational Taxonomic Units). Recognition of taxa is done based on similarities (affinities) by measurements of resemblances. These groups of OTUs are referred as clusters and the numerical methods employed are often referred as cluster analysis.

While clustering, objectively definable and exactly comparable limits can be drawn for all taxonomic groups within a particular study. The stability of classification based on numerical taxonomy depends upon the following:

• More information may accumulate; if the initial evaluation of resemblances is based on an adequate value and would change very little on the addition of further characters.

• New taxonomic entities may be included in subsequent studies. Although this will not change resemblance values among the old entities, application of the previous criteria for levels and number of transects that may result in new and different taxa. Similarities between taxa may be obtained from a matrix of resemblances with the help of multiple factor analysis for matrices based on correlation coefficient.

FACTOR ANALYSIS

To indicate the taxonomic relationship from a similarity expressed form of correlation coefficient, **factor analysis** is used. It is in fact a sort of statistical method used to describe the complex interrelationships among taxa in terms of the smallest number of factors. Thus the degree of similarity between an OTU and the average aspect of the taxon is represented by a factor and is given by the factor loadings. The higher the factor loading, the more typical is the OTU of taxon. In fact, each factor represents the 'type' of a taxon.

It is relevant to mention that the phylogenetic speculation does not hold well in numerical taxonomy. The numerical taxonomists keep strict separation of phylogenetic speculation from taxonomic procedure. Taxonomic relationships between taxa are evaluated purely based on the resemblances in the specimen at hand and the relationships are static or phenetic. They neither take into account the mode of origin of similarities found nor the rate at which the resemblances may have increased or decreased in the past. This is because of the following reasons:

• Until and unless methods are developed for assessing and quantifying the phylogenetic significance of character differences or affinities, the consideration of such information does not match with the objectivity and repeatability for the taxonomic process.

• There is no classificatory scheme, graphic or otherwise, which can provide information on degree of resemblances, descent and rate of evolutionary progress. Thus these aspects are treated separately, establishing the phenetic resemblances of taxonomic entities and basing the numerical classification on this alone.

In numerical taxonomy there is separation of phenetic and phyletic considerations. This is based on following reasons:

• The available fossil record is so fragmentary that the phylogeny of vast majority of taxa is unknown.

• Even when the fossil record is available this evidence has to be interpreted in a strictly phenetic manner.

It is very difficult to decide phylogeny on one set of character particularly those that are homologous (derived from a common ancestor).

Representation of Clustering

Similarity matrices have been represented by simple shading method. The dendrogram represent the most simple and convenient method to describe the results of numerical taxonomy. The dendrograms are based on phenetic evidence and show following features:

• Except their rectangular nature, they appear very much like the phylogenetic tree strictly based on phenetic evidence.

• The **abscissa** of such dendrogram does not convey any meaningful interpretation and serves to separate the OTUs.

• The **ordinate** represents the similarity coefficient in the scale ranging from zero (0) to one (1) and frequently multiplied by 100 or by 1000 in order to avoid decimal places.

• Junction points between stem along such scale represents the resemblances between the two stems at the similarity coefficient value shown at the ordinate.

• The dendrogram may be drawn differently.

• The tips point upward and the final joined stem pointing downward on the page.

• By lying on its side with the tip of the OTUs pointing to the right. In this case, the nomenclature of the classification can be listed in the same diagram.

• The similarity value along the ordinate can be divided into suitable number of equal class intervals by drawing phenon lines as class limits across the dendrogram. The number of classes into which the variation should be divided will depend upon the number of OTUs being classified.

• Once the class intervals along ordinate have been established, each class mark should be coded on a scale starting with unity at the low end (the end having the lowest similarity to value) and increasing by unit steps. Thus, starting with low classes, the highest class should be coded as 10. These values will then be proportional to the similarity value.

• A comparison of dendrograms is made simply by calculating an ordinary product moment-the correlation coefficient between the corresponding elements of the two matrices of co-phenetic values to be compared. These coefficients have been called co-phenetic correlation coefficient. The magnitude of co-phenetic correlation coefficient is a measure of extent of similarities between two dendrograms.

LIMITATIONS OF NUMERICAL TAXONOMY

Numerical taxonomy is down with convergence and **parallelism**. Phenetists attempt to classify unrelated organisms because it is based on overall morphological similarity and does not distinguish between **analogous** and **homologous** features; since analogous features are usually weighed down by large number of homologous features as supported by most evolutionary biologists. Numerical taxonomy produces a classification that reflects 'phenetic distances' i.e., degrees of similarity. Phenetic classification appears to be vague as there is more than one way of measuring similarity and the results may not be consistent.

SIGNIFICANCE OF NUMERICAL TAXONOMY

Numerical methods have enabled the exact measurement of evolutionary rates and phylogenetic analysis. Recent molecular techniques have made it possible to exactly measure the homologies of deoxyribonucleic acid (DNA) and ribonucleic acid (RNA) between different individuals. It has helped in the estimation of **hybridization** between DNAs of different taxa and therefore enabled the establishment of evolutionary relationships. Phenetics provide numerical tools, which help to examine overall pattern of variation and thus helps in the identification of discrete groups that can be classified as species. It is relevant to mention that the principles of numerical taxonomy are being applied in various fields of human activities. They marvel at the rapidity with which this knowledge is spreading throughout the biological, medical, geological and social sciences, as well as humanities.

Chapter 11

POPULATION TAXONOMY

A population is a group of individuals of a particular **species** living close together occupying a specific area at the same time, so that each member of a particular sex has an equal chance of mating with the member of opposite sex. The population may consist of 'single specie' or more than one species. The study of **population** growth and its mechanism is referred as **population dynamics.** A characteristic population has its own features like birth rate, growth rate, reproductive potential and finally death rate. The growth of the population is affected by several factors including the environmental factors; however, the favorable environmental factors do not persist for long period. The nature also keeps a check on the growth of the population. The populations of two successive generations are balanced by mortality rate which varies from species to species and is directly proportional to the fecundity.

CHARACTERISTICS OF POPULATION

Natural populations are composed of many interbreeding individuals, each with a unique combination of genes and alleles. A local interbreeding population is known as **deme** to distinguish it from a geographical population. However such populations share a common **gene pool**. The gene pool of adjacent populations may be partially or completely isolated from each other. Thus a population may be defined as the sum of individuals sharing a common gene pool. When two populations interbreed together, there is **gene flow** between them. The distribution patterns of the population show remarkable variations in the whole **biosphere.** For example, subdivision of species into discrete populations is illustrated in cases where geographical barriers restrict populations that are genetically separated in a geographical area. Even if there are no geographical barriers, distance alone tends to reduce gene flow between organisms of a species that live in different parts of same geographical range. Populations are characterized by following features:

Population Density

It refers to the number of individuals in a particular area in relation to available space and time. The population tends to remain relatively stable in spite of their remarkable power to reproduce. The fluctuation in the populations is an attribute of much significance. This variation may be due to different factors like, weather, temperature, humidity, water, light, wind, availability of food, competition (intra-specific or interspecific), **predation, parasitism, pathogens** and **mortality**. Since population is a changing entity, the rate of growth of a population is generally obtained by dividing the change by the period of time elapsed during the change. Thus the growth rate of a population is the number of organisms added to population per unit time. It is generally represented as under:

$$\Delta N/\Delta t$$

Where Δ (delta) denotes change
N is the initial number of individuals in a population
t is the time.
It is expressed by the following equation:

$$rN = dN/dt$$

Where, N is the number of individuals present in population
r is the rate of change per individuals
d is substituted for the entity that is changing

Population Dynamics

It refers to the study of changes in population size and its composition over a period of time. **Natality** (birth rate), growth rate and mortality (death rate) form the basis of population study. As we know, some animals may have great **reproductive potential** at certain part of their life cycle, which begins at the onset of sexual maturation and ends in later part of life or finally terminates with death. Birth rate may remain constant for a single parent during this time or may change. Growth is initially very rapid, then gets slower. Death rate is often high during early life, it often increases again later either abruptly (as in human beings) or gradually (as in many invertebrates).

According to Simpson (1961) populations not individuals are the units of systematics. There is great deal of variation in the populations which are dynamic systems that evolve both by progressive changes within them and by diversification and separation into multiple distinct new systems. Local population or series of populations with genetically unique features adapted to particular conditions are referred as **ecotype**.

It is relevant to mention here that the fundamental units of evolution are species which are sub-divided into discrete populations. Geographical barriers segregate species into natural populations and the distance acting as barrier alone tend to reduce gene flow between the members of a species that live in different parts of geographical range. A perusal of population as a unit requires a good understanding of the basic principles of taxonomy, without which it would not be possible to study the individuals constituting species. Initially, taxonomists considered species as an invariant unit. However it was observed that populations from different geographical range of a species show variation ranging from small to large. This has resulted in the replacement of certain animals groups - the typologically defined species by the **polytypic** species. The polytypic species is composed of different populations in time and space.

According to Harrison et al., (2004) a species that is ecologically linked to a specialized, patchy habitat may likely assume the patchy distribution of the habitat itself, with several different populations distributed at different distances from each other. This is the case, for example, for species that live in wetlands, alpine zones on mountaintops, particular soil types or forest types, springs and many other comparable situations. Individual organisms may periodically disperse from one population to another, facilitating genetic exchange between the populations. This group of different but interlinked populations, with each different population located in its own, discrete patch of habitat, is called a **metapopulation**.

There may be quite different levels of dispersal between the constituent populations of a metapopulation. For example, a large or overcrowded population patch is unlikely to be able to support much immigration from neighbouring populations; it can, however, act as a source of dispersing individuals that will move away to join other populations or create new ones. In contrast, a small population is unlikely to have a high degree of emigration; instead, it can receive a high degree of immigration. A population that requires net immigration in order to sustain itself acts as a sink. The extent of genetic exchange between source and sink populations depends, therefore, on the size of the populations, the carrying capacity of the habitats where the populations are found and the ability of individuals to move between habitats. Consequently, understanding how the patches and their constituent populations are arranged within the metapopulation and the ease with which individuals are able to move among them is key to describing the population diversity and conserving the species (Harrison et al., 2004).

Biotic Potential

It refers to the ability of an animal or population to reproduce among themselves under optimal conditions and represents the maximum reproductive power of the population (Chapman, 1928). In other words, it is the maximum reproductive capacity of an organism under optimum environmental conditions. It is often expressed as a proportional or percentage increase per year. The maximum rate at which a population can increase when resources are unlimited and environmental conditions are ideal is termed as the population's biotic potential. Each species will have a different biotic potential due to variations in its reproductive span and the frequency of reproduction. The population size is limited by different factors like **interaction, competition, parasitism, predation** and **biotic potential.** However, populations rarely reproduce at their biotic potential because of limiting factors such as disease, predation, and restricted food resources. Populations also vary in their capacity to grow.

There are always limits to population growth in nature. Populations cannot grow exponentially indefinitely. Exploding populations always reach a size limit imposed by the shortage of one or more factors such as water, space, and nutrients or by adverse conditions such as disease, drought and temperature extremes. The factors which act jointly to limit a population's growth are termed the environmental resistance. The interplay of biotic potential and density-dependent environmental resistance keeps the population in balance.

Carrying Capacity

For a given region, carrying capacity is the maximum number of individuals of a given species that an area's resources can sustain indefinitely without significantly depleting or degrading those resources. Determining the carrying capacity for most organisms is quite clear. The definition is expanded to include not degrading our cultural and social environments and not harming the physical environment in ways that would adversely affect future generations.

For populations, which grow exponentially, growth starts slowly, enters a period of rapid growth phase and then levels-off when the carrying capacity for that species has been attained. The size of the population then fluctuates slightly above or below the carrying capacity. Reproductive lag time may cause the population to overshoot the carrying capacity temporarily. Reproductive lag time is the time required for the birth rate to decline and the death rate to increase in response to resource limits. In this case, the population will suffer a crash or dieback to a lower level near the carrying capacity unless a large number of individuals can immigrate to an area with more favorable conditions. An area's carrying capacity is not static. The carrying capacity may be lowered by resource destruction and degradation, during an overshoot period or extended through technological and social changes.

INTERACTION AND VARIABILITY IN POPULATIONS

Populations of animals interact with each other and the environment in different ways. The interaction may be positive and negative amongst the members of the population. The size of the population may be limited by competition which may be inter-specific or intra-specific. The inter-specific competition occurs between the members of the same population for food, shelter and finding mate. On the other hand, intraspecific competition involves interaction between members of different population. Some of the interactions between the populations are discussed as under:

Positive Interaction

In positive interaction, the members in association serve to help each other.

Mutualism

In this type of interactions, there is an intimate association between members of the population which are in close contact with each other. Mutual relationship may be found between plants and animals. Many insect species like bees, moths and butterflies serve as an excellent pollinating agent and they also collect **nectar** in doing so. Termites feed on wood but are unable to digest **cellulose**. It is digested by a protozoan *Trichonympha* living in the gut of termite and in return, it gets food and shelter from termite. Many animals feed on seeds and fruits of different plants and help in their dispersal by dropping their excrements. Some animals like sponges, coelenterates, molluscs and worms show close association with unicellular algae like *Zoochlorellae, Chlorella* and *Zooxanthellae.*

Commensalism

It is a class of relationship between two organisms where one organism benefits but the other is neutral (there is no harm or benefit). In this type of interaction, two or more population live together without any physiological exchange and only one of them is benefited, while the other (**host**) is not affected. Some of the examples include a small tropical fish *Fierasfer* found in the cloacal cavity of **sea cucumber**, the sucker fish, *Echeneis* is found attached to the outer surface of sharks, whales and turtles with the help of sucker which is the modification of **dorsal fin**, oyster crab,

Pinnothers ostreum lives in the mantle cavity of **oysters** and *Epizoanthus* found attached to the root spicules of *Hyalonema* and shrimp of the genus *Spongicolla* in the **paragastric cavity** of the sponge *Euplectella.*

Further, cattle egrets found sitting over the back of the cattle or other livestock. As cattle, horses, and other livestock graze on the field, they cause movements that stir up various insects. As the insects are stirred up, the cattle egrets following the livestock catch the insects and feed upon them. The egrets benefit from this relationship because the livestock have helped them to find their meals, while the livestock are typically unaffected by it.

Proto-cooperation

It is a positive interaction where both the species are benefited but they can live equally well without this association. It is the ecological interaction in which both participants benefit and that is not obligatory for their survival. It is a harmonious (positive) interspecific ecological interaction. For example, the sea anemone gets attached to the molluscan shell having hermit crab (*Eupagurus*) inside. Sea anemone (*Adamsia*) gives protection to the crab and the crab in turn carries the anemone to the new feeding ground.

Negative Interaction

In negative interaction one or both the members of the association are harmed. It may include exploitation, **antibiosis** or competition.

Exploitation

In this type of association, one of the species is harmed by making direct or indirect use of the other for support, shelter or food.

Shelter

In this case, certain animals use another species for shelter. For example, the bird cuckoo never makes its own nest instead lays her eggs in the nest of crow.

Food

The animals may utilize some other animals to obtain food. This may range from **parasitism** to **predation**.

Parasitism

It refers to the phenomenon in which one species (**parasite**) benefits more than the other species (**host**). The phenomenon of parasitism is well developed in case of parasitic protozoan and helminthes. The parasite may be **ectoparasite** like *Hirudinaria* or **endoparasites** belonging to group protozoa and helminthes.

Predation

Populations of animals that feed on other individuals are known as predators. There is well developed interaction between prey and predator. When prey resources are abundant, predator number increases until the prey resources decrease in size. On the other hand, when the prey number drops, predators also show a declining trend. If ecological conditions are favorable for prey their number may increase again, thus the prey-predator cycle continues.

It is relevant to mention here that when two interacting species, e g., **prey** and **predator** evolve together, they can influence the evolution of the other. This is referred as **co-evolution**. Sometimes co-evolution results in two species that influence (both positively and negatively) from each other in a relationship referred as **symbiosis**. It would not be out of place to mention that the prey-predator relationship is very important in maintaining balance among different animal species. Adaptations such as chemical and physical defenses that are beneficial to prey, ensure that the species will survive. At the same time predators must undergo certain adaptive changes to make finding and capturing prey less difficult. Without predation, certain species of prey would drive other species to extermination through competition. Without prey, there will be no predators therefore, this relationship is vital for the sustenance of ecosystem.

Defence Mechanism

There are several ways animal avoid falling prey to a predator e g., running and camouflaging provides **cryptic** or protective coloration to the animal so that it can blend with its habitat to avoid being detected. In addition, **mimicry** and physical structures like sharp quills of porcupine also help it to escapes from enemies. Tough shell of turtles provides protective covering and certain animals including insects release **venom** or **alarm pheromone** from the specialized glands.

SOURCES OF VARIATIONS IN POPULATIONS

It is a fact universally acknowledged that evolution is not possible without variation and the later is the basis of evolution and classification. The populations may show following different types of variations:

Developmental Variations

This is also referred as **ontogenic** variations. The developmental stages help to understand the morphology, **phenotype** and **genotype** of an individual. The populations may vary depending upon the breeding season, thus the age differences alone can cause variation in the population. The average age in the population may affect variation in characteristics other than size. The complex life cycle pattern of certain groups may result in different developmental variations. For example, in **Foraminifera** the sexually reproducing generation regularly alternates with asexual generation, consequently many species are **dimorphic**. The developmental differences are significant to understand morphology and evolution.

Non-genetic Variations

Environmental factors like light, water, temperature, nutrients etc. may result in changes in phenotypic appearance. For example, in **social insects** like honey bees the larvae may develop differently depending on the type of food they get during a particular phase of development. The shells of many invertebrate species that live in crevices and cavities in hard substrata show continuous variations. Sometimes non-genetic effects of the environment are related to function and may even provide some advantage to the animal. The development of **callus** on the hand and feet of primates is the result of action of the environment over the physical tissue. It protects the animal from abrasion and bruising.

Age Variation

Sometimes the developing stages are so different from the adults that it becomes difficult to identify them. The task becomes even more difficult in case of free-flcating forms. In many species like fish, **meristic** characters e g., number of scales or fin rays varies and in insects the number of antennal segment also vary.

Seasonal Variation

Seasonal variations result in the developmental of special features in different groups of animals. In birds, the gonadotrophic cycle is correlated with seasonal variation. Many insects produce several generations in a year, e g., aphids. *Daphnia* in response to changes in temperature, turbulence and other **physico-chemical** parameters of water, show regular morphological changes through the season. Further, during certain periods of year, populations of some insects may be found in abundance. The annual fluctuations in population may be controlled by extrinsic factors like differences in physical factors. On the other hand cyclic changes in the population density are controlled by intrinsic factors.

Cyclomorphosis

In certain freshwater organisms, e g., rotifers and cladocerans regular morphological changes occur during the life cycle in relation to changes in temperature, turbulence and other physico-chemical parameters of water.

Social Variation

In social insects like bees, wasps, ants and termites, there is well organized division of labour. The honeybee colonies comprise three adult phenotypes, males (drones), reproductive females (queens) and sterile females (workers). The castes are modified females and are genetically identical and which sex would develop depends upon the type of food which is fed to the developing larvae. Larvae are initially fed with royal jelly produced by worker bees, later switching to **honey** and **pollen**. Whereas, the larva fed solely on **royal jelly** will develop into a queen bee specialized to lay eggs, while other females are sterile workers. All workers begin as nurses, tending to the eggs. After a few weeks some nurses switch to become foragers leaving the hive to search for nectar. Drones (males) are produced from unfertilized eggs and are haploid therefore represent only the DNA of the queen that laid the eggs i.e., have only a mother. Workers and queens (both female) result from fertilized eggs and therefore have both a mother and a father. A modified form of parthenogenesis controls sex differentiation. Honeybees detect and destroy diploid drones after the eggs hatch. This is a classical example of social organization in insects.

Genetic Variation

The genetic variation occurs due to the differences in genotype. It is relevant to mention that genetically controlled variations are of greater importance. For example, in human populations, the hair color, body size and fingerprints show wide variations. The genetically variant individuals of a population differ from each other on the basis of certain features associated with one or the other sex. The variation may occur due to structural differences in the sex organs. In most groups of animals sexual dimorphism occurs where males and females differ strikingly from each other. Further in some animal populations **gynandromorphs** and **intersexes** may also occur. Thus, the variation may be explained by discrete and notable differences or by a complete range of intermediate forms that overlap each other. In this case, changes in the genetic material like mutations, gene flow and recombination result in variation. The existence of genetic variability in the populations is maintained in different ways:

- There is time lag in the occurrence of mutation and its establishment or rejection by natural selection.
- The effect of chance variations in gene frequency in populations that have passed through **bottlenecks** of small size.
- The frequency dependent selection results mainly from small variations in the habitat occupied by a single population.
- Retention of both adaptive and less adaptive alleles at a locus occurs because certain gene combinations in the **heterozygous** conditions confer a particular

adaptive advantage as compared to **homozygous** conditions. In addition variation may be of following types:

Discontinuous Variation

The differences between individuals of a population may be polymorphic. In certain cases, a single gene may control the appearance of a single character e g., in insects of order **Hemiptera, Coleoptera** and **Lepidoptera.**

Continuous Variation

Each character may show different degree of variability within a single population. Further, the related species may show different degree of variability. It is suggested that no two individuals except in a population of sexually reproducing individuals are exactly alike genetically or morphologically.

Environmental Variability due to Phenotypic Plasticity

Phenotypic variation in the population occurs as the population is subjected to environmental factors. Changes in different environmental conditions force the population to develop new structures which enables the populations to adapt itself. The changed structures result in phenotypic variability produced as a result of environmental factors. Many evolutionists after Lamarck believed that such adaptive modifications become incorporated into **genome** and thus contribute to the adaptive evolution of species. It is observed that population samples drawn from different portions of geographic range on comparison show smaller or greater differences.

It is suggested that all modern species and races of organisms have existed as successful populations well adapted to their environment for thousands or millions of generations. It is believed that all of the potentially useful **mutations** would have occurred at least once during the evolutionary history of species and incorporated by **natural selection** into the gene pool. Further, the mutations could only be favored by the natural selection if the environment were changing relative to the needs of the animal. It is also relevant to mention, that a single mutation may not improve the overall adaptability of the organisms but may enable the organisms to adapt itself to a particular environment.

Genetic Variability in Populations

A population is a group of individuals of the same species that share aspects of their genetics or demography more closely with each other than with other groups of individuals of that species. The populations contain a hidden store of variability in the form of recessive **genes** and gene combinations which can be brought to the surface by inbreeding and selection. It can be explained by the following facts:

- If mutation is recessive, it is eliminated by natural selection.

- It is also submitted that some genes confer a greater adaptive advantage to the individuals when they exist in the population at low frequencies than when they are at high or medium frequencies. This results a gene to become less frequent until it is reduced to a low frequency, after which it may reach equilibrium at that frequency.

- The gene pool of a population can be enriched with hidden genetic variability. In a population, random mating may temporarily increase the frequency of a particular allele (which may be homozygous for one character). The fluctuations in the gene frequency do not produce any significant effect in large population, since any deviation in one direction is compensated by a random deviation in the opposite direction.

- The variability in genome of a population can also accumulate through heterozygote superiority. Thus individuals heterozygous for two alleles at a particular locus or for alternative chromosome or chromosome segments which contain adaptive gene clusters are adaptively superior to individuals homozygous for either of the two alleles. The superiority of heterozygotes for either genes or chromosomal segments is important because it maintains an extra delivery of genetic diversity in the population. In many natural populations the genes which form a balanced combination are located close together on a segment of chromosome and are known as **super genes**.

DIFFERENTIATION OF POPULATION

From the beginning of life, genetic differences between individuals in populations have existed and later on due to random mating these variations get organized into systems of adaptive gene combination. Changes in the environmental conditions either resulted in the loss of populations or made the populations to adapt themselves in accordance with the changed environmental conditions. It is also suggested that in response to different environmental conditions, related populations get modified variously by adapting new gene combinations. This further leads to diversity in populations. It is known that genes control the development of characters in an organism and there is inherent variation in the gene pool of a population, further sudden and abrupt changes in genes result in mutation. For example, changes in base pairs and chromosomal mutation may occur by **inversion**, rearrangement, addition or deletion of long sequences of DNA. **Transposons** or jumping genes may also result in recombination. It is also known that mutation is the primary source of variation and can spread in a population only through the effect of **diploidy**.

Population diversity may be measured in terms of the variation in genetic and morphological features that define the different populations. The diversity may also be measured in terms of the populations' demographics, such as numbers of individuals present, and the proportional representation of different age classes and sexes. However, it can be difficult to measure demography and genetics (e g., allele frequencies) for all species. Therefore, a more practical way of defining a population and measuring its diversity is by the space it occupies. Accordingly, a population is a group of individuals of the same species occupying a defined area at the same time (Hunter, 2002). The area occupied by a population is most effectively defined by the ecological boundaries that are important to the population.

The geographic range and distribution of populations (i.e., their spatial structure) represent key factors in analyzing population diversity because they give an indication of likelihood of movement of organisms between populations and subsequent genetic and demographic interchange. Similarly, an estimate of the overall population size provides a measure of the potential genetic diversity within the population; large populations usually represent larger gene pools and hence greater potential diversity.

Populations tend to be subdivided into small units referred as **gamodemes**, which remain isolated from other such gamodemes. Each gamodeme has its own gene pool and there is possibility of gene exchange between such gamodemes. Thus, within such small natural populations, favourable mutations occur and allow the individuals to adapt themselves. The selective advantage of the mutation soon spread to other gamodemes thereby increasing the effective range of useful mutant conditions. Such useful mutation taking place in a particular gamodeme or small cluster of gamodeme on the periphery of comparatively large population results in the formation of new and distinct species. Accumulation of few more mutations occurring in the favorable direction may isolate such population completely from the neighboring populations and in course of time, such species get differentiated from its neighbor. This kind of species formation is referred as **allopatric speciation**. In course of time, species so formed become entirely distinct from the parent stock; hence for speciation some sort of isolation is necessary.

Isolated populations with very low levels of interchange show high levels of genetic divergence (Hunter, 2002) and exhibit unique adaptations to the biotic and abiotic characteristics of their habitat. The genetic diversity of some groups that generally do not disperse well - such as amphibians, mollusks and some herbaceous plants - may be mostly restricted to local populations (Avise, 1990). For this reason, range retractions of species can lead to loss of local populations and the genetic diversity they hold. Loss of isolated populations along with their unique component of genetic variation is considered by some scientists to be one of the greatest but most overlooked tragedies of the biodiversity crisis (Ehrlich and Raven, 1969).

Natural selection is a process by which species adapt to their environment. It leads to evolutionary change when individuals with certain features develop greater survival or reproductive rate than other individuals in a population and pass on these heritable traits to their offspring. It must be made clear that natural selection is not possible without genetic variation. Thus, mutation and genetic recombination generate different genotypes which may or may not enable the organisms to survive and reproduce.

GENETIC DRIFT AND GENE POOL

It is also made clear that if a population is very small, it will contain only limited and random sample of total genetic variability within the whole species i.e., only a fraction of the overall **gene pool** of that species. The smaller the population, the fewer genes would be handed down to future generations. The frequency of alleles can change from generation to generation as a result of chance or random events in a small gene pool resulting in genetic drift. It causes gene pools of two isolated populations to become different as some alleles are lost and others are fixed.

Genetic drift occurs when founders establish a new population after genetic bottleneck and resulting interbreeding. A bottleneck effect is genetic drift in which population size is greatly reduced as a result of natural disaster, predation or habitat reduction. This results in severe reduction of the total genetic diversity of the original gene pool. Thus population bottleneck occur, if the population size is reduced for at least one generation as genetic drift acts more quickly to reduce genetic variation in small population.

Genetic drift has an important evolutionary consequence when new population becomes established by only a few individuals - a phenomenon known as **founder's principle**. Genetic drift can have profound effect on population. In population bottleneck, where the population suddenly contracts to a small size, genetic drift can result in sudden and dramatic changes in **allele** frequencies. Due to genetic drift, the variation between **demes** (small subpopulations) increases over time i.e., demes will diverge, especially if they become isolated. Genetic drift along with natural selection, mutation and migration is one of the basic mechanisms of evolution. Genetic drift affects the genetic makeup of the population but unlike natural selection through an entirely random process. Thus genetic drift results in:

- Loss of genetic material
- Genetic divergence between the populations, and
- Evolution

POPULATION GENETICS

A population is a group of interbreeding individuals of same species occupying a particular area. The members of population vary from one another. This variation provides raw material for natural selection to operate. It is a known fact that the **gene mutation** results in new alleles and is the source of variation within the populations. Rate of mutation varies greatly among individuals also. Mutations are sudden and abrupt changes in the chromosomal DNA. Natural selection acts on individuals not on individual genes. Sexual reproduction increases the chances of variation by reshuffling the genetic information from parents to their offspring.

MUTATION, NATURAL SELECTION AND ADAPTATION IN POPULATION

Mutation coupled with recombination is the primary source of variation and is regarded as the ultimate source of new and different genetic material appearing in the populations. It is the recombination which spreads the mutants through the population and results in the development of new gene combinations. Evolution based upon genetic variation, changes in gene frequency and natural selection acts as a guiding force of evolution; since it favors and encourages efficient gene combinations. Natural selection works on every stage in the life history of organisms. Further as the environment is subjected to constant change, the natural selection also fluctuates. Systematics thus helps to analyze wide variety of associations and interactions among vast array of populations. Positive natural selection or the tendency of beneficial traits to increase in frequency in a population is the driving force behind adaptive evolution. For a trait to undergo positive selection, it must have two characteristics. First, the trait must be beneficial; in other words, it must increase the organism's probability of surviving and reproducing. Second, the trait must be heritable so that it can be passed to an organism's offspring. Darwin and Wallace (1858) proposed that positive selection could explain the many marvelous adaptations that suit organisms to their environments and lifestyles, and this simple process remains the central explanation for all evolutionary adaptations yet today. It is submitted that the adaptations are random genetic changes which increase fitness for their carriers. When enough of these changes accumulate in one segment of the population or another population such that the differences between the two groups are distinctive enough to be catalogued then they diverge into two species.

Population taxonomy is now regarded as new systematics as it led to the proper understanding of the species concept and has given a biological approach to the study of taxonomy. It is relevant to mention that all organisms occur in nature as members of a population and specimens can be identified and classified only when they are treated as samples of natural populations. It aims to study and compare the differences between intraspecific populations and this is made possible by taking taxa as populations or aggregates of populations to study the variations and delimitation of lower taxa and categories.

With the rise of taxonomy, biological approach to taxonomy got fortified. The morphological characters of individuals were better analyzed with the help of different aspects of behavior, ecological parameters, physiology and biochemistry. Experimental analysis of molecular events, chromosomal pattern, amino acid analysis etc., also assisted taxonomy to a great extent. Further, with the development of population systematics, ideas regarding evolutionary biology are being analyzed with accuracy and this has resulted in the development of concept of population genetics. The population genetics has provided a sound basis for the development of population taxonomy.

Chapter 12

TAXONOMIC CHARACTERS

A character in terms of taxonomic studies is an observable trait or feature of a taxon, which helps to distinguish one type of individual from another. It is an expression of body organization, structure and function of an individual which can be used, measured or counted. The characters provide basic information or the primary data for taxonomic studies. Different authors have defined **character** from their own view point. Some of the definitions are given herein as under:

- According to Mayr et al., (1953) a taxonomic character is defined as "any attribute of an organism belonging to a different taxonomic category or resembles an organism belonging to same category."

- Character is a feature which varies from one kind of organism to another (Michener and Sokal, 1957a, b).

- Anything that can be considered as a variable independent of any other thing considered at the same time (Cain and Harrison, 1958).

- A character is an attribute by which a member of a taxon differs or may differ from another taxon (Mayr, 1991).

- Sokal and Michener (1958) defined taxonomic characters on the basis of which the different taxa can be worked out. Hence character in its commonest taxonomic usage is any feature of one kind of organism that differentiates it from another kind. For example, the grey body color of a particular *Drosophila* fly is a character which distinguishes it from another fly having black body color, in this case body color is a character, which occurs in two 'states' either 'grey' or 'black'.

- Ashlock (1985) has introduced a new term **'signifier'** for traits which varies from one organism to another. Thus, in insects antenna would be a signifier and if it is 'pectinate' or 'serrated' it would be a character.

A character is trait which refers to form, structure or behavior on the basis of which we can differentiate between different **taxon.** It thus helps the taxonomist to separate one taxa from the other, i.e., allows differentiation or potential differentiation from others. Further, the alternative forms of a character is referred as

character state, e g., different types of antennae in insects (**serrated, monoliform, filiform** etc.) help to differentiate one group from another. The distinguishing features of individuals are referred as **attribute**. For example, grasshoppers of family **Acrididae** possess short antennae in comparison to the members of family **Tettigonidae**, which have long antennae. In this case, the length of the antenna is its 'state' (Fig. 12.1 a, b). The alternative character state of antennae may either be long or absent. Similarly, different types of tail in fish viz., **homocercal, heterocercal** or **diphycercal** also serve as an example of character state (Fig. 12.2 a to c).

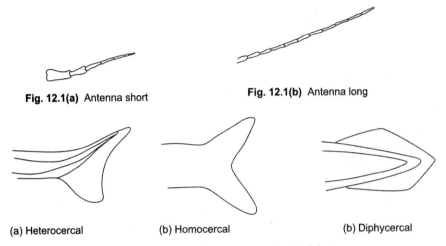

Fig. 12.1(a) Antenna short   **Fig. 12.1(b)** Antenna long

(a) Heterocercal   (b) Homocercal   (b) Diphycercal

Fig. 12.2 Different types of tail in fishes

Presence of a particular character is more important than its absence. It is on the basis of a particular feature that one can differentiate between different groups of insects or other animals. It has been shown that the appearance of a particular character is dependent on combined action of sets of genes. If some mutational changes occur in the **genome**, a particular character may be lost completely and not all characters may help in classifying individuals.

CHARACTERS AND SYSTEMATICS

In each group of organisms there are different taxonomic characters which help in the identification of individuals. Study of character for comparison is much easier when sufficient number of specimens is available. Database of characters, viz., morphological, molecular and others may be reduced to the matrix with records (data on the species) and characters recorded in the field study. Characters are the rows of the states having the adaptive significance, the most important are those characters advanced states of which are typical for the majority of species. Similarities in the advanced states may not reflect the relationship but it may arise independently because of the adaptive usefulness for the species selection. While using the matrix for the identification and phylogenetic analysis, it is necessary to consider that the significance of taxonomic character may have at least three different aspects:

Diagnostic significance

It emphasizes the importance for identification, capability to reach the final identification by using minimum number of steps.

Evolutionary significance

Study of characters is important to understand the origin and evolution of the taxon.

Phylogenetic significance

In this case, the characters can be ordered or unordered in the evolutionary row but with the minimal possibility of the controversial changes (reversions) and with the minimal variability within species. Unlike the diagnostic characters, it is not recommended to split the complex characters to simple components.

The value of character increases with its complexity like number of states, absence of the intra-species variability and reversions, accordance of the tree based characters and independent biological data like geographical distribution, life cycle and taxonomic position of host. The identification process includes the character sorting using the above mentioned criteria of significance in deliberate or concealed form. Complex characters (such as the head or tail or lateral **sensilla** patterns in nematodes combined with the shape of the sensilla) help in the identification of taxa easily.

KINDS OF TAXONOMIC CHARACTERS

Characters are used in classifying animals making keys in description and diagnosis of a taxon. The choice of characters and their states depends upon their use. Morphological characters help to measure the quantitative variations in organisms. It describes variation in size of structures or counts of meristic characters. Thus, the characters may be categorized into following two different states:

Diagnostic Characters

A taxon can be distinguished from other taxon on the basis of diagnostic characters. These characters should be clear and specific to that particular taxon and help in identifying and distinguishing one group of organisms from the other. For example, natures of wings in insects serve as an important character in distinguishing one group of insect from the other.

Key Characters

These characters are used for identification and can be observed easily in a particular taxon. Based on these characters, the taxon can be distinguished easily. It is also relevant to mention that the characters should not be too variable within taxa; otherwise it would be difficult to identify and classify individuals and moreover, the diagnosis of taxa and consideration of **keys** would become difficult. Characters which appear due to the variation of environmental factors or availability of nutrients may not be reliable than the features which are controlled by genes. For example, in insects, size may vary depending upon nutritional status. It is also important to note that the reliability of characters is not fixed between groups; it depends on the fact that how reliable a particular character is, when it is matched with another character - a phenomenon known as **congruence**.

MEASUREMENT OF CHARACTERS

It is known fact, that animals show relationships with other individuals due to similarities between them. It may be mentioned that characters have alternative states and they are transmitted from ancestor to the descendants without being altered or they can change in one species and transmitted in a new form to the descendants.

Coding of Characters

The characters and character states are represented by conventional style. It consists of list of characters used and a table showing the distribution of characters.

List of Characters
The characters are examined and the character states found in the organisms are represented in tabular form.

Distribution of Character States
The characters present in an animal are usually given in a matrix with the number corresponding to the characters is listed across the top and the individuals are mentioned down the side. The state for a species is given where the name of the organism and the character intersect.

Missing Characters

In some cases, the specimen itself or the relevant parts of some specimens are missing or the characters are inappropriate i.e., not relevant for some of the taxa. Such missing characters are represented by a question mark (?) or a dash (-). This indicates that the missing characters should not be used as an evidence for classifying the individual. It is further stated that the characters might be missing for a number of reasons including inadequate sampling of all life cycle stages or gender for some taxa or poorly preserved or fragmentary specimens (e g., as in fossils). Sometimes, the missing characters are indicated by symbol (0) and if it is present, it is indicated by number (1).

Inapplicable Characters

Certain characters are relevant for one part of the **cladogram** but are not applicable to another part. For example, variation in **molars** is relevant for grouping different kinds of mammals. Since, molars are absent in birds, dinosaurs, lizards and snakes, such characters become irrelevant or inapplicable for these groups and are indicated with dash (-) so as to distinguish them from missing characters. The absence of a feature may be a **synapomorphy** e g., absence of tails in the great ape is a derived feature and should be recorded in data matrix. Characters can be classified according to the scales of measurement and ordering of their states. It can be of following types:

Qualitative Characters

The characters which relate to form and structure are referred as qualitative characters. The qualitative characters are described in words for example, antennae may be short or long, wings may be present or absent. These characters can be of following types:

Binary Characters

The characters which occur in two states, such as the presence or absence of single feature like wings, **halteres**, **spurs** on tibiae and in wing venation, particular vein like **median** or **cubitus** which may be present or absent. These are referred as character states.

Multiseriate Characters

These characters have more than two discrete states such as antennae being **clavate** or **aristate** (Fig. 12.3 a and b).

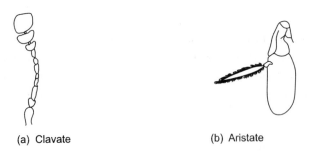

(a) Clavate   (b) Aristate

Fig. 12.3 Different types of antenna in insects

Quantitative Characters

The quantitative characters can be described using numbers which may consider the relative size or shape of structure, i.e., **morphometric** characters or a count of serially homologous traits i.e., **meristic** characters such as the number of teeth, limbs or vertebrae. In other words these characters can be counted or measured and are of following types:

Meristic Characters

Such characters represent parts that are consistent within taxa and vary discontinuously between the taxa e g., number of spots on the **elytra** of a beetle (Fig. 12.4 a, b and c).

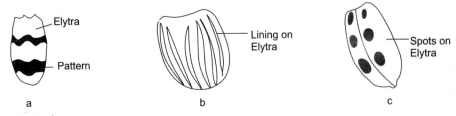

Fig. 12.4 Varying pattern on the elytra in beetle

Continuous Quantitative Those characters which vary continuously such as length and width. These characters can be divided into states by arbitrary or statistical gap coding e g., length 0-10 mm, 10-20 mm and 20-30 mm.

Ordered Characters These characters show states or conditions which can be connected by possible series of steps or leaps referred as transformations e g., lobed fins, forelegs or wings.

Unordered Characters These characters cannot be connected in any logical sequence.

HOMOLOGY

Homology is fundamental to systematics. The comparative biology aims to distinguish between homology and analogy. Characters are compared across and amongst taxa to estimate whether they are **homologous** or **analogous**. Homology refers to concept of 'similarity' or 'natural relationship'. Thus, two characters are homologous, if built on a common basic plan, but have different functions.

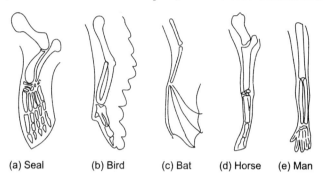

(a) Seal   (b) Bird   (c) Bat   (d) Horse   (e) Man

Fig. 12.5 Serial homology in the forelimbs of vetebrates

The paddle of dolphin, wing of a bird and bat, fore limb of horse and man are homologous as they arose by the modification of the forelimb present in the first tetrapod (Fig. 12.5 a to e). On the other hand wings of flying reptiles, birds, bat and insect all help in flying, thus apparently they have similar function but are built on different structural plan and are referred as analogous organs. It is relevant to mention here that wings arose separately in ancestors of birds and bats as well as in insects (Fig. 12.6 a to d).

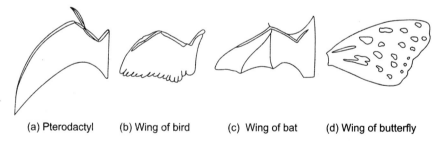

(a) Pterodactyl   (b) Wing of bird   (c) Wing of bat   (d) Wing of butterfly

Fig. 12.6 Analogous structure in different groups of animals

Tests for Homology

According to Patterson (1988) homology can be tested on the basis of following three grounds:

Similarity

It refers to appearance, development, composition and position of different structures.

Conjunction

Two homologous characters cannot occur simultaneously in the same position in the same animal.

Congruence with other homologies

In a natural scheme of classification, homologous characters should not appear to originate independently in more than one natural group.

Serial Homology

In segmented animals such as annelids and arthropods, there is linear repetition of structures on successive segments. For example, **abdominal gills** in **Ephemeroptera** (mayflies) occupy corresponding position on each abdominal segment. These structures are different from conjunction tests and form a special case of homology.

Homoplasy

It occurs when the characters are similar but are not derived from a common ancestor. In cladistics, a **homoplasy** or a **homoplastic character state** is a trait (genetic, morphological etc.) that is shared by two or more taxa due to convergence, parallelism or reversal. Homoplastic character states require extra steps to explain their distribution on a most parsimonious cladogram. Homoplasy is only recognizable when other characters imply an alternative hypothesis of grouping because in the absence of such evidence, shared features are always interpreted as similarity due to common descent. It is divided into **convergence** and **parallelism.**

Convergence

It refers for characters that come up independently from unrelated characters which have little genetic similarity. The convergent structures are fundamentally different in structure and shape and evolve in response to same adaptive response. For example, evolution of eye indicates that it has originated independently in many different species by convergence.

Parallelism

It refers to characters that arise independently from the same original character, i.e., they arise in solely related lineages, e g., **fussorial** legs in mole cricket *Gryllotalpa*. It is relevant to mention the difference between parallelism and convergence. If the phylogenetic relationship between characters is distant, then it is more likely to be referred as convergence, on the other hand parallelism refers to the close phylogenetic relationship.

WEIGHTING OF CHARACTERS

The term character weighting is referring to a procedure that is allowing a choice between conflicting hypotheses of homology and monophyly, according to certain weighting criteria, by assigning higher weight to some characters than to other conflicting characters. The most important criteria for an a priori weighting of characters (a weighting that is preceding the phylogenetic analysis) are the compatibility and structural complexity of the characters. Compatibility refers to the number of conflicting characters i.e., those characters that conflict with fewer other characters are regarded as stronger evidence, than characters that conflict with more other characters.

Weighting on the basis of structural complexity means that simple structures that might easily evolve by convergence or superficial similarities that might be based on an insufficient analysis are regarded as weaker evidence than characters that are so complex that they could hardly be non-homologous and that are so well-investigated that the proposed similarity is not just superficial. It is relevant to mention that all available weighting criteria should be used to estimate the relative weight of a character.

In order to identify and describe a particular taxon, it is essential to look for those characters which provide a correct identification and description of a particular species. This is referred as character weighting and is very helpful in describing a taxon in following manner:

- The characters can easily be observed.
- The characters show important phylogenetic correlation.
- The characters serve in the construction of a natural group.
- The important characters show a highest correlation with others in natural group. This is also referred as **posteriori weighting** or correlation weighting of characters. This weighting of characters takes several forms. This can be illustrated with the help of following illustrations:

Suppose there are ten species of a particular individual having ten distinct morphological characters. The character may refer to size, which may be large and small, the color which may be red or white or the teeth may be present or absent. Consider these ten species divided into two groups occupying the different branches of the phylogenetic tree and the morphological information is used to categorize them. There are 637 possible divisions of ten species in to two groups. Some of the possible divisions are more reasonable than the others. The three alternatives have been shown as under:

- The first subdivision produces two groups, that are each heterogeneous in all morphological characters.

• The second produces groups homogenous for one of the characters but heterogeneous for all others.

• The third produces groups homogenous for three of the ten characters (say, 1, 7 and 8).

The weighting of characters is done hypothetically. For example, in vertebrate classification, presence or absence of feather may serve as distinguishing character in separating birds from other groups of vertebrates. The presence or absence of feathers is thus analogous to characters 1, 7 and 8 in the given example. In vertebrates teeth is an important character but it is absent in birds, though *Archaeopteryx* had teeth. The presence or absence of teeth as a character is analogous to character '6' in the example. An additional character (6) almost supports the classification: one group is nearly homogeneous (all but one species are minus). The characters truly supporting the grouping may be referred as 'excellent' taxonomic characters because they each suggest the same classification. The character '6' is called a 'usable' taxonomic character because it suggests a classification very similar to that supported by the other three. On the basis of preliminary study of ten species, it would be clear that character 1, 7 and 8 are significant which help in dividing the larger group of species. Character 6 could also be added tentatively because of the 90% agreement between its distribution and that of selected three. These attributes (viz. 1, 7, 6, 9 and 8) could be used to test large number of species. If their separate use produced the same or nearly the same division of the larger group and if the divisions are geographically, ecologically or by the strata analysis appear reasonable, it can be concluded that these four characters (1, 6, 7, 8 and 9) are critical in determining a major evolutionary division.

Further subdivision of each group is done; the original group of ten species may constitute a family. The initial division would produce subgenera and so on. The character used to determine genera may not be used in the determination of subgenera because the genera are homogeneous for those characters. Thus the characters are weighted in taxonomic studies for clear understanding the concept of category however, weighting of characters is not possible at lower levels. The grouping of insects into subclasses, superorders, orders and families is based on the nature of wings and its presence or absence, its modifications, venation, nature of mouthparts, mode of development and the type of metamorphosis which they undergo.

CHARACTER ANALYSIS

The task of the systematist is to consider certain criteria for selection of characters and methods of character analysis. Taxonomic characters may be drawn from various sources. For example, morphological character analysis requires many methodological decisions and inherent assumptions at every step in the process. In fact different taxonomists have used different criteria which are being discussed herein as under:

Morphological Characters

Morphology i.e., shapes and outward appearance of a particular taxon is the usual feature for recognizing particular taxa and it is easily observable. Many morphological characters are considered relative to each other and they describe variation in size, shape or counts of serially homologous structures. Further, these quantitative traits can be codified. This helps to unravel three basic problems of character analysis viz., character state definition, delimitation and ordering.

Body Parts and Coloration
The morphological characters of the individuals draw immediate attention. Among protozoans, teeth shells, **thecal plates**, cysts and other hard parts are vital for classification in groups like **foraminifera, radiolarians, rhizopods,** flagellates and other individuals. Structure of **genitalia** in different groups of insects vary in shape and size and thus are specific to particular species. In molluscs, the **shell** pattern forms the basis of classification. In extinct groups, fossil evidences provide some basis for classification.

Sometimes variation may exist in a single species or even the two related species show genital structures, which may not be distinguishable. Varying color pattern of fishes, plumage of birds, amazing coloration in insects and reptiles serve as a basis of classification or help in differentiating them to large extent. In trichuroid worms, *Trichuris susi* and *Trichuris trichura,* the length and width of the **spicules** and shape of the body is the best diagnostic characters for the identification of trichurid males.

Developmental Characters
Life cycle stages like eggs, larvae and pupae of insects are of great help to distinguish the different species as well as help in the identification of adults also.

Genetic Characters

The expression of phenotype of an individual depends upon the expression of genes. It is also known that closely related species can mate successfully. Hybridization in such species results in production of sterile individuals. The genetic characters help in correct identification of certain individuals.

Cytological Characters

Cytological study of chromosome helps in the establishment of relationship between different groups of animals. The chromosome may be **acrocentric** (terminal position of **centromere**), **metacentric** (both arms are equal and centromere is in central position) or **holocentric** (diffuse centromere). Sometimes species and higher taxa of animals may differ by an extra ordinary number of chromosomal patterns including **Robertsonian changes**. Banding pattern and **karyotype** studies also help in tracing the phylogeny of different groups of *Equus*. Chromosomal studies also help in the comparison of closely related species which often differ markedly in chromosomal structure than in external appearance. Studies of the **giant chromosome** of certain larval dipterans help in the construction of precise **dendrograms** of related species.

Physiological Characters

Certain physiological functions of living organisms, like the ability to tolerate temperature and salinity are characteristics features of particular groups of animals. These characters can be observed in living individuals only and help to differentiate dissimilar groups of animals.

Molecular Characters

Molecular characters are becoming important in taxonomy especially in **prokaryotes** which have very simple organization. The most primitive prokaryote utilized the same **macromolecules** and made use of common metabolic pathways, which are present in the highest animals. Molecular techniques utilizing sequencing of proteins, nucleic acids and amino acids help in establishing relationship between different species.

The similarity and relationship between humans (*Homo*) and chimpanzee (*Pan*) is indicated in the similarity of **hemoglobin** and **fibinopeptides**. It has been found that typical human and chimpanzee homologs of proteins differ in only an average of two amino acids. About 30% of all human proteins are identical in sequence to the corresponding chimp protein. DNA-DNA hybridization has also helped in working the extent of similarities and differences between two species and in the construction of phylogenetic trees. Comparison of hemoglobin molecules of human beings with chimpanzee has shown that there has been little evolutionary variation in the hemoglobin molecules since the groups have diverged from each other. It has been found that the chimpanzee and human **genomes** are more than 98% identical.

It is relevant to mention one very important aspect of molecular evolution, viz., the relationship is not determined by the presence or absence of shared characters but it enables to estimate the degree of difference between the two taxa.

Behavioural Characters

Behaviour is one of the very fundamental criteria for determining taxonomic status. Comparative study of behaviour has helped in the classification of different groups of animals like birds, insects, frogs and fishes. Feeding, courtship and mating behaviour of insects and different animals groups are very important and specific to particular group, e g., social organization and nest building is at its climax in hymenopterans. Behaviour is also correlated with **isolation**, since it results the population to face an entirely new ecological conditions and the isolated populations in order to survive, develop new behavioural trends. Courtship and **acoustic** behaviour in different groups of animals also serve as a basis of taxonomy. Webbing pattern of **mites** and **caterpillars** also help in classification to some extent. Use of extraneous materials for construction of nests or larval-pupal covering provides characters at different levels of classification in insects. Different species of *Mantis* show unique egg case specific for a particular species.

Effect of Ecological Factors

Distribution of different groups of animals depends upon ecological factors. Selection of suitable habitat, ability to tolerate different ecological factors, varied distribution of organisms and resistance to predators serve as an important character during classification of animal groups. Intra and inter species competition provide enough support about the innate behavior of animals to survive and reproduce. Some species prefer dry habitat whereas, many others may be aquatic and still others are primarily terrestrial. These attributes enable organisms to survive and get well adapted to its ecological niche.

It is also observed that when two clearly related species co-exist in the same general area, they avoid fatal competition by means of species-specific **niche** characteristics. Sometimes populations of some species show ecological variations due to the effect of different ecological factors. Further animals show host-specificity because of niche specialization. Closely related species may show variation in many aspects of life cycle such as life span, **fecundity** and time or duration of breeding season. For example, the type of habitat and feeding habit can characterize each species of **Galapagos finches**, as given in the Table 12.1.

Parasites and Symbionts

Host-parasite relationship is of much interest in taxonomy. **Parasites** play an important role in interpreting relationship of higher taxa. Parasites in order to establish themselves always face an attempt of being expelled by the host. Similarity between parasites indicates relationship between the hosts also, e g., common external and internal parasites are found in human beings (*Homo sapiens*) and in the African apes (*Pan* and *Gorilla*). This shows close relationship between *Homo* and *Pan*. Termites also show very interesting relationship with protozoan parasite, *Trichonympha*.

Table 12.1 Habit and habitat of Darwin's Finches

S No.	Genus	Habitat	Chief Food Material
1.	*Geospiza*	Ground Finch	Seeds
2.	*Camarhynchus*	Tree Finch	Insects
3.	*Certhides*	Warbler Finch	Small insects

Geographical Characters

The distribution of different groups of animals provides a useful taxonomic character. Geographical characters are important at two levels in the identification of taxon. In **microtaxonomy** the **sympatric-allopatric** relationship of populations is often decisive in the determination of species status. Two sympatric populations when reproducing can never be **conspecific**. The mapping of populations, subspecies and allopatric species is indispensable in the delimitation of **polytypic** species and **super species**. In **macrotaxonomy** relationship of higher taxa is analyzed by the distribution of nearest relative (sister group) of the taxon. If the taxon has a disjunctive range, the probable cause of the range disjunction arises due to:

- Primary isolation i.e., establishment of **founder population** through dispersal.
- Secondary isolation i.e., the fragmentation of a previously continuous range by a new distributional barrier.

It is relevant to mention here, that the knowledge of earth's history and geological records helps to work out the nature of isolation. Geological events like mountain building, plate movements or climatic variation shift such as range disjunction caused by the Pleistocene ice caps or by post Pleistocene habitat shift result in different distributional pattern of population. One very interesting question arises that, how can distribution pattern help in determining the nearest relative of a higher taxon. It is generally observed that the nearest relative is found in the same geographic region or continent, e g., Australian song birds are found in Australian continent; however, there may be some exception to it also.

UNRELATED CHARACTERS

The characters may be non-independent as they sometimes represent the same underlying features then it should be discarded e g., larval color 'red' and 'hemoglobin present'. There are certain characters such as distribution and habitat which represent biogeography, ecology and evolutionary processes based on phylogeny and classification.

Character Displacement

It refers to the phenomenon where differences among similar species whose distributions overlap geographically are accentuated in regions where the species co-occur but are minimized or lost where the species distribution do not overlap. Character displacement was first explicitly explained by Brown and Wilson (1956) as two closely related species have overlapping ranges. In the parts of the ranges where one species occurs alone, the populations of that species are similar to the other species and may even be very difficult to distinguish from it. In the area of overlap, where the two species occur together, the populations are more divergent and easily distinguished i.e., they displace one another in one or more characters. The characters involved can be morphological, ecological, behavioral, or physiological and they are assumed to be genetically based.

Brown and Wilson (1956) used the term character displacement to refer to instances of both reproductive character displacement, or reinforcement of reproductive barriers and ecological character displacement driven by competition. As the term character displacement is commonly used, it generally refers to morphological differences due to competition. This pattern results from evolutionary change driven by competition among species for a limited resource (e g., food). The rationale for character displacement stems from the competitive exclusion principle, also called **Gauses' Law**, which explains that to coexist in a stable environment two competing species must differ in their respective ecological niche; without differentiation, one species will eliminate or exclude the other through competition.

CLASSIFICATION AND SOURCE OF DATA

It is often suggested that the classification drawn from different sets of observation should be the same or nearly the same. The observations drawn from different sources should indicate common evolutionary pattern i.e., the phylogeny.

SIGNIFICANCE OF CHARACTERS IN BIODIVERSITY ASSESSMENT

Biodiversity is an important feature of the ecosystem that is receiving quite attention now, as such correct identification of organisms is necessary to monitor biodiversity at any level. This requires extensive examination of specimens from different localities including the type localities. Thus, wide range of materials examined help for proper identification of specimens as the groups distinguished by just one character may not accurately reflect phylogenetic subdivisions. The significance of character analysis can be summarized herein as under:

• The taxonomic characters which evolve slowly are most useful in the recognition of higher taxa whereas, those which changes rapidly are most useful in lower taxa.

• Taxonomic characters are the expression of biology of taxon, as such its knowledge is also essential.

• The same phenotypic character may vary in value and constancy from taxon and even within single phyletic series.

• The taxonomic characters, which are subject to parallel evolution, i.e., involving loss or reduction, should be used with utmost care.

Chapter 13

THE SPECIES CONCEPT

The species is the principal unit of evolution and it refers to the concrete phenomenon of nature and various taxonomists have proposed different definitions for species. Species is an assemblage descended from one another or from a common parent and of those who resemble each other (Cuvier, 1829). According to Mayr (1957), "a species is an array of populations which are actually or potentially interbreeding and are reproductively isolated from other such arrays under natural conditions." These characters describe variation in quantitative traits like differences in size, shape or counts of serially homologous structure. The species is the smallest aggregation of populations (sexual) or lineage (asexual) diagnosable by a unique combination of characters states in comparable individuals. Thus, a species is an ancestral descendant sequence of populations evolving separately from others with its own evolutionary role and tendencies. It differs from all other species not only morphologically but also in physiological manifestations and its 'psychic behavior'. Species may also be explained as a group of individuals of common descent with certain constant specific characters in common which are represented in the nucleus of each cell by characteristics sets of chromosomes carrying **homozygous** genes causing **intrafertility** and **intersterility**.

According to Godfray and Marks (1991) the species is a fundamental unit of biology. It marks the boundary between macroevolutionary and microevolutionary processes. The component organisms of species are genetically similar to one another and are visually similar to one another. A species also occupies a particular niche, or has a unique way of life in the natural world. Finally, a species is an enduring entity through time, whose longevity transcends that of any of its component organisms. The term species refers to two distinct entities in nature, the species **taxon** and the species **category**.

The Species Taxon

It refers to concrete zoological objects having a classifiable population or group of populations. Species taxa are particular individual **biopopulations**. The common crow (*Corvus splendens splendens*) and the house cricket (*Acheta domesticus*) are species taxa. They can be described and delimited against other species taxa.

The Species Category

It contains all taxa of species rank. The word species indicates the rank in Linnaean hierarchy. The species category is the class that contains all taxa of a given species rank.

DIFFERENT TYPES OF SPECIES CONCEPT

The different types of species are discussed herein as under:

Typological Species

According to this concept, the observed diversity of the universe reflects the existence of a limited number of underlying 'universals' or types. Individuals do not have any special relation to each other as they are merely expression of the same type.

The typological species concept is sometimes referred as essentialist species concept (Mayr 1982 c). It is well accepted that morphological differences between individuals serve as criteria for species status. The variation between the individuals is the result of imperfect manifestations of the idea implicit in each species.

All taxonomists for practical purposes use morphology as the basis of defining different species; however, morphology alone cannot be the sole criterion for this proposes. The concept of species accepted by all the taxonomists is based on the following:

- The species is completely constant through time.
- It is separated from the other species by a sharp discontinuity.
- There are strict limits to the possible variation within any one species.
- The species consists of similar individuals sharing the same fundamental nature.

However, the typological species concept has been rejected on the following two grounds:

i. Individuals are frequently found in nature that are clearly **conspecific** with other individuals in spite of striking differences due to sexual dimorphism, age difference, polymorphism and other related individual variation. The members may have been described as species as soon as they are found to be the members of the same breeding population.

ii. There are **sibling species**, which hardly differ at all morphologically yet they are members of same biological species. Thus the degree of difference cannot be the sole criterion in ranking taxa as species.

Nominalistic Species

According to Darwin (1859) "I look at the term species, as one arbitrarily given for the sake of convenience to a set of individuals closely resembling each other......." The species have been invented in order that we may refer to great numbers of individuals collectively (Bessey, 1908). It is suggested that nature produces individuals and nothing more; species have no actual existence in nature, they are just mental concept. In fact species of animals are not human constructs but are formed because of evolution and variation. Members of a species taxon show some common attributes due to common heritage.

Biological Species

The biological species concept states that if two organisms can and do breed and produce fertile offspring in the wild, then they are the same species. This doesn't take into account morphological similarity. The organisms may look quite different; an example of this is sexual dimorphism. According to this concept, species have independent reality and are typified by the statistics of population of individuals. It emphasizes on population and genetic cohesion of the species and the traits which the species receives from its **gene pool**. In other words, biological species are groups of interbreeding natural populations that are reproductively isolated from other such groups. The biological method classifies organisms according to their ability to reproduce successfully and produce fertile offspring of both sexes. This method is useful for and limited to classifying organisms that are living and very closely related. It cannot be used to classify fossils (as we could not test to see if fossils can reproduce successfully) or distinguish between the relatedness of organisms that cannot reproduce successfully. The biological species has following characteristics features:

- The members of the species form reproductive community where the members are capable of perpetuating their race.
- The biological species is an ecological unit regardless of the individuals that constitute it and interacts as a unit with other species with which it shares its habitat.

• The biological species is a genetic unit having large intercommunicating gene pool, whereas the individual organisms are merely a temporary vessel holding a small portion of the content of the gene pool for a short period.

It is relevant to mention here that population of different species are reproductively isolated from each other under 'natural conditions'. Most species are divided geographically into subunits or breeding populations. It is implied that such breeding populations are actually or potentially interbreeding with each other. In other words, the segregation of the total genetic variability of nature into discrete packages known as species occurs, which are separated from each other by reproductive barriers that prevent the production of too great numbers of disharmonious incompatible gene combinations. This is the basic concept of biological species which holds species as a reproductive community. DuRietz (1930) defined species as the "smallest natural populations permanently separated from each other by a distinct discontinuity in the series of biotypes". However species hybridize readily or under domestication interbreed in captivity.

Disadvantage of Biological Species

Biological species concept suffers from following drawbacks:

Asexual or Apomictic Species It fails to meet concept of interbreeding which is the main characteristic features of biological species. In **parthenogenesis,** species is able to reproduce without sexual reproduction. Some characters of sexually reproducing species are present in the **apomictic** group.

Sibling and Cryptic Species It includes groups of similar closely related species which are reproductively isolated but morphologically identical, such species pose greatest problem to the taxonomists, as they are weakly established or not at all separated morphologically. A large number of biological species do not differ morphologically or only very slightly. They occur at lesser or greater frequency in almost all groups of organisms. Many sibling species are genetically different from each other as morphologically highly distinct species. Further, morphological species concept sometimes fails as there are numerous morphological types within a biological species, either due to different life cycle stages viz., male, female and immature stages quite different from each other than to the corresponding morphological types in different species.

Gradual Speciation Unless differentiation between individuals and the intermediate stages like **biotypes**, races, subspecies, **ecotypes** or **semi-species** are clear, the biological species concept may not hold good.

Rings of Races The species that overlap can't inbreed, this means that breeding occurs all through the area but at either end of the range, the more extreme versions of the species are so different that they are unable to interact and breed. When extreme races are **intersterile**, they fail to satisfy the definition of biological species. If interbreeding populations are taken into account, each such population irrespective of noticeable differences sometimes interbreeds with the members of own population.

Hybrid Complex The biological concept of species also fails in case of hybrid complexes. Such **hybrid complexes** reflect the sum total of species or semi species, which are linked by hybridization and such species may be either fertile or sterile.

Evolutionary Species

According to Simpson (1961), "an evolutionary species is a single lineage (an ancestral descendant sequence of populations) evolving separately from others and with its own unitary evolutionary role and tendencies ". Wiley, (1978; 1980) defined species as "an evolutionary species is a single lineage of ancestral descendant populations of organisms, which maintains its identity from other such lineages and which has its own evolutionary tendencies and historical fate".

The existence of separate evolutionary lineage means that an important factor affecting their success and survival may be the success and survival of other such lineages. The evolutionary species does not consider the criterion of interbreeding. It is observed that the various species have different roles in their **niche**. This role cannot be observed in series of dead specimens, recent or fossils, kept in a museum. Valid and sufficient evidence of separation and unity in roles can however be obtained from observations on such specimens. Morphological resemblances and differences (as reflected in populations, not individuals) are related to roles, if they are adaptive in nature.

The distinction between species may be difficult to observe in practice due to the fact that often a few features are observed to distinguish between them. It seems clear that the evolutionary species concept justifies bringing to bear ecological, behavioural, genetic and morphological evidence as a reflection on evolutionary separation. It is the evolutionary distance between the populations, which accounts for species differences and of course, without evolution, speciation is not possible at all.

The foregoing discussion makes it clear, that the different species concept overlap each other. It is difficult to estimate the populations of species in the living world. In any case the biological distinction is primary and the morphological distinction is secondary. The morphological differences are not an essential feature of a species as this can be acquired with or without the simultaneous delayed acquisition of differential morphological characters. The evolutionary species concept is also not very significant. The biological species may also appear to have similar evolutionary role.

Phylogenetic Species

This states that a species is a group of organisms at the tip of a phylogeny has the same ancestor and can be clearly distinguished from other branches. It consists of an irreducible (basal) cluster of organisms, diagnosable distinct from other such clusters and within which there is a parental pattern of ancestry and descent. It distinguishes traits (variable features within a group) from characters (invariant features within a group). Each phylogenetic species is defined by a unique character or unique combination of characters. The phylogenetic method classifies organisms according to their evolutionary predecessors and how closely related they are, using methods such as DNA profiling and sequencing. The limitation of this method is that it has a significant reliance on hypotheses of the order of evolutionary events rather than hard evidence, which is not available because fossil records are often too poor to be of good help and because the extended evolution process cannot be observed directly.

Recognition Species

It is because most inclusive population of individual biparental organisms shares common fertilization system. Thus speciation is based on the development of new mate recognition system.

Other Kinds of Species

In addition to types of species described earlier, there are some other kinds of species, being discussed herein as under:

Taxonomic Species
It refers to any taxon that has been referred as species and given a species name following the provisions of the Code:

Morphospecies
It states that if the organisms look similar then they are from the same species. Further, animals from the same species show strong morphological differences across the sexes due to sexual dimorphism. It includes features that come from convergent evolution e g., development of eye in vertebrates and also the traits derived from mimicry e g., many species of butterfly appear similar due to morphological similarities, irrespective of other attributes.

Palaeospecies

It has successive species in a single lineage.

Agamospecies

A species of organism in which sexual reproduction does not occur represent typically a collection of clones. Species that do not require fertilization of male and female gametes to produce offspring. A term used for groups of primarily or secondarily uniparental (agamic, apomiktic, asexual) organisms that are morphologically so similar that they have been classified as a single species. Examples include many bacteria and some plants and fungi. The absence of sexual reproduction means that the biological species concept cannot be applied and instead taxonomists must rely on identifying certain diagnostic traits to distinguish between closely related asexual lineages. Consequently, the boundaries of agamospecies are often difficult to define.

Phenetic Species

It represents dense regions within a hyper dimensional environmental space. In other words, it is the smallest (most homogenous) cluster that can be recognized upon some given criterion as being distinct from other clusters (Sneath and Sokal, 1973).

Chronospecies

It includes the species that we can see developing from the fossil record over time. However, it is very difficult to decide where one species could have interbred with another and where to put names.

Cohesion Species

Templeton (1989) defined the cohesion species concept as the largest delimited population that functions as an internal mechanism ensuring mutual phenotype cohesion of its members. Phenotype cohesion of a population is understood to mean maintenance of mutual similarity of its members even when the average appearance of individuals in the population changes in time and the population develops as a whole.

It is apparent that in a sexually reproducing species the exchange of genetic information between members of the population taking place during sexual reproduction functions as a mechanism capable of maintaining mutual similarity of the members of the population. In other cases, cohesion is ensured by the existence of a common species-specific mechanism of recognition of sexual partners. Thus, the cohesion species concept encompasses the biological species concept and also the ethological species concept.

Ring Species

It arises from a peculiar form of allopatric speciation that takes place when the centre of a species range is effectively unoccupied. Adjacent races all around the ring will interbreed, however, the races at the termini are so divergent, that there is no interbreeding and they can exist in sympatry without interbreeding. The species range of sea gulls represents classical example of ring species.

Umbrella Species

Certain species play a crucial role in the maintenance of ecosystem processes; beyond which additional species have no influence. The role of a greater diversity of species in this case may ensure the presence of those species fundamental to the maintenance of an ecosystem during that time. A species with large area requirements when protected offers protection to other species that share the same habitat (Simberloff, 1998). The umbrella species are typically large and require a lot of habitat. By protecting this larger area, other species are protected as well. Umbrella species generally have some features like, their biology is well known, easily observed or sampled, have large home ranges, migratory and have a long lifespan.

Thus, protecting an umbrella species should theoretically save an entire suite of sympatric species with similar habitat requirements. However, for an umbrella species to function effectively the area selected for protection under an umbrella species approach should more efficiently protect other species than an equivalent area not selected using such an approach (Caro et al., 2004) or selected on the basis of random procedures (Fleishman et al., 2001).

Keystone Species

These are the species that enrich ecosystem function in a unique and significant manner through different activities and the effect is disproportionate to their numerical abundance. Their removal initiates changes in ecosystem structure and often loss of diversity. These keystones may be habitat modifiers, keystone predators or keystone herbivores.

Indicator species

Landres et al., (1988) defined an indicator species as "an organism whose characteristics (e g., presence or absence, population density, dispersion, reproductive success) are used as an index of attributes too difficult, inconvenient, or expensive to measure for other species or environmental conditions of interest." A species that is particularly sensitive to environmental conditions, therefore can give early warning signals about ecosystem health. Because they are so sensitive a decline in indicator species health can serve as an indication of air and water pollution, soil contamination and climate change or habitat fragmentation. Indicator species are often threatened or endemic (native) species.

Biospecies

The biospecies concept takes the resemblances among organisms as epiphenomenal and regards the fundamental nature of a species to be its exclusive gene pool

(Dobzhansky, 1937; Mayr, 1942). The process of speciation involves formation of reproductive isolating mechanisms. However, Peterson (1978, 1985) is the of the opinion that parts of species look similar and participate in an exclusive genetic community, but the process of speciation does not involve isolating one gene pool from another, but rather generating new fertilization system. It is a manner in which a population of individuals identifies potential mates and establishes genetic succession (Lambert et al., 1987).

Flagship Species

Flagship species are used to attract the attention of the public (Western, 1987). According to Johnsingh and Joshua (1994) "by focusing on one flagship species and its conservation needs, large areas of habitat can be managed not only for the species in question but for other less charismatic taxa." Flagship species are frequently chosen after a species has suffered from exploitation or habitat destruction; consequently, they may be species that are sensitive to disturbances. Flagship species can earn concern for nature at a global level, as in the case of the giant panda (*Ailuropoda melanoleuca*). Successful flagship species are sometimes chosen on the basis of their dwindling population size or endangered status (Dietz et al., 1994). Flagship species may be most effective, if they are endemic to one country (Kleiman and Mallinson , 1998).

SUBSPECIES

All the species vary and this variation may be cyclic or random in the long trend of evolutionary march. The species also vary in space at any one time, within local population (or demes) and geographically between non-isolated, incompletely isolated or temporary isolated populations. In other words, when two or more populations of a given species are geographically isolated and show consistent differences between them, they are given sub-specific names.

Subspecies is a term usually applied to geographically isolated populations or groups of populations that are genetically different but are not sufficiently different to be reproductively isolated. Subspecies was considered as 'varieties' by the earlier authors. The term subspecies in zoological nomenclature gave rise to trivial nomenclature. This was later approved in the International Rules of Zoological Nomenclature. A subspecies is defined as "an aggregate of phenotypically similar populations of a species inhabiting a geographic subdivision of the range of that species and differing taxonomically from other populations of that species". Many authors have applied the term and 'subspecies' to the individual variants or **sibling species**. Based on the above discussion, following deductions may be drawn:

• A single species may consist of many local populations all of which though very similar are slightly different from each other genetically or phenotypically. A subspecies is therefore a collective category.

• Each local population is slightly different from every other local population.

• Any population is assigned the status of a subspecies on the basis of phenotype alone. This does not hold well for all the members of that population, because the members share similar traits specific to the populations.

• Subspecies should be named only if they differ taxonomically i.e., by sufficient diagnostic morphological characters.

Demerits of Subspecies Concepts

Various aspects of geographical variation cause difficulties for taxonomists. Different authors have named insignificantly dissimilar local populations as subspecies. On the other hand every subspecies has been considered as unit of evolution instead of recognizing it merely as an object to facilitate intraspecific classification.

Based on above discussion, Wilson and Brown (1953) and Inger (1961) have described the subspecies. Subspecies concept often fails on following grounds:

• Different characters show independent trends of geographic variations.

• The independent occurrence of similar or phenotypically indistinguishable populations in geographically separated area as **polytypic** subspecies.

• The occurrence of micro-geographic races within formally recognized subspecies.

• The degree of distinction between subspecies should not be arbitrary.

Temporal Subspecies

In paleontology, it is difficult to apply commonly accepted biological species definition (based on reproductive isolation) and at the same time recognize species as discrete, non-arbitrary entities. These are referred as the 'species problem' in **paleontology**. There are different opinions regarding recognition and definition of **fossil species**.

• Number of workers believe that the biological species concept fails as it cannot be applied to an evolutionary group and moreover, fossils cannot be tested for reproductive isolation.

• A fossil can be recognized on the basis of morphological differences and these differences can serve to distinguish different species.

Most paleontologists favor in applying the commonly accepted biological definition of species in an indirect and imperfect way. A paleontologist can include specimens in a fossil species by considering them to belong to a single biological species. The biological species concept defines species based on genetic and morphological criteria. This sort of morphological differences between biological species can also be used as reference for living groups closely related to the fossil group being studied. Once, the quantity of morphological variation between the recognized species is made out, it becomes difficult to place within lineage.

It we construct a three dimensional model of evolution of life with time represented by the vertical dimension and evolutionary differences between lineages by distance of horizontal separation, it would take the form of a complex tree branching towards top. The biologist for all practical purposes observes only the horizontal two-dimensional upper surface of the tree, the tops of uppermost branches. The evolutionary lineages remain obscured. Biological species tend to remain distinct because most of them belong to the lineages that have been reproductively isolated from the other lineages for a considerable period.

OTHER INTERSPECIES GROUPS

In addition to subspecies, there are certain other species categories discussed here as under:

Deme

A breeding population or local population that occurs in nature and which consists of similar organisms that interbreeds more or less at random. A small locally interbreeding group of individuals within a larger population is referred as deme. Demes are isolated reproductively from other members of their species, although the isolation may only be partial and is not necessarily permanent. Because they share a somewhat restricted gene pool members of a deme generally differ morphologically to some degree from members of other demes. Since a **deme** is a population in which there is random mating, it is also referred to as a panmictic unit. A deme may correspond with a subspecies or even in extremely exceptional cases with whole species. Some demes represent independent status for population on small islands, in separate grooves of trees, in anthills etc. Demes constitute basic population units yet they are not classified, as they do not have long evolutionary roles. Further, adjacent deme does not show any observable differentiation and they may merge, split or fade away in course of time.

Race

According to Mayr (1969) race is a subspecies and is an aggregate of phenotypically similar populations of a species, inhabiting a geographic subdivision of the range of a species and differing taxonomically from other populations of the species. As the gene flow between the races is reduced, such races are equivalent of geographic races in free-living animals. As the environment of two localities is never identical; every subspecies also tends to be an ecological race.

According to Dobzhansky (1970) races are genetically distinct Mendelian populations. They are neither individuals nor particular genotypes; they consist of individuals who differ genetically among themselves.

Templeton (1989) defined race as a subspecies (race) as a distinct evolutionary lineage within a species. This definition requires that a subspecies be genetically differentiated due to barriers to genetic exchange that have persisted for long periods of time i.e., the subspecies must have historical continuity in addition to current genetic differentiation.

Cline

A gradual change in a character or feature across the distributional range of a species or population usually correlated with an environmental or geographic transition. It refers to the gradual change in certain characteristics exhibited by members of a series of adjacent populations of organisms of the same species. In other words, it shows a continuous variation in form between members of a species having a wide variable geographical or ecological range. Huxley (1939) described cline as a character gradient. A single population may belong to different clines as it has attributes. It is not a taxonomic category. A cline is formed by a series of continuous populations in which a given character changes gradually. In clines at right angle, are the lines of equal expression of characters representing parts of phenotype referred as **isophenes**. Clines may be smooth or they may be 'step clines' with rather sudden changes of values.

Variety

It is somewhat controversial and not well treated term in taxonomy. On typological principles, each species has a fixed pattern. Any species that do not fit under this pattern is named as **variety**. Varieties have frequently been used in classification. The variety may represent the following:

- An individual variant.

- A group of such variant conceptually associated by the variation alone and not forming a population.

- A distinguishable population within a species which are analogous to or perhaps identical with the subspecies.

Morphotype

It refers to aggregate of particular variations within populations rather than to all varying organisms forming a population. The morphotypes can be defined as distinguishable sympatric and synchronic interbreeding population of a single species.

Super Species

Mayr (1931) introduced the term 'super species' which is a monophyletic group of closely related and largely or internal allopathic species. Super species represent groups of populations that seem to have passed beyond the point of potential interbreeding and have acquired separate evolutionary role.

Sylvester-Bradely (1951, 1956) has suggested that the term super species should be applied not only to contemporary geographical super species but it should also cover the ancestral species through a chronological super species lineage. It is believed that the super species are still near the critical part of speciation that of definitive isolation and it can not be quite certain whether they are really past that point and are not just below it. They are **nascent species** that will, if survive, collectively form subgenera or eventually a genus but have hardly reached the degree of divergence and expansion. They are not given any special names.

HOW DO NEW SPECIES OCCUR?

Although geographical isolation is the common means of establishing discontinuity between the populations, changes in the environmental factors and reproductive patterns also contribute to the process of speciation. Many species have specific differences from others, however, with closely related species which are very similar in appearance, the first thing is to examine the original specimen to make it sure that this newly discovered specimen does not show any resemblance or does not fit in to the variations of named species to the specimens already described. Then the species is described, which establishes a constant and clear difference between it and related species. According to Mayr (1957) there are two factors responsible for speciation:

• Development of diversity.

• The establishment of discontinuity between diverging forms.

The entire course of evolution depends upon the origin of new populations, which possess greater adaptive efficiency than their ancestors. To understand evolution, it is essential to know the divergence in population i.e., the process of **speciation**. Therefore, evolutionary divergence and differentiation of a formerly **homologous** population system into two or more distinct entities occurs.

Initially it was thought by the evolutionists that the process of speciation is similar to evolution. However, it must be made clear that evolution is not a simple process involving only the divergence of species. Further, the factors responsible for the formation of **race** and **species** are considerably different from those that result in **macroevolution** and evolution above the species level also requires a different complex processes.

It is relevant to mention, that the species may originate in different way under the influence of variety of factors. Populations belonging to different species on the other hand very rarely hybridize successfully when they meet in the wild. There are several patterns of random and non-random variations within a species. This shows that it has occurred through divergence. New populations are formed when a single deme becomes established in a new habitat. The new demes may arise by **migration** and **fragmentation**. The original population tends to expand and spread out from the center of introduction into the surrounding areas. At first the gene flow between all the portions of expanded population is continuous but later localized populations develop in most suitable habitats. When these subpopulations become partially isolated from one another, as the immediate habitat gets ecologically incapable of supporting the local population, fragmentation occurs. At this juncture the partially isolated demes may still retain the same genetic constitution as the original ancestral population but the ecological barrier between them provides a base for genetic divergence. Thus, isolation becomes a decisive factor in evolutionary divergence, without isolation, speciation is not possible.

Due to demic fragmentation, each segment of the population will be under slightly different pressures of mutation, natural selection and genetic drift. If the fragment deme occurs in a small but suitable portion of the habitat, the primary evolutionary forces will modify it from the time of origin. Occasional gene flow results in variation which is acted upon by natural selection, independent of other demes in the population. Over a period of time, each deme diverge from the other due to the impact of evolutionary force, however all the demes remain interconnected by a web of congenial gene flow.

If the impact of natural selection on variation is different in each demes' habitat or if the selection pressure is low and the populations are small, random population variation results. This **microevolution** apparently takes place independently in each deme. If the selection pressure is moderately strong for certain degree of adaptive genotypic combination, it affects a considerable number of demes, this produces variation. Genetic divergence also occurs when populations undergo genetic differentiation during a period, when fragments of original population gets completely isolated from one another in space. It may be added that before differentiation, partial isolation must be established. When fragments of original homogeneous ancestral population are isolated, variation, natural selection and genetic drift operate differentially upon the isolated population, this results in genetic divergence. Each population fragment tends to develop its own variation pattern due to evolutionary forces operating upon it. Thus each restricted population tends to develop its own adaptive features.

The natural selection and **genetic drift** operates synergistically, the selection provides a platform for the genotype to adapt, while drift produces fixed neutral and non-adaptive gene combination in the population. It is also relevant to mention the relative role of natural selection and genetic drift in the divergence of population especially at the demic and racial level. The evolutionary forces viz., natural selection and genetic drift operate in the direction of reduced population variability. Here the natural selection encourages adaptive genotypes, while drift produces fixed neutral and non-adaptive gene combinations in the population.

PROCESS OF SPECIATION

The process of speciation takes a long time and occurs on a very large geographical scale. There are following two process of species formation:

Anagensis

It is also referred as phyletic evolution, which refers to a gradual change over evolutionary time, until the resulting descendants can be recognized and accorded status of new species.

Cladogenesis

It is known as branching evolution, where an ancestral species gives rise to two or more daughter species without reticulation event. According to Mayr (1963) spatial distance is a pre-requisite for speciation. Before spatial isolation there must be natural selection pressure to diverge. This divergence or separation must be sufficient enough to prevent gene exchange with adjacent populations. Speciation occurs following divergence due to isolation. Mechanisms that produce reproductive isolation resulting in the origin of new species can be categorized as under:

Pre-mating Isolating Mechanisms
These prevent union of gametes and can occur in one of the following ways.
   • Individuals are not able to mate due to seasonal or habitat isolation; it implies that population live in different localities and do not meet.

- Individuals meet but do not mate (ethological or behavioural isolation) thus, species recognition occur but the courtship signals are ineffective.
- Individuals mate but sperm transfer does not occur this is because structural differences in genitalia prevent copulation or sperm transfer.

Post-mating Isolating Mechanisms

These produce varying degree of **hybrid** sterility. It prevents hybrid offspring from developing or breeding when mating occurs. It may occur due to following mechanisms:

Hybrid Inviability In this case, either the hybrids fail to develop or fail to reach sexual maturity. Thus, mortality acts as isolating mechanism when hybrids do not survive and die before reaching to maturity.

Hybrid Sterility Here the hybrid is viable but partially or completely sterile and thus fails to produce functional gametes. Thus, sterility of hybrids can act as isolating mechanism.

Behavioural Sterility In this case, hybrids reach sexual maturity but fail to produce appropriate courtship displays or mating behavior. Finally a reduction in fitness of the hybrid offspring can isolate two populations. This happens when the F_1 hybrid is fertile but the F_2 hybrid has lower fitness as compared to the parent.

ROLE OF REPRODUCTIVE ISOLATION IN SPECIATION

Reproductive isolation occurs when segments of same population are unable to interbreed; as a result, two reproductively isolated populations are formed. If one or both populations move out of the original habitat the species would be **allopatric**. If the two populations remain in the same area, the population is referred **sympatric**. Once reproductive isolation is established, each population follows its own evolutionary course. Allopatric speciation is the primary method of speciation.

PATTERN OF SPECIATION

The entire course of evolution depends upon the origin of new populations having better adaptive efficiency in comparison to their ancestors. Therefore the pattern of speciation is very crucial in understanding the variation and assemblage of animals. The speciation can be categorized as under:

Allopatric Speciation

Here speciation occurs due to geographical isolation. In fact geographical barrier prevents gene flow between two spatially isolated populations descended from a common ancestor. Once gene flow is prevented due to isolation then drift or natural selection leads to divergent characteristics. If the population remains separated for a long period and interacting forces of evolution operate to produce divergence, allopatric speciation results. Thus, allopatric speciation occurs when a population splits into two because of some kind of barrier (environmental and geographical) and the two groups in due course of time become differentiated into two different species due to lack of gene flow between them. It can occur quickly, if it is accompanied by **genetic bottleneck**. Most animal speciation is allopatric. It is relevant to mention here that the genetic drift is simply the process that causes small fluctuations in the frequency of an **allele** simply due to limited size. It is just by chance alone, that the frequency of alleles will fluctuate up and down in the population. This type of speciation occurs under following conditions:

- When a colony is isolated (e g., island and mainland).
- Division of range by geographical barrier.
- Isolation due to geographical distance.

It is relevant to mention that all of the above referred factors severely restrict gene flow and allow populations of a species on one side of the range to start differentiating from the species on the other side of the range. Over a period of time, the genetic differences become large enough in individuals of populations to be categorized into a new species.

Interestingly, **ring species** arise from a peculiar form of allopatric speciation that takes place when the center of a species range is effectively unoccupied. Adjacent races all around the ring will interbreed, however, the races at the termini are also so divergent that there is no interbreeding and they can exist in **sympatry** without interbreeding e g., the species range of sea gulls in the Arctic Ocean represents a classical case of ring species. The polar ice cap limits the species range of sea gulls to a ring that circumnavigate the globe. Thus, races from America freely interbreed with races from Europe. Races from Siberia freely interbreed with races from America and Caucuses. However, Central European races do not interbreed with those from Western Europe.

Parapatric Speciation

For a species to arise by parapatric speciation, the environmental conditions should show a very strong change. It is the natural selection that favors one set of characters compared to other traits favored on the other side. Animals that do not interbreed at such hybrid zones, tend to show under-dominance in fitness i.e., their progeny (hybrids) have lower fitness.

Allo-parapatric Speciation

This type of speciation follows micro-evolutionary pathways and it depends largely upon spatial isolation, time and environmental factors. The population is initially separated, but then secondarily gets exposed to subsequent parapatric speciation. The genes in **heterozygous** conditions have lower fitness than **homozygous** ones because of the strong environmental differences on either side of the contact zone.

Sympatric Speciation

In this case reproductive isolation develops without geographical barriers. The two populations remain in the same area but are not able to mate due to reproductive incompatibility. In sympatric speciation, the origin of species with the two forms occurs in the same place i.e., within a **panmictic** population. Thus, sympatric populations have the same or overlapping geographical range. Two populations are said to be reproductively isolated, if interbreeding does not occur even if they both lived in the same areas.

Thus, the crux of the species definition lies in the fact that the assemblage of populations constituting one species is reproductively isolated from other without interbreeding. However, there are certain vexed issues to be resolved, detailed herein as under:

- Can two species live sympatrically without interbreeding?
- Are the species completely isolated without gene exchange?
- If two populations are found to hybridize successfully and able to produce offspring, whether they constitute one species?

Sympatric speciation is run-away process and **sexual selection**, mutation for female choosiness arises in the population and the female opt for males with some kind of specialized trait. It is suggested that the sexual selection operates on the whole individual. It is relevant to mention, that as the genetic drift is an entirely random process, each sub-population would fix for different selection or male trait. When such phenomenon occurs, strong differential mating mechanism leads to isolation. The end result of sub-mechanism of run-away is reproductive isolation by mating preferences, the key to speciation. In other words, the divergence without geographical isolation results in differentiation of closely related species that co-inhabit a single geographical area.

Thus, the present species are the survivors of longtime of continuous evolution influenced by periods of environmental changes, where the organisms are involved in continuous struggle to survive. Only the fittest survive to produce greatest number of offspring which after many generations become distinct from its ancestors.

MECHANISM OF ISOLATION

Isolating mechanisms operates differently in allopatric speciation discussed herein as under:

- If two populations are ecologically isolated for a considerable period of time, differential mutation, drift and selection will ultimately lead to gene combinations producing reproductive isolation.
- If two formerly ecologically isolated populations become sympatric, selection will operate against any hybrids produced by accidental interbreeding; favor and reinforce all reproductive isolating mechanism.
- If the genetic control of hybrid sterility is correlated with genetic features with a positive selection value, then genetic incapability will result from any crosses.
- If sterility producing genes are neutral or even non-adaptive, they may become fixed by genetic drift in small populations.

It still remains a crucial issue as how isolating mechanism arises in allopatric populations and prevents interbreeding when populations become sympatric.

NATURAL SELECTION, ENVIRONMENT AND SPECIATION

Evolution implies changes in the gene frequency from generation to generation and as there are many genes and many individuals within most species, evolution is almost taking place. It is believed that most evolution is due to the phenomena of natural selection. There are three important factors governing the rates of evolution:

- Rate of evolution depends on the intensity of selection.
- Variability within populations, which provides raw material for evolution.
- Generation time, i.e., the faster the turnover of lifetimes, faster would be the process.

This implies the extent of genetic and morphological change per generations. However, every change does not occur through natural selection. There are directional changes in the gene frequency that arise by chance but that persist without being favoured by natural selection. However, genetic drift is much less likely in large populations where chance changes in the gene frequency tend to average out than in small populations.

It is also relevant to mention here that the rate of evolution occurs rapidly within large populations divided into many partially isolated subgroups or demes. This is because of the fact, that variety of adaptations will tend to arise in these populations and those that turn out to be of general adaptive value to the species, will spread by the small degree of gene flow that occurs among the demes. However, according to Mayr (1982 a), the evolution actually tends to proceed slowly, and well-established species evolve due to the unity of the genotype and well-integrated structure of an adaptively successful arrangement of genes.

Further, environment plays an important role in the evolution of species. Following features are attributed to this:

- Dramatic environmental changes trigger extinction as well as speciation.
- Species arises after splitting of the ancestral species, when they acquire new adaptations to the changed environmental conditions. Species get stabilized for millions of years and may disappear abruptly due to imbalance in the ecosystem.

The origin of new species from existing ones largely depends through changing interactions between the animal and the environment, this is referred as microevolution. Evolution is a hierarchical phenomenon where the changes of evolutionary importance may take place at any level viz., the genome, the individual and the species or above. Further, natural selection and genetic drift may operate at all these levels. Further, evolution at higher level that causes diagnostic differences of major taxonomic rank is referred as macroevolution. It is relevant to mention, that individuals are the units of selection at the micro-evolutionary levels and the species are the corresponding selection units at higher levels.

ESTABLISHMENT OF NEW SPECIES

One of the trivial problem remain to be discussed i.e., how to distinguish two populations as a separate species. It is argued, that the two populations cannot live sympatrically unless they belong to different species. Some differences can be observed in all allopatric populations, which may be morphological, physiological, behavioral or ecological. The extent of differences between species is reasonably constant within single evolutionary group. Most living species are quite different from one another because they have been reproductively isolated for long enough to have developed several noticeable differences. The geographical differences of allopatric populations can also be used as a helpful criterion in making distinction of species. There is different opinion with respect to species and speciation.

i. Darwin's concept of *Origin of Species* is based on the fact that:

- Species arise through development and further modifications of adaptations under the guidance of natural selection.

- Evolutionary change is a slow, steady and gradual process.

- Species are temporary stages in the continuous evolution of life.

- Species are reproductive communities having members capable of interbreeding among themselves.

• It is because of different factors, the ancestral species splits into descendant species, where interbreeding is not possible. It follows that physical and geographic isolation is a precursor to speciation.

ii. According to Eldredge and Gould (1972):

• Species arise by a process of splitting.

• This phenomenon may occur relatively quickly (around 5-50,000) years, compared with the quite longer period in the process of speciation.

iii. It has been suggested that:

• Instead of prompting adaptive changes through natural selection, environmental changes causes organism to seek familiar habitats to which they are already adapted through the phenomenon referred as **habitat tracking**. Species also remain stable due to the specialized nature of their internal and structural organization; all species are broken into local populations that are integrated into local system. There has been long period of stability, followed by abrupt extinction of species. The established pattern of speciation is given as under:

• When physical and environmental events start triggering regional species level extinction, then evolutionary changes predominantly occur through speciation.

• When environmental factors remain stable for comparatively longer period, speciation and species wise evolutionary changes are comparatively rare.

• It is the natural selection which shapes most evolutionary adaptive changes nearly simultaneously in genetically independent lineages, as speciation is triggered by extinction in turnover events.

SPECIATION VERSUS BIODIVERSITY

According to Mora et al., (2011) the diversity of life is one of the most striking aspects of our planet; hence knowing how many species inhabit earth is among the most fundamental questions in science. They have shown that the higher taxonomic classification of species (i.e., the assignment of species to phylum, class, order, family and genus) follows a consistent and predictable pattern from which the total number of species in a taxonomic group can be estimated. This approach was validated against well-known taxa, and when applied to all domains of life, it predicts ~8.7 million (± 1.3 million SE) eukaryotic species globally, of which ~2.2 million (± 0.18 million SE) are marine.

In spite of 250 years of taxonomic classification and over 1.8 million species already catalogued in a central database, it has been estimated that some 86% of existing species on earth and 91% of species in the ocean still await description.

Documentation of biodiversity is possible only when the pattern and event of speciation is properly known. In order to conserve biodiversity, it is essential to know the different species, their habit and habitat, the different ecological factors that govern their life cycle and the distribution pattern. If different factors are favorable for a particular species, it multiplies, adapts and establishes itself as a dominant species of the locality in due course of time.

According to Kopp (2010) the neutral theory of biodiversity advocates that patterns in the distribution and abundance of species do not depend on adaptive differences between species (i.e., niche differentiation) but solely on random fluctuations in population size (ecological drift), along with dispersal and speciation. In this framework, the ultimate driver of biodiversity is speciation. However, the original neutral theory made strongly simplifying assumptions about the mechanisms of speciation, which has led to some clearly unrealistic predictions.

Chapter 14

ZOOLOGICAL TYPES

In taxonomic procedures, it often becomes difficult to identify the taxa carrying the name of particular genus. The type specimen serves as a standard of reference, helps to recognize and define particular taxa. The standard specimen based on whose characters a particular genus is described is referred as 'type' specimen. It in fact represents the kind of specimen or taxon. The method of describing a particular genus is referred as **typification**. Type specimens are the objective standard of reference for the application of zoological names. When a new species or subspecies is described, the specimen(s) on which the author based his description become the type(s) (Article 72.1). In this way names are linked to type specimens, which can be referred to later if there is doubt over the interpretation of that name. The type specimens thus become the name bearer of the species and used to define a species group. However, there has been tremendous modification in the principles and practice of typification since the time of Linnaeus. It is known that the species consists of variable populations. Further, variation is one of the characteristic features of the species and the variability of population can not be described by just one single specimen. Further, no single specimen can sub-serve to describe the entire range of variation of a population. Type specimens are selected by the taxonomists to designate individuals which typify the described species.

TYPE SPECIES

Each genus must have a type species as the description of a genus is usually based primarily on its type species, modified and expanded by the features of other included species. The generic name is permanently associated with the name bearing type of its species. A type species describes the essential features of the genus to which it belongs. If closer examination of the type species shows that it belongs to a pre-existing genus or disassociated from the original type species and given a new generic name; the old generic name becomes the **synonym** and is abandoned, unless there is pressing need to make an exception decided case by case, by filing a petition to the International Commission of Zoological Nomenclature.

TYPE GENUS

A type genus helps to provide the name of a **family** or **subfamily**. As with type species, the type **genus** is not necessarily the most described representative but is usually earlier described largest or best known genus. It is not uncommon for the name of family to be based upon the name of a type genus which has passed into synonymy; the family name does not need to be changed in such cases.

FEATURES OF TYPE SPECIMEN

The type specimen has been assigned following distinguishing features:
- A type specimen serves the basis of description.
- A type is a zoological specimen; it does not represent the name of the genus.
- In most cases, the type is capable of showing only part of the basic traits of the species.
- A type serves as a standard for comparison with newly discovered specimens of a particular species.
- The type of a species is a specimen superficially chosen as such by the original taxonomist or by the later author.
- The type is often the principal source of the features given in the description. However, 'type' does not serve as the exclusive or primary basis of description. Simpson (1961) has introduced a term *hypodigm* for entire sample of specimens known at a given time.

PRINCIPLES OF TYPIFICATION

Article 61 of the International Code of Zoological Nomenclature provides for typification of taxa. Some of the essential concepts of typification are outlined herein as under:
- The valid name of a taxon is determined from the name bearing type (s) of a particular locality.

• Each nominal taxon in the family genus or species group has actually or potentially a name bearing type.

• The type specimen helps to serve as standard reference for the application of the name it bears.

• A new species is described by examining all the specimens including the type specimen. The type does not serve as an exclusive basis of description.

• The objectivity provided by the typification is continuous through the hierarchy of names. It extends in ascending order from the species group to the family group.

• Once a type specimen is designated, it can not be changed even by the author of the taxon except by the exercise of plenary powers of the Commission (Article- 79) through the designation of a **neotype** (Article-75).

CORRECTING MISIDENTIFICATION OF TYPES

Sometimes original authors of new taxa misidentify type species of new genera and the type genera of new taxa of the family group. The errors committed by the original authors can be corrected under the provisions of Article 41, 49, 65 b, 67 I and 70 b. Before initiating any correction, it is to be made clear that the type of a taxon is not a 'name' but a 'zoological entity'. The species or genus is the specific individual which the original author had come across while designating it as 'type' specimen and not the name, which the present author might have erroneously attached to the specimen in question. There are provisions under Article 65 a, 70 a; for correct identification of type specimens. In case of misidentification, the matter is referred to the Commission under the provisions of Article 41. The Commission then brings out certain rulings to maintain the stability and continuity. It is to be made clear that during re-examination of the referred case, if the Commission comes to conclusion that the type actually belongs to a different species then the Commission under the plenary powers vested with it under Article-79, suppresses the original type and designate a neotype, which conforms to the accepted concept of the species.

DESIGNATION OF TYPE SPECIMEN

The type specimen serves as official reference for the description of a species. A taxon of the species group (species or subspecies) can have only a single type. This can either be a specimen designated or marked as the type referred as **holotype** by the original author at the time of publication of original description or designated from type series i.e., **lectotype** or neotype.

KINDS OF TYPE SPECIMENS

Different names have been applied to type specimens such as holotype, lectotype, neotype, syntype, plastotype and topotype etc. The type specimens have been classified under following two main groups:

Primary Type

It is the original specimen used by the author in describing or illustrating a new species. It includes the following:

Holotype

When a single specimen is designated in the original description as the type for a species, it is referred as holotype. In other words, it is the single specimen chosen as the type by the original author at the time of publication of the original description. The specimen which is designated as holotype must bear the identification label having following information:

- Date of collection.
- Locality from where the specimen has been collected.
- Name of the person who has collected the specimens.
- Sex of the specimens (male or female).
- Size and developmental stages of the collected specimens-mature (adult) or immature (larval, nymphal or pupal) stages.
- Host from which the specimen has been collected and in case of parasites, whether it is **ectoparasite** or **endoparasite**.
- The geographical locality of the collected specimens and the height or depth of the locality from where the specimen has been collected.
- The fossil specimens should also have a record of zoological sphere.
- The name of the museum, where the holotype has been deposited and the serial number assigned to it.

Thus, the type specimen should bear as much as possible information for taxonomic purposes. Further, following points must be given due consideration in describing type.

- Type designation should always be completed before publication.
- Type designation should be clear and without any confusion.
- It is not advisable to change or remove type label.
- Only an expert should carry out revision work of type study.
- The details regarding deposition of 'type' with some museum should also be recorded. Before publication, type of species should not be disturbed.

Neotype

A specimen selected as type, subsequent to the original description in cases where the original type gets badly damaged, is known as **neotype**. If the holotype, lectotype or syntype gets lost due to damage to the specimen or otherwise, there is provision under Article 75 of the ICZN, for designation of neotypes. The neotype should be designated only in connection with the revision work and only in exceptional circumstances, when a neotype is necessary in the interest of stability of nomenclature (Article 75 b). A neotype is designated for a species whose name is not in general use, either as a valid name or as a synonym or for any name, not in use viz., *nomen dubium*. The neotype should be designated strictly in accordance to the provisions laid down in Article 75 of ICZN.

Further, the Commission has the power under Article 79, to suppress an existing type in the interest of stability of nomenclature and to designate a neotype to conform to the traditional usage of a name. This is done in order to avoid confusion in typification. In case the type designation appears to be doubtful, then only the taxonomist can submit proposal to the Commission for suppression of type and request for designation of neotype in accordance with the traditional use of name. However, such suppression and designation of neotype should be carried out only under exceptional cases. It implies from this ruling that the taxonomist can not go against the provisions of the Code at his own.

Secondary Type

These include the following:

Syntype

When several specimens are used to define species as a standard for comparison, they are referred as **syntype**. Further, syntype may include specimens labeled as **cotypes** or 'types'; these are the specimens without any identification labels. One of two or more specimen cited by the author of the species, when no holotype was designated or it is any one of two or more specimen originally designated as types.

Isotype

A duplicate specimen of the holotype collected at the same place and time as the holotype.

Lectotype

A lectotype is a specimen originally designated as a syntype but later on marked as definitive type specimen for a species. The lectotype may be designated from syntype to become the name bearer of a nominal species. If a nominal species has no holotype, one of the syntype may be designated under the provisions of Article 74 and

Recommendations 74A, 74B, 74C, 74D and 74E. Lectotype for all practical purposes are holotype selected by the reviser of that species, chosen from the syntype series. The provision of the Code however cautions the selection of lectotype. It should not be done just to add a type specimen to the collection. If the description of a species is clearly based on a particular specimen, that specimen should be made the lectotype. Due consideration must be given in the selection of lectotype. Sometimes, a single specimen is designated as 'type' without citing such a specimen as 'the type' in published description.

Paralectotype
Any additional specimen from among a set of syntypes after a lectotype has been designated from among them; these are not name-bearing types.

Hapantotype
In Protistans, the type consists of two or more specimens of directly related individuals representing distinct stages in the 'life cycle' and these are collectively treated as a single entity.

Tertiary Type
These include paratype or **allotype**.

Paratype
Every specimen in a type series other than holotype or lectotype is known as **paratype**. It refers to a specimen other than holotype formally designated by the author of a species as having been used in the description of the species. In fact paratype belongs to the same population from which holotype has been derived. For convenience the author considers other species as paratype. In case of any doubt, the final comparison is made with the paratype. Thus, when an author designates a holotype (Article 73.1), then other specimens of the type series are referred as paratype. It is relevant to mention, that the holotype remains as the name bearer but the paratype serves to express more precisely the author's concept of species.

Specimens of Special Origin
It includes **topotype** which is not a part of original type material but that has been collected at the type locality. If the holotype and paratype has been lost, then topotype is considered for describing the species.

SIGNIFICANCE OF TYPE SPECIMENS

It is relevant to mention that the validity of species name can rest upon the availability of original specimens or if the type cannot be found or it has never existed upon the clarity of description. A complex nomenclature is applied to the different sorts of type specimens. Type specimen serves as basis for description and definition of taxa. It is the principal source of the features given in the description. The type also serves for later examination to add some further data to the original description. The provisions of Code help in declaration of a name bearing type by a process of fixation. In taxonomy, typification serves following important functions:

- Type carries the name along with it wherever the type may go.

- Type also subserves as a source of data not shown by the type. In most cases, the type is capable of showing only part of the basic attributes of that species.

- The type is used as a standard for comparison with the newly discovered specimens of that species.

It is relevant to mention that the determination of the exact organism designated by a particular name usually requires more than the mere reading of the description or the definition of the taxon to which the name applies. New forms, which may have become known since the description was written, may differ in characteristics not originally considered; or later workers may discover, by inspection of the original material, that the original author inadvertently confused two or more forms. No description can be guaranteed to be exhaustive for all time. Validation of the use of a name requires examination of the original specimen. It must, therefore, be unambiguously designated.

At one time an author might have taken his description from a series of specimens or partly (or even wholly) from other authors descriptions or figures, as Linnaeus often did. Much of the controversy over the validity of certain names in current use, especially those dating from the late 18th century, stems from the difficulty in determining the identity of the material used by the original authors. In modern practice, a single type specimen must be designated for a new species or subspecies name. The type should always be placed in a reliable public institution, where it can be properly cared for and made available to taxonomists.

Chapter 15

TAXONOMIC CATEGORIES

Taxonomy refers to the rules and methods of biological nomenclature, whereas, classification involves grouping of organisms into a formal system based on similarities in external and internal anatomy, physiological functions, genetic makeup or evolutionary history. Classification thus determines the method of organizing the diversity of life on earth according to presumptive phylogenetic relationship. It is a dynamic process that reflects the very nature of organisms which are subject of modification and change over generations during the process of evolution.

Classification is based on grouping of animals into those having similar characteristic features, i.e., **phena** and placing them in definite named taxonomic **categories**, i.e., taxa (pl.- taxon). To make it clear, if we come across a collection of animals, it would be sorted on the basis of similarity and the specimens would be known as phena which would be assigned a place at a particular level of taxonomic categories, i.e., taxa in the hierarchical scheme of classification. In other words, each category may be defined as subdivision of the next higher category.

Linnaeus (1758) created a hierarchical classification system using only six taxonomic categories: kingdom, class, order, genus, species and variety. These categories are based on shared physical characteristics or phenotypes within each group. Linnaeus presented a system for organizing different groups of animals into a series of nested groups based on their similarities and differences. This system uses clearly defined shared characteristics to classify organisms into each group represented by these different levels. It is interesting to mention that organisms are not classified as individual but as groups of organisms.

Since the kingdom as an assemblage of different individuals was used by Linnaeus to separate plant and animals, thus only three categories **class, order** and **genus** were considered to be the **higher categories**. Further, as the number of species known at the time of Linnaeus was relatively small, the categories used by him could easily manage the different animal groups known at that time. Later on, **phylum** and

family were also added. The scheme became **kingdom,** phylum, class, order, family and genus. Further, categories like **subphylum, subclass, superorder, suborder, superfamily, subfamily** and **subgenus** have been added. Of this, kingdom, class, order and genus are referred as higher categories.

In addition to ordering organisms, taxonomists give a scientific name to new species, to distinguish it from similar organisms. This is the standard way of knowing the same organisms the entire world over and also to avoid confusion. Thus, taxonomic categories utilize the knowledge of similarity and shared properties of taxon.

HIERARCHY

Classification is the practical application of taxonomic principles so that the species can be identified and placed at an appropriate place in hierarchical system. This hierarchical system moves upward from a base containing large number of organisms with very specific characteristics. This base taxon is a part of large taxon. Each successive set of taxon is characterized by a broader set of characteristics. In zoological classification a hierarchical system of categories is used in a sequence of decreasing size. Terms such as species, genus, family and order designate categories. A category is an abstract term. The taxa are assigned an appropriate categorical rank. It is an attempt to express similarity i.e., individuals having similar characters and common descent are grouped together. The most closely related species are combined into genera, groups of related genera into subfamilies and families, these into orders, classes and phyla. In this procedure, there is drastic difference between the species, taxon and the higher taxa. Intrinsic characters define higher taxa e g., mammalia is a class consisting of animals having 'dermal mammary glands'.

Detailed subdivisions within these primary categories, outlined as above are often employed. Following is the detailed account of different levels of category used (the ranks italicized are obligatory):

- *Kingdom*
- Subkingdom
- *Phylum*
- Subphylum
- Superclass
- *Class*
- Subclass
- Infraclass
- Cohort

- Superorder
- *Order*
- Suborder
- Superfamily
- *Family*
- Subfamily
- Tribe
- *Genus*
- Subgenus
- *Species*
- Subspecies

TYPES OF CATEGORY

The categories designate a rank or level in hierarchical scheme of classification to which groups are assigned. According to Blackwelder (1962) category are arbitrarily chosen before groups are assigned to them. The categories may be of following types:

- The lower category i.e., the species category.
- The infraspecific category.
- The Higher categories i.e., category above species level.

The Lower Category

The lower category includes species which is the fundamental unit of taxonomy and is defined as a group of organisms which can interbreed freely to produce fertile offspring but cannot breed with individuals from other species (i.e., reproductively isolated from other such group). Species is the true biological unit which represents the relationship between two populations. It may or may not be morphologically distinct. In a given locality a species is usually separated from other sympatric species by a complete reproductive gap. When two populations co-exist together they do not interbreed. Species is a stable and distinct population of interbreeding forms of life characterized by similar morphology.

Species form a genetic unit with a shared gene pool; it also entails an ecological unit as a member of a species interacting with similar environmental factors. Species may also be considered as populations (not an individual) allowing for variation and evolutionary changes. Taxonomic research at the species level is known as microtaxonomy whereas; macrotaxonomy involves the process of classifying genera. The species category is characterized by singularity, distinctness and differences. A prefix is used to indicate a ranking of lesser importance. The prefix 'super' indicates a rank above and the prefix 'sub' indicates a rank below, e g., superclass, class, subclass and infraclass.

Table 15.1 Category level of honeybees and humans

Taxonomic Category	Taxon	
	Honeybee	Human Beings
Kingdom	Animalia	Animalia
Phylum	Arthropoda	Chordata
Class	Insecta	Mammalia
Order	Hymenoptera	Primates
Family	Apidae	Hominidae
Genus	*Apis*	*Homo*
species	*indica*	*sapiens*

The Infraspecies Category

Any category below the rank of species is an infraspecific category. It includes subspecies (abbreviated as sub sp.), variety (var.), **subvariety** (sub var.), **forma** (f.), **forma biologica**, **forma specialis** and **individuum**. Initially, binomial nomenclature served the purpose of describing genus and species, however due to advancement in the field of biosystematics, the species is considered to be represented by the initial binomial and all subsequently described variants attributed to that binomial. When more than one infraspecific taxon is present, a trinomial nomenclature is adapted. For example, hooded crow is known as *Corvus corvus cornix.* In extreme cases, the nomenclature may be polynomial e g., *Amaranthus hybridus* ssp. *Cruentus* var. *paniculatus.*

In biological classification, **subspecies** (abbreviated '**subsp.**' or '**ssp.**'; *plural:* 'subspecies') is either a taxonomic rank subordinate to species or a taxonomic unit in that rank. A subspecies cannot be recognized in isolation: a species will either be recognized as having no subspecies at all or two or more (including any that are extinct), never just one. While the scientific name of species is a binomen, the scientific name of a subspecies is a trinomen - a binomen followed by a subspecific name. A tigers' binomen is *Panthera tigris*, so for a Sumatran tiger the trinomen is *Panthera tigris sumatrae.*

Polytypic Species

It has been found, that many species consists of populations which are different from each other. Such populations which differ from previously named populations of a species are known as new sub-species. The species that contain two or more subspecies are known as polytypic species and the species which are not subdivided into subspecies are known as **monotypic species.**

The polytypic species are composed of allopatric or allochronic populations which differ from one another. Recognition of subspecies often creates problems, as all populations of sexually reproducing organisms differ slightly. Closely related species with similar ecological requirements occasionally replace geographically and it is difficult to decide whether they are complete species or constitute subspecies.

Variety

Linnaeus had proposed the term 'varietas' as the only subdivision of the species. However, due to technical reasons the taxonomists are not using the term variety now. Different views have been expressed regarding variety:

- It has been considered as a morphological variant of the species, irrespective of distribution.

- It is a morphological variation of the species having its own geographical distribution.

- It may be variation of species due to acquiring a particular color or habit phase.

- It is a morphological variant of species which share a particular habitat also shared by one or more varieties of same species.

Subspecies

The subspecies is the lowest category recognized in the International Code of Zoological Nomenclature (Article 45e). A subspecies is a major morphological variation of species that has distinct geographical distribution other than occupied by other subspecies of the same species. The term subspecies refers to the variation between species in sense of 'geographic race'. Subspecies is a category quite different from species. A subspecies can be defined as 'an aggregate of phenotypically similar populations of a species inhabiting a geographic subdivision of the range of that species and differing taxonomically from other populations of that species'.

In other words, subspecies can be defined as 'species within species'. It is relevant to mention that subspecies may consist of many local populations which are quite similar, yet are slightly different from each other genetically and phenotypically. A subspecies is therefore a collective category. Subspecies are to be named only when they differ by enough morphological characters. In brief, a subspecies is a small group of a primary species that has made evolutionary change in order to adapt to geographical difference that is not present, where the majority of the primary species inhabit.

It should also be kept in mind, that phenotypic differences alone are not sufficient to assign populations the status of subspecies. For example, breeding range of two species may overlap geographically, but the breeding range of two sub-species does not. If two discrete breeding populations co-exist together in the same locality they are designated as full species except in cases of circular overlap. It is also worth mentioning that when two subspecies occupy the same locality, intermediate or hybrid populations may combine to represent the characters of both species.

Super Species

A super species is a monophyletic group of closely related largely or entirely allopatric species. The components of super species were originally designated as semi species.

Race

The term is applied to local populations within subspecies. Since the environment of localities is never identical; every subspecies is an ecological race. However, some populations differ in their ecological requirements without acquiring taxonomically significant differences.

The Higher Category

A higher category is class into which are placed all the higher taxa that are ranked at the same level in a hierarchical classification. The category selected for a given taxon indicates its rank in the hierarchy. It is relevant to mention here that taxa are based on zoological realities, while categories are based on concepts. In Linnaean hierarchy there is no difference between species category and higher categories.

The taxa ranked in these highest categories represent the main branches of the phylogenetic tree and they are characterized by a basic structural pattern laid down early in the evolutionary history. The taxa in the higher categories are definable in terms of basic structural pattern but except for certain highly specialized groups, such as order Siphonaptera (fleas), Chiroptera (bats) and Impennes (Penguins), the higher taxa are not primarily or even predominantly distinguished by special adaptations. The taxa contained in the higher categories are in most cases widely distributed in space and time.

The International Code of Zoological Nomenclature entails many procedures used in dealing with next higher taxonomic categories. The Code itself rules out for formation of categories up to the superfamily level. The recommendations of the International Commission attached to the Code cover the entire range of higher categories. The rules are almost parallel to those for species. A new species must be assigned at least to the next higher category (the genus). Thus all the genera can be assigned to families, orders and so on.

The type concept also extends to the definition of higher categories, when a new genus is proposed; a type species designation must accompany the original description. The type species thus becomes the name bearer of the genus. Similarly a family must have a type genus and so on. The higher categories signify affinities among group of species. The different higher categories are discussed herein as under:

Genus

The generic name refers to the genus (taxon), having a group of species that are fairly closely related. A genus may contain a single species or a monophyletic group. For example, insect of genus *Dysdercus* may include different species, such as *Dysdercus koenigii* and *Dysdercus cingulatus*. Likewise, fish genus *Labeo* may include specimens like *Labeo rohita, Labeo calbasu, Labeo dero* etc. The generic name e g., *Dysdercus* or *Labeo* can be used alone to describe a genus; whereas, the species name is always used with the generic name, it does not convey any sense if written alone. Similarly genus *Paramecium* has different species e g., *Paramecium aurelia, Paramecium bursaria, Paramecium caudatum, Paramecium multimicronucleatum, Paramecium nephridium*, etc.

According to Mayr (1969) "a genus is the obligatory taxonomic category directly above that of the species in Linnaean hierarchy." A genus is a monophyletic group consisting of one or more species that are separated from other generic taxa by a decided gap (behaviour, morphology or some other characteristics). It is recommended that the size of the gap be in inverse ratio to the size of the taxon. In other words, the gap needed to recognize that group as a separate genus the smaller the species group, the larger the gap needed to recognize it.

Features of Genus The genus has the following attributes:

- A genus should not be expanded unnecessarily.

- A genus must be different from other taxa of the same rank.

- A genus taxon is a phylogenetic unit; it implies that the included species are descended from the nearest common ancestors. In other words, all descendants of an ancestral taxon are grouped together.

- Each genus must have a type species. The generic name is permanently associated with the type species. If this specimen is assigned to another genus, the name of the genus linked to it becomes a junior **synonym** and the remaining taxa in invalid genus need to be reassessed.

- A genus may also be considered as an ecological unit consisting of species adapted for a particular mode of life.

- The genus should occupy a distinct niche or adaptive zone. A genus niche is obviously broader than the species niche but both kinds of niche exist.

- The genus should show distinctness with regards to evolutionary criteria. The genus must possess considerable genetic identity. Species of same genus sometimes are able to produce hybrids.

Sometimes, many genera are divided into subgenera (singular-**subgenus**). In zoological nomenclature, the genus and species names are sometimes identical, e g., *Gorilla gorilla* and *Rattus rattus*. When a newly discovered species is named and described, every attempt is made to find a reasonable assignment in existing higher categories. If it is impossible, then new categories are sometimes proposed. Further, a higher category may be monotypic.

Family

Genera are grouped into families. A family can be defined as a taxonomic category which includes single genus or a group of genera of common phylogenetic origin, which is separated from other families by a decided gap i.e., by important characteristic differences like behaviour, morphology or some other characters. For example, family **Canidae** includes dogs, wolves and foxes whereas, the cats like animals ranging from domestic cats, tiger, lion and cheetah come under family **Felidae**. The size of the gap is in inverse ratio to the size of the family. The name of the family always ends with the letters '*ae*', but is written in normal face.

Features of Family Like genus, the family also possesses certain adaptive characters which help it to fit in particular adaptive zone. The family represents different animals but all of them possess certain features specific to that particular family. In Linnaean hierarchical classification, family has not been recognized as category but significantly most of his genera have been elevated to the rank of families. Linnaean generic concept was compatible with modern family concept with the difference between the genus and the family being minor. Linnaeus did not consider having an intermediate category between genus and order. With the increase in animal types after 1758, naturalists universally applied the family concept to designate an intermediate level between genus and order. It is relevant to mention that if the adaptive zones are distinct, the wider is the gaps between the families i.e., the members show much difference further, there is increase in number of recognized taxa above the family rank. The family has following attributes:

• Family is older than genera they contain and more often have a worldwide distribution.

• Family is a very useful category.

• At a given locality, the various families are generally distinct.

• Families are often found to form different distinct groups on each continent and types that bridge the gap between them.

Some of the important family and the major group they represent are being given herein as under:

• Apidae   - Honey bees
• Picidae   - Woodpecker
• Canidae   - Dogs, foxes and wolves
• Bovidae   - Cattle, sheep, goat
• Hominidae   - Human beings

Subfamily A subfamily is a division of family.

Superfamily In case of insects, superfamily category is also used.

Order

Families are grouped into orders, whose individuals may vary in many ways. For example, order **Carnivora** includes cats, dogs and weasels, which differ from each other. The members of the family may vary significantly in morphology e g., sheep, cattle, deer, goat, pig coming under order **Artiodactyla** differ morphologically to a great extent. An order may further be subdivided into superorder, suborder and infraorder. Many phyla include distinct orders whereas; in some phyla they are less well known than the classes.

Class

The class is a major division of animal kingdom and it includes superclass and infraclass. The class forms the basis on which most fossil study is based. A class is generally subdivision of a phylum. For example, class Mammalia includes lion, cat, tiger, cow and sheep etc. The vertebrates can be grouped under five classes as mentioned below:

- Pisces   - Cold blooded animals having gills for gaseous exchange.
- Amphibia - Cold blooded animals laying eggs near water.
- Reptiles   - Cold blooded animals laying eggs on dry land.
- Aves   - Warm blooded animals having wings and lay eggs.
- Mammalia - Warm blooded animals having mammary glands to nourish the young ones.

Phylum

In biological taxonomy a phylum is a taxon in the rank below kingdom and above class sharing certain common traits. At this level, animals are grouped on the basis of similarities in basic body plan or organization. All the members of the phylum share this pattern; however; there might be structural difference during the course of evolution either due to **divergence** or **convergence**. It is assumed that all the members of the phylum have a common ancestry. Thus, classes are grouped into phyla which are further grouped into kingdom. The chordates possess following distinguishing traits:

- Pharyngeal gill slits to obtain oxygen.
- Notochord is present at some stages in their life cycle.
- Single hollow nerve chord on the top of notochord.

It is on the basis of above features that there animals have been grouped into chordates and non-chordates.

Subphylum/Super phylum
The Phylum may be divided into subphylum and superphylum.

Kingdom
It is the highest taxonomic category in Linnaean hierarchy. The Kingdom Animalia is divided into two subkingdoms, viz., **Protozoa** and **Metazoa**. Protozoa includes solitary or colonial acellular microscopic animals. At this level, organisms are considered on the basis of following:

- Cellular organization.
- Mode of nutrition.

ORIGIN OF CATEGORIES

According to Darwin (1859) the natural system is genealogical (family lineage and evolutionary history) in its arrangement like a pedigree but the degree of unification which the different groups have undergone have to be expressed by ranking them under genera, subfamilies, families, sections, orders and classes. During the course of geological history, the descendants of an aberrant species may evolve into a different genus, the genus into a different family and so on. Higher categorical rank of taxa evolves through evolution from a lower rank, whereas, lower does not evolve through subdivision of higher categories. Higher categories are poorly defined. They are quite artificial and help in the sorting of species. A large number of species, each having several characters of taxonomic importance may be divided into many quite different systems of groups. For each higher category there are criteria for characterization that are comparative and relate to the member taxa. Sometimes, few characters are singled out on the basis of priori grounds as being of greater evolutionary significance than the rest and the division is based on these characters in preference to others.

DEMERITS OF HIGHER CATEGORIES GROUPING

The higher categories assigned just on the basis of single character for grouping of species does not work well. For example,

- On the basis of shell microstructure or morphology of **lophophore**, it is difficult to divide Brachiopod families into orders.
- Classification of trilobites into orders on the basis of facial suture proved failure.

- In classification of bivalve, two schemes were followed. The configuration of the shell and the hinge teeth of the shell served as an important character (several teeth types are known and include cardinal teeth, pseudocardinal teeth and lateral teeth, these are typically missing or very much reduced).

Thus two different classifications are developed. At present several characters including dentition and gill type are considered. Modern taxonomists use single character approach in higher-category taxonomy, for example:

- Feathers in birds represent a definitive taxonomic character.
- Structure of pelvic girdle is good criteria to classify dinosaur families into two major orders.

From the foregoing discussion it would be clear that the classification involves grouping or ordering of millions of species into some well-defined system based on morphological characters and similarities. This diversity is immense, therefore; for organizing this vast assemblage of animal life some set of ordering or categorization is needed. Since, life first appeared on earth around 3.5 billion years ago, variety of animals have evolved. Many of these organisms have become extinct, the remaining have developed into recent day fauna and flora of the world. Extinction and diversification is continuing and taxonomists are frequently encountering fluctuations that may affect the way organisms are classified.

Chapter 16

MERISTIC AND NON-MERISTIC DATA ANALYSIS

Taxonomic characters may be defined as an attribute by which a member of a **taxon** differs from an individual or group of another taxon and these traits help in the establishment of relationship between different groups of organisms. Further greater the number of **attributes**, greater would be the parameters for comparison. For example, **antenna** between two different groups of insects may be **pectinate** or **plumose**. In fishes, the anal fin rays may be 8, 9 or 10, it may have smooth skin or it may be covered with scales. The morphology of any individual has been the primary source of taxonomic studies. It is relevant to mention that the vast majority of classification for identification is based on morphological characters.

MORPHOLOGICAL VARIABILITY IN NATURAL POPULATION

Variability exists in natural population of a given species not only on the basis of morphology, physiology and behavior as well. The cause for such variability is attributed at genetic level as well as due to environmental factors. It is crucial to understand this variability as it provides complete information about the population of the individuals that consists of species under study. **Morphometrics** is the measurement and analysis of body form and acts as a powerful tool when used in the context of adequate biological knowledge.

MERISTIC CHARACTERS

A **meristic** character refers to measurable or any countable structure of the body. Meristic approach to taxonomy includes certain characters like body length, fin length, head length, eye diameter, **fin rays, gill rakers, dermal denticles, claspers, scales** of fish or number of segments in the **antennae** and **tarsi** of insects or the ratios between such measurements. Examples of internal meristic characters include **vertebrae, pyloric caeca, pterygiophores, branchiostegial rays** etc.

Some meristic characters are measured empirically, such as percent moisture and age of the individuals. Other meristic characters are written in the form of codified words. For example, if a particular sex has to be mentioned, terms like M- for male and F- for females may be used. The meristic characters must be recorded very carefully in the meristic table. It is relevant to mention here, that counting of vertebrae is a discontinuous variable, whereas, measurement of body parts is treated as continuous variable. Genera, species groups, species, subspecies, population or groups with in species and individuals can be differentiated on the basis of meristic characters.

Fish Meristic Characters

Fish meristic characters include features that can be counted, such as scales, **fins** and fin rays and vertebrae. Fish **morphometry** can be analyzed by linear measurement of body or its parts. In addition to meristic characters, certain other characters like dermal denticles, claspers, fin spines and counts of serially repeated elements such as vertebrae, teeth and rays in the **spiral valve** may be used to distinguish between different fish stock Criteria for some of the morphometric analysis in fish are detailed herein as under (Fig.16.1).

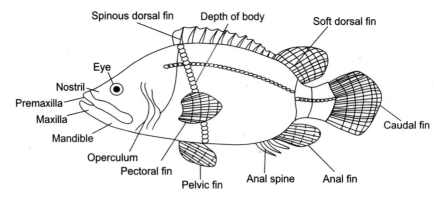

Fig. 16.1 Structure of fish showing different body parts

Standard Length
It refers to the distance between the tip of the **snout** to the base of caudal fin in fish.

Total Length
It is measured from the tip of the snout to the tip of the caudal fin.

Length of the Head
It refers to the distance in a straight line between anterior most part of the upper lip and posterior most edge of the **opercular** bone. In case of **percoids**, it includes opercular spine also.

Snout Length
It is measured from the tip of the snout to the anterior margin of the orbit.

Upper Jaw Length
It refers to the greatest length of the upper jaw.

Lower Jaw Length
It refers to the greatest length of the lower jaw.

Eye Diameter
It is measured from margin to margin of the bony orbit.

Girth Length
It is measured covering the circumference of the body at its deepest part.

Height or Depth of the Body
It is measured along the vertical line at the deepest part.

Length of the Caudal Peduncle
It is measured from the posterior base of the caudal fin to the base of the **caudal fin**

Fin Formula
The fin ray formula represents the count of fin rays in different fins of the body of a fish species. Dorsal fin is represented by the symbol D, anal fin by A, Pectoral fin by P, Ventral fin by V and caudal fin by C.

- Spines are the rays that are single shafted. Spine counts are expressed by **roman** numerals (like I, II, III IV), while the soft ray counts are expressed by **arabic** numerals (1, 2, 3, 4). Soft rays are the rays, which are bilaterally paired, segmented and flexible.
- In cases where both spines and soft rays co-exist, the roman and arabic numerals are used to give their counts in the same order.
- Sometimes capital and small numerals are used to distinguish between strong and weak spines. This can be explained as under:
  - i. D II 18 means that the dorsal fin has two strong spines and eighteen soft rays.
  - ii. D ii 18 means that the dorsal fin has two weak spines and eighteen soft rays.
  - iii. D_1 I 4 means that the first dorsal fin has one strong spine and four soft rays.

iv. D_2 5 means that the second dorsal fin has five soft rays alone.

Fin formula of fish, *Clarias batrachus* can be expressed as under:

$$D\ 62\text{-}76,\ P\ 1/8\text{-}11,\ V_6,\ A\ 45\text{-}58,\ C\ 15\text{-}17$$

Scale Counts

Scales show varying pattern, which may be regularly arranged, smooth, moderate/small in size. Some examples of scale counts are given herein as under:

- L 1 30 means, that the number of scales on the lateral line of trunk is 30.
- Ltr 16 indicates that the number of scales on the lateral line of trunk is 16.
- $85^5\ l_7$ indicates that there are eighty-five scales in the lateral line, five scales above it and seven scales below it, which can be distinguished by their perforated nature.

Tiwari et al., (2004) have compared various body proportion ratios between **diploid** and **triploid** fish. Many significant differences of morphometric ratio between two **ploidy** groups have been observed. However, only the ratio between standard length and body length (SL/BD) was found to be a precise indicator of triplody in Indian catfish, *Heteropneustes fossilis*. This ratio is useful in segregating triploids in aquaculture farm practices.

Hermida et al., (2002) while working in *Gasterosteus aculeatus* have observed that phenotypic variability was smaller in laboratory-reared fish than in the wild fish.

Pawson and Ellis (2005) have also discussed different methods for the stock identity of elasmobranchs in the northeast Atlantic Ocean in relation to assessment and management. This signifies the importance of meristic traits in segregating fish populations.

The inferior pharyngeal teeth counts and five morphometric characters of diagnostic significance were opted for statistically evaluating the populations of *Schizothorax richardsonii* and *S. plagiostomus* inhabiting the river Alaknanda in Srinagar (Garhwal). The relationship of AFL with A-CFBL (anal to caudal fin base length) in the species under investigation was found to be an important feature. The relatively longer AFL and smaller A-CFBL in *S. plagiostomus* served to distinguish it from *S. richardsonii* (Pandey and Nautiyal, 1997).

Various morphometric measurements have also been studied in case of other animals also. For example, in house lizard *Hemidactylus*, **snout** vent length (SVL), tail length (TL), and inter limb distance (ILD), and hind limb length (HLL) from groin of straightened limb to the tip of longest toe. The different meristic character studied in lizards, are number of **supralabial** and **infralabial** (SUPL) and (INFL), number of longitudinal rows of tubercles on the dorsum and number of adhesive lamellae on the fifth finger and fourth toe (Vences et al., 2004).

Insect Meristic Characters

Many insect species can be differentiated based on differences in general shape and appearance. Daly (1985) has given a very detailed account regarding applications and interpretations of morphometrics in insect biology. Different types of antennae, legs, nature of mouth parts and wings serve as an important taxonomic feature (Fig. 16.2 to 16.5). For example, the number of segments in antenna or tarsi varies as described with the help of following examples:

- In longicorn beetles, the number of segments in antennae is generally constant being 11.

- In *Prionus imbricornia,* the females have 18 segments in antennae, while 18 to 20 segments are present in the antennae of males.

- In *Forficula auricularia* (earwig), the number of segment in antennae differs; different species may possess 11, 12, 13 and 14 antennal segments, which are treated as normal.

- In families of insect order **Orthoptera**, the number of tarsal joints differs. For example, in **Forficulidae, Acrididae** and **Gryllidae** the number of tarsal joints is 3, in **Blattidae, Mantidae** and **Phasmidae** it is 5, whereas 4 tarsal joints are found in the members of family **Locustidae**.

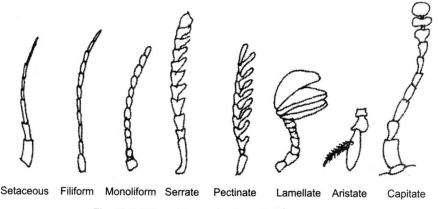

Setaceous   Filiform   Monoliform   Serrate   Pectinate   Lamellate   Aristate   Capitate

Fig. 16.2 Different forms of antenna in insects

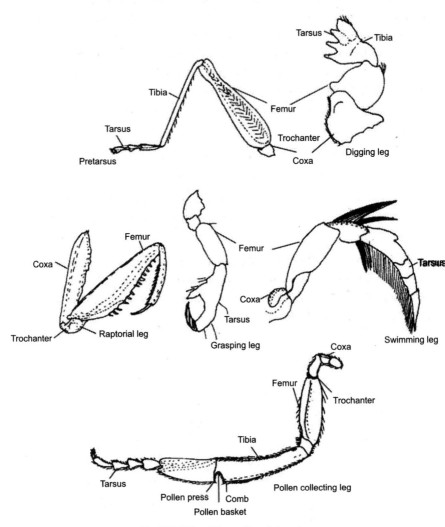

Fig. 16.3 Different types of legs in insects

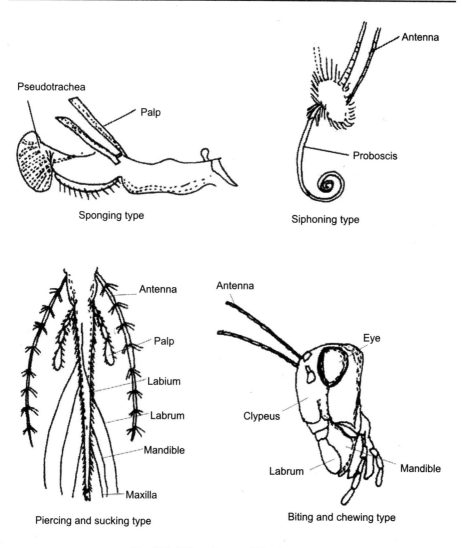

Fig. 16.4 Different types of Mouthparts

NON-MERISTIC CHARACTERS

These features include those traits which do not have serial repetition and can not be counted. Such characters in fishes include shape of the body which may be stream lined or dorso-ventrally compressed, head may be elongated or flattened, mouth may be large or narrow, body size may be small or large, texture of the skin may be leathery or scaly, **basi-caudal** spots may be present or absent. An account of different non-meristic characters in fishes and insects has been out lined in Table 16.1 and 16.2.

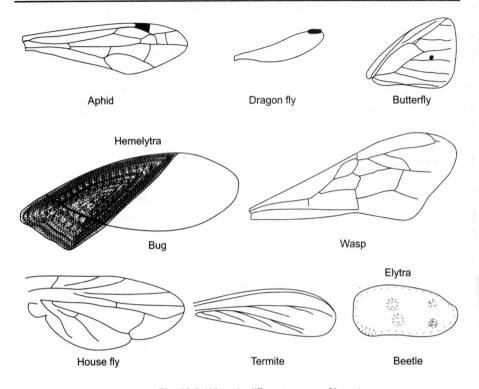

Fig. 16.5 Wings in different groups of insects

OTHER CHARACTERS

In addition to meristic and non-meristic features, certain other features are also used to segregate animal populations they are discussed herein as under:

Chemical Marks in Otoliths

Certain chemicals like **alizarine, strontium** and **calcein** are used to develop certain marks in the otoliths or other body structure of the fish. The chemical is mixed with the feed which produces a spot in the particular part of the body which can be observed later on. However, this process is time taking.

Thermal Marks

These are produced by exposing the animals like fish to different temperature regimens due to which distinct and recognizable patterns appear in the otolith. This method requires prolonged experimental period. The Alaska Department of Fish and Game, U.S., have been using otolith thermal marking of hatchery-raised salmonids to distinguish stocks and assist with management of mixed-stock fisheries. In addition, thermal marked otoliths have given clues regarding high sea distribution and movements of salmonids in the North Pacific Ocean and the Bering Sea (Agler et al., 2011). Similarly effective management of Pacific salmon species depends on accurate and timely analysis for mark status and mark identification (Williams and Moffitt, 2011).

Elastomeres

These are the internal marks, visible externally and may be of various colors produced by injecting colored and/or fluorescent plastic paint placed between the fin rays or at the base of the fins.

Table 16.1 Alternative forms of non-meristic characters in fishes

Character	Alternative forms of Non-meristic Characters	
Skeleton	Bony	Cartilaginous
Body shape	May be elongated, Depressed, ribbon like, fusiform etc.	Neither elongate nor cylindrical
Gill openings	Single	Paired
Dorsal spine	Present	Absent
Symmetry	Body asymmetrical	Body asymmetrical
Jaws	May project in to beak (as in *Mastacembelus)*	Jaws normal; absent
Barbels	Present	Absent
Head	Scaly	Not scaly
Upper jaws	Prominent	Not prominent
Abdomen	Serrated	Not serrated
Teeth	Absent (as in example *Gonialosa*)	Present
Anal Fin	Long (with more than 35 anal rays) as in *Notopterus*	Moderate (less than 23 rays)
Dorsal Fin	Normal, adipose, may occur as single elongated single fin as in *Clarias*	Modified to form sucker as in *Echeneis* May even be absent as in *Notopterus*
Tail	Homocercal	Heterocercal or diphycercal
Fins	May be simple or provided with fin rays, horizontal e g., *Bagarius sp.* Dorsal & pectoral fins spines strong and hollow	May be without spines or fin rays, not horizontal, Dorsal and pectoral fins spines neither strong nor hollow

Biological Marks

A trait that is measured and evaluated as an indicator of normal biological or pathogenic processes or pharmacological responses, is referred as biological mark. Biological markers can reflect a variety of disease characteristics, including the level of exposure to an environmental or genetic trigger. The detection and exploitation of naturally occurring DNA sequence polymorphisms are among the most significant developments in molecular biology. Polymorphic genetic markers have wide potential applications in animal and plant improvement programmes as a means for varietal and parentage identification, evaluation of polymorphic genetic loci affecting quantitative economic traits and genetic mapping (Nagaraju et al., 2001).

Recent progress in DNA marker technology particularly PCR based markers such as random amplified polymorphic DNA (**RAPD**), amplified fragment length polymorphisms (**AFLP**), and inter simple sequence repeat markers (**ISSR**) have augmented the marker resources for genetic analyses of a wide variety of genomes. Among the PCR based technology the RAPD is a simple and easy method to detect polymorphisms based on the amplification of random DNA segments with single primer of arbitrary nucleotide sequence (Williams et al., 1990). Genomic DNA is subjected to the PCR adding only a single short oligonucleotide of random sequence in each reaction.

Table 16.2 Alternative forms of non-meristic characters in insects

Insect Body Part	Characters/variation observed
Antenna may be modified into different types	Setaceous - Cockroaches
	Filiform - Grasshoppers
	Monoliform - Termites
	Serrated - Beetles
	Pectinate/comb like - Moths
	Clubbed/clavate - Butterflies
	Plumose/feathery - Mosquitoes
	Aristate - Houseflies
Mouth parts adapted for feeding	Biting and Chewing type e g., Grasshoppers
	Piercing and sucking type e g., Mosquitoes
Legs modified for different purposes	Saltatorial –*Acrida* species
	Raptorial - Praying *Mantis*
	Natatorial - *Hydrophilus*
	Fossorial - *Gryllotalpa*
	Pollen collection – *Apis*
Nature of Head	Prognathous, Hypognathous or Opisthognathous
Wings show profound variation	Wings covered by scales - Lepidoptera
	Fore wings narrow & thickened to form tegmina – Grasshoppers
	Proximal area well sclerotized & distal area membranous - Hemiptera
	Forewings toughened to form horny or leathery elytra - Coleoptera

The RAPD method has been demonstrated to be useful for the studies of taxonomic problems (Vilatersana et al., 2005), systematic relationships (Pharmawati et al., 2004), phylogeny reconstruction (Poczail et al., 2008), population genetic structure (Sales et al., 2001), species hybridization (Caraway et al., 2001), assessment of genetic diversity (Souframanien and Gopalakrishna, 2004). Galal (2009) has worked out the intraspecific variation within the three ant species viz., *C.*

maculatus, M. pharoensis and *C. bicolor* on the basis of RAPD and PCR-RFLP marker system. Naturally, occurring parasites have been used as an **indicator** of various aspects of fish biology and of other aquatic invertebrates.

SIGNIFICANCE OF CHARACTER STUDY

Taxonomic characters are animal's traits used as an evidence for locating a particular species. Characters indifferent animal groups can be studied as under:

- The animals show varied morphological and biological organization, behavioural pattern, ecological adaptations, biochemical, and molecular adaptations as well.
- They can be qualitative or quantitative to describe variation that is discrete or continuous.
- They can have fixed states or polymorphic within species but with states distributed in different frequencies across species,.
- By the evolutionary processes that produced different groups of animals (sexual or natural selection, or genetic drift) and by the role, they play in the speciation process.

Taxonomists need to resort to different taxonomic characters to conform to the biological peculiarities of particular taxa. For example, behavioral characters-especially those genetically fixed without ontogenetic learning - such as call patterns of insects, bats and frogs are routinely used by taxonomists working on these groups and their use has led to the discovery of many cryptic species in some groups of organisms. In addition, the ecology of organisms can be an important source of evidence in some cases. For example, changes in ecological conditions of a locality may alter the taxonomic character that can help to distinguish between closely related species. In bacteria, the lack of a conspicuous morphology coupled with extensive gene transfer has forced taxonomists to develop a model-based strategy that combines data on ecology and on genetic diversity to delimit species. Identification of fish population being integral components of fisheries stock management is essential for aquaculture practices and for conservation of endangered species. The identification of fish stock helps to determine the appropriate strategies for the development of fishery and harvest strategies. Character study in case of insects helps to distinguish between harmful and useful insects and their use as biological agents for control of different insect pests.

Chapter 17

BIODIVERSITY ASSESSMENT

Biodiversity observed on earth today is the result of nearly 4 billion years of evolution. Conservation scientists like Jenkins and Lovejoy (1980) used the term 'biological diversity'. Rosen (1985 coined the term **biodiversity** and Wilson (1992) used it for the first time in a publication. The total heritable variation of differences in the characteristics exists in individuals and their species in various ecosystems of different parts of the earth. Biodiversity is an attribute of ecosystem it forms the basis of evolution and is a universal phenomenon occurring in nature. Biodiversity can be explained at three levels, viz., **gene**, **species** and the **ecosystems** because of cellular and molecular, taxonomic and geographic criteria respectively.

Biodiversity can be defined as the "variability among all living organisms from all sources, including *interalia* terrestrial, marine and other aquatic ecosystems, and the ecological complexes of which they are part: this includes diversity within species, between species and of the ecosystem. This is in fact the legally accepted definition of biodiversity, since it is adopted by the United Nations Convention on Biological Diversity held at Rio de Janeiro in 1992". After the Rio Earth's summit 'Biodiversity' has become the focus of attraction of people by and large. According to Wilson (1992), the real biodiversity is the genetic diversity, i.e., the diversity of genes and **genome** dynamics occurring at the **DNA** level resulting in evolution. The term biodiversity is defined in different way as under:

- Biodiversity refers to variation of life at all levels of biological organization.

- Biodiversity is a measure of the relative diversity among organisms present in different ecosystem. This includes diversity within species, among species and comparative diversity among ecosystem.

- Biodiversity refers to the 'totality of genes, species and ecosystems of a region. In order to describe the 'diversity of life' following three components of diversity are detailed herein as under:

Genetic Diversity

It is also referred diversity within species and is important for the conservation of species. It is particularly important for natural resources such as aquaculture and fisheries. Distribution pattern of different species can be assessed by population survey studies. More detailed analysis of reproductive patterns, physiology and genetic variation of species help to workout the variation between the species. It should be made clear that the measurement of within species biodiversity do not necessarily require costly genetic analysis as genetic diversity is closely related with phenotypic life history traits. Within species, diversity help to assess the species whose populations are of particular interest as they are resource, endangered or a past species.

The measure of biodiversity within subspecies can thus be described by attributes of domestication such as growth rate and body shape. It is relevant to mention that the genetic diversity used in aquaculture is fundamental for breeding strains to produce economically important species. Use of **genetic code** is the accurate method of biodiversity measurement as it draws similarities at the molecular level. It is relevant to mention that genetic variation between the species is not because of differences in the number of functional genes but due to percentage of non-coding DNA. It is also significant to mention that there are many **universal gene segments** in a wide range of organisms suggesting the existence of an ancient minimal set of DNA segment that all cells must have. The variation between the species helps to assess the level of populations which are either well established or threatened. Measuring within species biodiversity should also include phenotypic traits and genetic markers. The habitability of these traits shows that they have genetic components which can be identified by molecular analysis like DNA, mRNA and protein polymorphism.

Diversity between Species

It refers to the diversity of genes within a species. Accordingly, there is genetic variability among populations and the individuals of the same species. Most ecological indices of biodiversity relates to groups of species. The relative abundance indices tend to increase with the number of species (i.e., species richness) and the **evenness** of the distribution of individuals among the species. Each index is differently weighted for richness and evenness, so that they can be ranked by their sensitivity to these variables. The relative abundances of species can be described with a number of different indices. These include the following:

- Single number indices (univariable) of which species richness and Shannon Index are best known.
- Graphic techniques for visual comparison of relative abundances.

Diversity of Ecosystem

It is the diversity at higher levels of organization e g., the different types of ecosystem on earth. Species richness is widely used in **ecology** as a measure of species diversity. The species richness of a particular habitat is correlated with the general status of the habitat in relation with different ecological variables. In the species richness index, all species that exist in an ecosystem count equally. There are two schools of thought in this regard: according to one stand, different species are weighted according to their relative abundance in the system. However, this is criticized by the observation for not taking into account the differences between the species. In other words, the functional role of species may vary with their abundance in the system. On the other hand, another group advocates that different species should be given different weight in the index on the basis of the characteristic features they possess.

Further, climate change, migration, deforestation and aquatic biota can break down ecological barriers resulting in species with an allopatric distribution becoming sympatric. A question may arise that what happens when different species overlap? Depending upon the phylogenetic distances between species, crossing may lead to parthenogenesis or hybridization and impact on disease epidemiology. The measurement of biodiversity enables to plan conservation of important species with more accuracy. Sometimes, indicator species are used for measuring biodiversity. The measurement of biodiversity comprises of following two components:

- The degree of differences between the species.
- The number of individuals, species and different habitats etc.

DISTRIBUTION OF BIODIVERSITY

Biodiversity is not evenly distributed on earth, it is consistently richer in the tropics and towards Polar Regions there are fewer species. The floral and faunal diversity depends on climate, altitude, soil and the presence of other species. The IUCN has categorized Earth's species as rare, endangered or threatened. A biodiversity **hotspot** is a region with a high-level of **endemic** species. Myers (1988, 1990) has identified these biodiversity hotspots and is of the opinion that these hotspots tend to occur near areas of dense human habitation leading threats to many endemic species. Following pressures of rapidly growing human population, developmental activities in many of these areas is increasing.

ASSESSMENT OF BIODIVERSITY

Biodiversity is the diversity of interaction among species. It encompasses the species and their immediate environment. Biodiversity a simple contraction of term biological diversity, which is the sum total of all biotic variation from the levels of genes to ecosystems. Biodiversity has a multistage content ranging from genes, species, **habitat** and ecosystems. The biodiversity needs to be studied because of the following:

- Several countries are signatories to the convention on biological diversity.
- Diversity of animals and plants including dense forms of life need to be measured for their conservation.

Biodiversity conservation has become imperative in view of growing degradation of natural ecosystem. Effective conservation depends upon ways to measure and monitor biodiversity changes. Biodiversity also contains structural and functional attributes at the following levels:

- The number of different species in the ecosystem.
- The characteristic features of different species, i.e., their functional traits.
- The relative abundances with which the individuals are distributed over different species.

It is suggested that a system is supposed to be more diverse than the other, if it contains a greater number of different species and the species in a particular habitat are quite distinct from each other and the individuals are more evenly distributed.

Consider two grassland ecosystems namely A and B, each of which contains eight insects. System A is characterized by three beetles, one dragonfly, one moth, two grasshoppers and one bug, whereas, system B has four grasshoppers and four bugs. According to first criterion, if we consider the number of species, system A has five different species, thus it has higher species diversity than system B (containing only two different species). According to second criteria i.e., characteristic features of species help to distinguish between members of different species. According to third criteria depending upon the evenness of relative abundances, it can be said that system B has a higher density than system A, because there is less chance in system B that two randomly chosen individuals will be of the same species (Fig. 17.1 and 17.2).

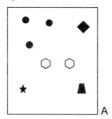

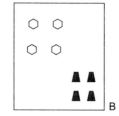

Fig. 17.1 Different species in an ecosystem

Fig. 17.2 Showing high species diversity

Where different symbols indicate diversity if different species like beetle (●), dragon fly (★), moth (◆), grasshopper (⬡), bugs (▲)

TYPES OF BIODIVERSITY

Species tend to evolve in a community by utilizing the available resources like habitat areas, light intensity, prey size etc. for its survival. Biodiversity of individuals refer to richness, variance, abundance, evenness, equitability and relative rarity (as in **Shannon-Weiner Index**). Further the diversity may be compared by taxa, areas, time, behaviour, trophic groups, sex and age, weight of the individuals, similarity and genetic make up. The species differ in the proportions of niche hyperspace they are able to occupy and the share of the community's resource they utilize. The quantity of resource utilized is expressed in species productivities and when species are ranked by relative productivities (or some other measurement) from most to least important, dominance diversity curves are formed. Diversity in strict sense is the richness of the species and is measured as the number of species in a sample of standard size. Whittaker (1972) described three terms for measuring biodiversity over spatial scales: alpha, beta, and gamma diversity:

Alpha diversity (α diversity)

It refers to diversity within a particular area, community or ecosystem and is measured by counting the number of taxa present within the ecosystem. It is concerned with identifying and describing the basic units of earth's diversity (the species) and in grouping these into preliminary classifications (genera). It is the number of species found in a small and homogeneous area. It is the same as species richness for a local community. In other words the complexity and richness of community is referred as alpha diversity.

Beta Diversity (β diversity)

It attempts to construct hierarchical classifications that incorporate evolutionary relationships. The extent of differentiation of communities along habitat gradients is known as beta diversity. Hunter (2002) defines gamma diversity as "geographic-scale species diversity". The beta diversity entails comparing one small and homogeneous community with another in the same general area. In other words, it is the number of species unique to each community in the pair of communities being compared. This involves comparing the number of taxa that are unique to each of the ecosystems, thus working out the species diversity between ecosystems along with the environmental gradients. The simplest of Whittaker's original definitions of beta diversity is as under:

$$\beta = \gamma/\alpha$$

      where α is alpha diversity
          β is beta diversity
          γ is gamma diversity

Gamma diversity (γ diversity)

It refers to measure of overall diversity for different ecosystems within a region. The position of species in a landscape of communities as described in terms of both habitat and niche relationships, may be termed as **ecotope**. It seeks to understand the process that is responsible for taxon formation and evolution in general. Thus, the total species diversity in a landscape (gamma diversity) is determined by two different things, the mean species diversity in sites or habitats at a more local scale (alpha diversity) and the differentiation among those habitats (beta diversity). According to this reasoning, alpha diversity and beta diversity constitute independent components of gamma diversity:

$$\gamma = \alpha \times \beta$$

         where α is alpha diversity
            β is beta diversity
            γ is gamma diversity

BIODIVERSITY INDICES

In an ecosystem the status of each species is considered by its relative abundance and the species diversity is measured by formulating **diversity indices**. It is stated that rare species should contribute less than the common species to the biodiversity, so far as species richness and distribution of relative abundances with community size and shape is considered.

     Biodiversity indexes are based on species differences but do not account correctly the abundance conditions or taxonomic differences. Biodiversity is usually plotted as taxonomic richness of a geographical area with some reference to the temporal scale. The basic idea of diversity index is to summaries the data regarding number of species and their proportional abundances into a single **numeric index** (Hill, 1973). The diverse index helps to discriminate between sites or samples size, species richness or **species evenness.** There are different indices to assess the biodiversity. Whittaker (1972) has described three common **matrices** used to measure species level biodiversity. These are discussed as under:

Species Richness

It refers to the number of species present in a given area. It is represented in equation form. The richness of species is measured by counting the total number of different species present in that ecosystem. This is the simplest of all the measures of biodiversity of ecosystems and is given by the following formula:

$$D^R (\Omega.) = n$$

Where D represents diversity index

Ω (Omega) =species richness

R = richness of species

n = number of different subspecies

The above method does not indicate the manner in which the diversity of the population is distributed among those particular subspecies. For example, if there are 4 different subspecies in zone I and zone 2, the species richness would be equal. It is not clear as to what percentage of the abundance of these subspecies would be.

Species richness also serves as **biodiversity indicator** used to evaluate the biodiversity loss *versus* species extinction. It does not take into account the proportion and distribution of each subspecies within a zone. This measure of species diversity does not take into account the way, diversity of the population is distributed or organized among those particular subspecies. For example, if there were four different subspecies in Zone 1 and Zone 2, the richness would be equal. However, this measure does not indicate the percentage of abundance of each subspecies. It is relevant to mention that the environmental factors like climate, soil type and disturbances strongly affect the functioning of the ecosystem and individuals can also affect the environment.

Darwin and Wallace (1858) gave the concept that biodiversity could affect the functioning of the ecosystem. Accordingly, a diverse combination of crops (polyculture) should be more productive than **monoculture**. They also highlighted the underlying biological mechanism, since coexisting species differ ecologically. Loss of species could result in vacant niche space and potential impact on the ecosystem.

It has also been observed that diverse communities have greater variety of positive and complimentary interactions and so outperform any single subspecies and can hold right species at the correct place and at accurate time. Such systems are generally more stable and function better than communities that have lost species. Relationships between biodiversity and ecosystem function are closely linked.

Factors Affecting Species Richness

There are many ecological factors which affect the species richness being discussed herein as under:

• There is inverse correlation between species richness and latitude. With the increase in latitude, the species richness decreases. This shows that there is an effect of area, available energy, isolation and /or zonation. The level of species richness increases rapidly from the northern region but decreases slowly from the equator to the southern region. Further the area at lower altitude is longer than at higher altitude, this result in high species richness at lower altitude.

• Different components of ecosystem also affect species richness through **abiotic** and **biotic** factors. For example, variation in the density of producers may affect **herbivore** populations, which in turn may affect carnivores and so on. In nature, high species diversity and high productivity are often not positively correlated since factors that increase productivity often lead to lower species richness because more productive species out compete less productive ones.

• It is just possible that larger areas are more diverse topographically and environmentally. Thus it provides opportunities for more species to set up their populations due to greater habitat diversity.

Simpson Diversity Index (D)

It is a measure of species richness as well as the percent of each subspecies from a biodiversity sample within a zone. The proportion of individuals in an area indicates their importance to diversity. Simpson (1949)has given this index and it is obtained by the abundance of a given subspecies in a zone, divided by the total number of subspecies observed in that zone. There are three different methods to calculate the abundance of a given sub species:

• Simpson Index: $D = \text{sum } (P_i^2)$ According to this concept, the probability that two randomly selected individuals in the zone belong to the same subspecies.

• Simpson's Index of Diversity: 1-D represents the probability that two randomly selected individuals in a zone belongs to different subspecies.

$$D = \frac{N\ (N-1)}{\sum n\ (n-1)}$$

where D = diversity index
N= total number of organisms of all species found
n = number of individuals of a particular species
$\sum$ = sum number of individuals

• Simpson's Reciprocal Index: 1/D refers to the number of equally common subspecies that will produce the observed Simpson's index.

In the above referred cases, diversity index is influenced by two parameters - the equitability of percent of each species present and their richness. For a given species richness, D will decrease as the percent of the species becomes more equitable. The 'D value' stands for the dominant index. As 'D' increases, diversity decreases. Thus, Simpson Index is used in Environment Impact Assessment (EIA) to identify the role of disturbances.

Shannon-Weiner Index

The Shannon index also known as **Shannon-Weaver index** and is used to measure diversity in definite data. It takes into account the number of species and the evenness of the species. It is simply the information - entropy of the distribution, treating species as a symbol and their relative population size as the probability. With the help of this index the number of individuals observed for each subspecies in the sample plot is calculated. It is based on the following assumption:

- Individuals are randomly sampled from independendently large populations.
- All species are represented in the sample.

In this method, the first step is to calculate the P_1 for each category of subspecies. The log of the number is multiplies this number. This number is then multiplied by the log of the number. Any base may be used; however, the natural log is commonly used. This index is computed from the negative sum of these numbers. The Shannon-Weiner index is represented as under:

$$H^`(P) = \sum_{i=0}^{s} P_i \log P_i$$

Where P_i = nl /N in which nl is the number of species present in the season.

N is the number of individuals

S denotes the number of season

H is the Shannon -Weiner index

When there are similar proportions of all subspecies then evenness is one but when the abundances are very dissimilar (some rare and some common species) then the value increases. It may be mentioned that Log^2 (0.3010) is often used for calculating Shannon-Weiner Index. The log base should be consistent while comparing diversity between samples or estimating evenness. It is suggested that the value of Shannon-diversity lies usually between 1.5 and 3.5 and only rarely it exceeds 4.5. Expected Shannon diversity is also used (Exp H`) as an alternative to H. Exp H` is equivalent to the number of equally common species required to produce the value of H` governed by the sample. In conditions, when all species are equally present, the observed diversity (H`) is always compared with maximum Shannon-diversity (H_{max}). It is also relevant to mention that Shannon-Weiner diversity is very widely used for comparing diversity between various habitats.

Sòrenson's Similarity Index

It is used for environmental assessments particularly in conjunction with or comparison to diversity and biotic indices. The **Sòrenson index** is a very simple measure of beta diversity ranging from a value of 0, where there is no species overlap between the communities to a value of 1 when the same species are found in both communities. It is expressed by the following formula:

$$\beta = \frac{2C}{S_1 + S_2}$$

Where S_1 represents the total number of species recorded in the first community and S_2 is the total number of species recorded in the second community, β is beta diversity and C is the number of species common to both communities.

Association Index

This is an important analysis tool in biological assessment (Johansson and Minns, 1987). In experimental studies analysis of variance, t-test and other statistical rules are applied to such indices, which help to evaluate the differences in community response. In aquatic biota as the environmental factor fluctuates off and on, the community goes on constantly changing. Thus, it is very difficult to compare different parameters in the samples taken either at different locations in the same year or at the same locations in different years. For example, there is seasonal variations in the populations of zooplanktons and phytoplankton in a community.

Dominance Index

This index is used to describe the dominant status of a particular trait in a community. This index can be calculated even for a brief period or over a season. Features like aggressive behaviour or territorial recognition can also be calculated with greater ease. Whittaker (1972) proposed a **dominance index** that measures relative index (y) of any given species (n) to the total species (N). It is expressed by the following formula:

$$\text{Whittaker Dominance Index} = \sum_{i=1}^{n} (Y/N)^2$$

It can be based on any type of measurement (numerical abundance, productivity, percent cover etc.). This index fails to explain the factors that affect species dominance.

Dissimilarity Index (DS)

This is based on the fact that degree of association or dissimilarity between pairs of samples can be determined. This type of index is applicable in cases of constantly changing communities.

Quadratic Entropy Index

Traditional ecological diversity indices are based on the abundances of species present. Such indices do not serve to work out taxonomic or similar differences. With equal species abundances, these indices measure the species richness only.

Biodiversity indices are based on species differences and do not account for abundances of species (Izsak and Papp, 2000).

Izsak and Papp (2000) has suggested that the **quadratic entropy index** (Q) is the only ecological diversity index used in the ecological practices that incorporates both species relative abundances and a measure of pair-wise taxonomic differences between the species in the analyzed data set.

Ricota (2002) has attempted to bridge the gap between ecological diversity index and measure of biodiversity with **Shannon's entropy**. He has shown that a number of traditional ecological diversity measures can be generalized to take into account a taxonomic weighting factor. Since, these new indices violate part of the mathematical properties that an index should meet to be termed an ecological diversity index. Ricotta has termed these new families of indices as 'weak diversity indices'.

MEASURES OF BIODIVERSITY

Biodiversity can be measured as species richness or species diversity (Magurran, 1988). It is not easy to measure biodiversity based on just one single measure and for the estimation of biodiversity, proper identification of species is necessary to monitor the biodiversity of a particular area. Measures of biodiversity are used as basis for making conservation programmes more effective. Often the **indicator species** are used as a parameter for measuring biodiversity. **Genetic diversity** is measured directly by looking at genes and chromosomes or by observing the phenotypes. Generally, multicellular organisms tend to have more DNA than single celled organisms leaving few exceptions. It has been suggested that genetic variation occurs due to the differences in non-coding DNA.

Phylogenetic Measurement of Biodiversity

Phylogeny provides an insight of relationship between species. In other words, phylogenetic species variability summarizes the degree to which different species in a community are phylogenetically related. As mentioned in the preceding paragraphs, three matrices based on species variability, species richness and species evenness have been proposed. These matrices are derived by considering the value of some unspecified neutral traits shared by all the species in a community. As the phylogenetic tree is constructed on the basis of this neutral trait, **speciation** occurs and from this point onwards, evolution proceeds independently along each phylogenetic lineage.

The **phylogenetic species variability** (PSV) helps to unravel the supposition that phylogenetic relatedness decreases the variance of these hypothetical traits shared by all species of the community.

- PSV is calculated on the basis of information about the phylogenetic relatedness of species in a community.
- The phylogenetic species richness (PSR) represents the number of species in a community multiplied by the PSV of a community. These matrices are directly comparable to the traditional matrices of species richness.
- The third matrices measure **phylogenetic species evenness** (PSE). It is the matrices of PSV modified to incorporate relative abundances. The maximum attainable value PSE occurs only if species abundances are equal and species phylogeny is like a star i.e., it depicts bursts of radiation with each species evolving independently from a common ancestor. Thus PSE is measure of both phylogenetic and species evenness. The phylogenetic species variability represents the degree to which different species in a community are phylogenetically related. When a community phylogeny is a star, the index is equal to 1, indicating maximum variability. As relatedness increases, the index approaches 0 (zero), this indicates reduced variability. It is assumed that branch length of all community phylogeny is proportional to the evolutionary divergence between the species. It is believed that community phylogeny represents the measure of PSV.

TAXONOMIC SURROGACY

During the last decade, measurement of biological diversity has emerged as a major discipline in biology, with an immense practical importance to minimize human-induced impoverishment of life on earth (Purvis and Hector, 2000). The taxonomic surrogacy approach is based on the idea that there are predictable relationship between species and higher taxonomic ranks such as genera and families. Taxonomic sufficiency concerns the use of higher taxon diversity as a surrogate for species diversity. Furthermore, to be of practical value, the higher taxon counts must be able to predict species richness with a reasonable amount of precision.

The majority of biodiversity assessments use species as the base unit. Recently, a series of studies have suggested replacing numbers of species with higher ranked taxa (genera, families, etc.); a process known as taxonomic surrogacy, a system that has an important potential to save time and resources in assessment of biological diversity (Bertrand et al., 2006). Conservation biologists have used surrogate species as a shortcut to monitor or solve conservation problems. Indicator species have been used to assess the magnitude of anthropogenic disturbance, to monitor population trends in other species, and to locate areas of high regional biodiversity. **Umbrella species** have been used to demarcate the type of habitat or size of area for protection, and **flagship species** have been employed to attract public attention. Unfortunately, there has been considerable confusion over these terms and several have been applied loosely and interchangeably.

Members of a taxonomic group served to be a surrogate for another taxonomic group if they possessed similar assemblage patterns. Surrogate taxa are used widely to represent attributes of other taxa for which data are sparse or absent. Surrogacy values of taxa have been evaluated in diverse contexts, yet broad trends in their effectiveness remain unclear. Several biodiversity indicator measures have been proposed. The species richness of selected taxa is often used as a surrogate to measure the richness of other taxa.

Conservation biologists often use one or a small number of species as surrogates to help them tackle conservation problems (Bibby et al., 1992). Surrogate species are employed to indicate the extent of various types of anthropogenic influence (Stolte and Mangis, 1992); or to track population hangs of other species (McKenzie et al., 1992). Surrogate species are also used proactively to locate areas of high biodiversity (Ricketts et al., 1999b) or to act as 'umbrellas' for the requirements of sympatric species (Berger, 1997); they thus can help in locating and designing reserves. Finally, surrogate species may be used as flagships in a sociopolitical context for attracting public attention and funding for a larger environmental issue (Dietz et al., 1994).

According Gaston (2000) to measure biodiversity, the species rank holds a central position and biodiversity is too complex to measure directly, so conservation planning must rely on surrogates to estimate the biodiversity of sites.

BIODIVERSITY HOTSPOTS

A biodiversity hotspot is a biogeographic region that is both a significant reservoir of biodiversity and is threatened with destruction. It is a region with a high level of endemic species that is under threat from humans. The hotspot region must be biologically diverse, with a high proportion of species that are not found anywhere else on earth and the security of the region must be threatened.

India is one of the 12 mega biodiversity countries in the world and is one of the richest countries in the world in terms of biodiversity. This natural variation in life is also reflected in the demography of the land. A biodiversity hotspot is a biogeographic region with a significant reservoir of biodiversity that is threatened with destruction. An area is designated as a hot spot when it contains at least 0.5% of plant species as endemic. There are 25 hot spots of biodiversity on a global level, out of which two are present in India. India has two hotspots-the the Indo-Burma region. (earlier the Eastern Himalaya), the Western Ghats also known as 'Sahyadri Hills'. These hotspots have numerous endemic species. These hot spots covering less than 2% of the world's land area are found to have about 50% of the terrestrial biodiversity.

GENETIC POLLUTION

It occurs when original set of naturally evolved (wild) region specific genes / gene pool of wild animals and plants become hybridized with domesticated and wild varieties or with the genes of other non-native wild species or subspecies from neighbouring or far away regions. It is also known as genetic contamination or genetic swamping.

Usually genetic pollution i.e., uncontrolled hybridization, introgression and genetic swamping happens because of human interaction with natural environment and a lack of foresight. However, in rare instances, it has also been observed to occur naturally more commonly in case of closely related subspecies of plants whose ranges overlap forming hybrid zones making it easier for insects to cross-pollinate them.

In most cases nature has its own interspecies genetic barriers to guard against genetic pollution to keep species distinct. The hybrids which are produced where the ranges of closely related wild species overlap they may display hybrid vigour (heterosis) in the first generation and are in the long run less fit than the two parent species which have evolved over hundreds of thousands of years specializing in exploiting their own particular niche in nature. It is extremely rare that the hybrids ever become fitter than the two wild parent species so that natural selection may then favor these individuals and it is even rarer that reproductive isolation is ever achieved to lead to the birth of a new species through the process known as hybrid speciation.

It is important to prevent genetic pollution so that we can conserve the naturally evolved region specific wild gene stock and genetic makeup of wild animals and wild plants for maintaining the health of natural ecosystems and the environment in general. Scientists now, also consider this naturally evolved biodiversity to be a valuable source of strong genes which in the future may be used on a continual basis to hybridize domesticated varieties, to make them even stronger and more resistant to climate and diseases, thus leading to an ongoing improvement in food security and medicines.

BIODIVERSITY AND ECOSYSTEM

Human beings depend on natural ecosystems for several essential ecological aspects such as productivity, maintenance of gaseous composition of atmosphere, preservation of soil fertility, cycling of mineral nutrients, control of pest outbreaks, preservation of **gene library** etc. Knowledge of biodiversity plays an important role in the study of ecosystem. Further, there is threshold of biodiversity, below which the present complex ecosystem will loose its identity. As such, the biodiversity needs to be studied as it is essential to the viability of ecological systems. Further, species diversity has been a key concept in understanding ecosystem. Biodiversity loss can affect ecosystem functions and services. Individual ecosystem functions generally show a positive **asymptotic** relationship (a method of describing limiting behaviour) with increasing biodiversity. This suggests that some species are redundant; ecosystems are managed and conserved for multiple functions which may require greater biodiversity. Species diversity in ecosystem is promoted by cyclic, non-hierarchical interactions among competing populations. Insects are by far the most valued in conservation for their ecological roles. They are key components in the composition, structure and function of ecosystem (Wilson, 1987).

The wealth of diverse species in an ecosystem is described by the relative abundance of individuals of different species. Human beings form an integral component of ecosystem since our many activities affect ecosystems in different ways, like ongoing anthropogenic development and indiscriminate use of agrochemicals, pesticides and deforestation affect ecosystem in one way or the other. Linking ecological and social process is crucial for understanding the relationship between biodiversity and ecosystem function. This helps to utilize the relationship for human welfare through sustainable development and the judicious management of natural resources. Biodiversity may help the ecosystem in the following manner:

- Production of renewable resources like food, wood and fresh water.
- Regulating services that decrease environmental change like climate regulation, pest/disease control.
- Biodiversity also plays a part in regulating the chemistry of our atmosphere and water supply. It is directly involved in water purification, recycling nutrients and providing fertile soils.
- Cultural services concerning human value and enjoyment including seeing panoramic beauty of landscape cultural heritage, outdoor recreation and spiritual significance.

It is an established fact that biodiversity loss reduces the efficiency by which ecological communities capture biologically essential resources, produce biomass, decompose and recycle biologically essential nutrients. There is sufficient evidence that biodiversity increases the stability of ecosystem function through time. Diverse communities have been found to be more productive as they contain **key species** that have large influence on productivity and differences in functional traits among organisms.

BIODIVERSITY AND CLIMATE CHANGE

Climate change is widely regarded as one of the most serious threats to global biodiversity. According to Thuiller (2007) the evidence for rapid climate change now seems overwhelming. Global temperatures are predicated to rise by up to 4° C by the year 2100, with associated alterations in precipitation patterns. The change in temperature and humidity is likely to delimit species boundaries. Higher temperatures are likely to be accompanied by more humid and wetter conditions but the geographical and seasonal distribution of precipitation will change. The ability of species to respond to climate change largely depends on their ability to track shifting climate through colonizing new territory or to modify their physiology and seasonal behaviour to adapt to the changed conditions where they are. Different groups of animals face the anthropogenic threats including changes to the habitat loss, fragmentation, impact by exotic species, pollution, climate change and indiscriminate use of pesticides.

BIODIVERSITY AND GENETIC RESOURCES

Ever since the dawn of civilization, herding, farming and management have resulted in vast range of useful variations in domesticated animals and crops. Changes in biodiversity occur through time in all communities and ecosystems. Some of these changes result from natural and others from human disturbances. Genetic diversity is the basis for speciation, as a species responds to natural selection and adapts itself within the community and the ecosystem. Conservation of animal genetic diversity is essential to global food security and to protect our ability to meet the challenges of the future. The biodiversity is threatened because of loss of habitat, over-exploitation and introduction of exotic species. Loss of biodiversity due to **genetic erosion** reduces the future options for global community and small farmers, as they depend on wild species and natural habitats.

LOSS OF BIODIVERSITY vis-à- vis FOOD SECURITY

Endemic species can be threatened with extinction through the process of genetic pollution, i.e., uncontrolled hybridization, introgression and genetic swamping. Genetic pollution leads to homogenization or replacement of local genomes as a result of either a numerical and/or fitness advantage of an introduced species. Hybridization and introgression are side-effects of introduction and invasion. These phenomena can be especially detrimental to rare species that come into contact with more abundant ones. The abundant species can interbreed with the rare species, swamping its gene pool. This problem is not always apparent from morphological (outward appearance) observations alone. Some degree of gene flow is normal adaptation and not all gene and genotype constellations can be preserved. Hybridization with or without introgression may, nevertheless, threaten a 'rare species' existence.

In agriculture and animal husbandry, the 'Green Revolution' popularized the use of conventional hybridization techniques to increase yield. Often hybridized breeds originated in developed countries and were further hybridized with local varieties in the developing world to create high yield strains resistant to local climate and diseases. Local governments and industry have been pushing hybridization. Formerly huge gene pools of various wild and indigenous breeds have collapsed causing widespread genetic erosion and genetic pollution. This has resulted in loss of genetic diversity and biodiversity as a whole. Genetically modified organisms (GMO) have genetic material altered by genetic engineering techniques such as recombinant DNA technology. Genetically modified (GM) crops have become a common source for genetic pollution not only of wild varieties but also of domesticated varieties derived from classical hybridization.

Genetic erosion coupled with genetic pollution is affecting unique genotypes thereby creating a hidden crisis, which could result in a severe threat to our food security. Diverse genetic material could cease to exist which would impact our ability to further hybridize food crops and livestock against more resistant diseases and climatic changes.

BIODIVERSITY, AGRICULTURE AND SUSTAINABLE DEVELOPMENT

Biodiversity forms the basis for sustainable development and continuance of ecosystem in dynamic state. Genetic diversity in agriculture provides crops with ability to adapt to ever changing needs and natural adversity. Biodiversity is a valuable asset for both the present and future generations. It serves human beings by providing food, raw materials, industrial chemicals and medicines as well as important social and cultural benefits. For example, genetic diversity may enable certain moth species to have particular genes to protect itself against pollution. If pollution increases, genes may become more active to camouflage them more effectively. The environment also affects the genes responsible for the diversity within the species and this could affect the ecological diversity. It has been suggested that proper identification of organisms is necessary to monitor biodiversity at any level (Veechine and Collete, 1996b). Some of the important features of biodiversity studies are as under:

- Biodiversity can act as a continuum within the biological world. It can help to link molecular studies and taxonomy in unraveling some of the most vexed problems in biology.

- It is the backbone of agriculture, animal husbandry and forestry as well as for **aquaculture**. Wild relatives of domestic animals and cultivated plant species are best suited for breeding new varieties and for biological inventions.

- It holds promise in the new era of industrial applications. A wide range of industrial products are derived directly from biological resources. These include building materials, useful insects, fishes, resins, dyes, gums, adhesives, rubber and oil.

- Many crops provide food, shelter and clothing. Although, about 80% of our food supply comes just from 20 kinds of plants, humans use at least 40,000 species of plants and animals.

- It also provides avenues for reintroduction of wild varieties of domesticated species to yield a better variety than the previous ones. Thus, economic contribution of biodiversity is enormous.

- It can play an important role to establish much needed bridge within social and cultural world.

- It also helps in the land use and regional development.

- Many drugs are derived from biological resources and micro-organisms.

- It also plays an important role in recycling nutrients and providing fertile soils.

Yet today's taxonomy stretches the surface of extant global biodiversity (Heywoods and Watson, 1995). Biodiversity thus represents totality of genes, species and ecosystem in a region and the species being the most fundamental unit of biodiversity. Variations in biodiversity represent variation in species richness. Further, along with the evolutionary process, the hierarchical organization of biodiversity reflects one of the fundamental organizing principles of modern biology.

BIODIVERSITY AND HUMAN HEALTH

Many of the anticipated health risks of climate change are associated with changes in populations and distribution of disease vectors, scarcity of fresh water and impacts on agricultural biodiversity and food resources. Availability of safe potable water is the greatest issue facing human health. Certain health issues which are influenced by the biodiversity include dietary health and nutrition security, infectious diseases, medical science and medicinal resources, social and psychological health. Biodiversity makes available the medicinal resources. A significant proportion of drugs are derived directly or indirectly from biological wealth. Different medicines are derived from plants, animal and natural resources.

BIODIVERSITY, TRADE AND COMMERCE

Many industrial materials are derived directly from biological sources. These include fibers, timber, dyes, rubber and oil. Biodiversity is also important to the security of resources such as timber, paper, fiber and food. As a result biodiversity loss is a significant risk factor in business development and a threat to long term economic sustainability.

BIODIVERSITY THE REFRESHING ASSET

- Biodiversity serves as a source of inspiration to musicians, painters, sculptors, writers and other artists. Popular activities such as gardening, fish keeping and specimen collecting strongly depend on biodiversity. Biodiversity has intrinsic aesthetic and spiritual value for us. It also provides means for recreational activities such as during leisure walk in parks, bird watching and animal safari.

- Biodiversity has intrinsic aesthetic and/or spiritual value to mankind and has inspired musicians, painters, sculptures, writers and artists alike. Followers of Linnaeus, natural historians and taxonomists have explained nature and methodically documented biodiversity world wide through the outset of New Millennium. The culmination of this effort is what we call **global biodiversity**.

BIODIVERSITY INFORMATICS

The term biodiversity informatics was coined in 1992 by John Whiting to cover the activities of an entity known as the Canadian Biodiversity Informatics Consortium, a group involved with fusing basic biodiversity information with environmental economics and geospatial information in the form of GPS and GIS. The term biodiversity informatics is generally used in the broad sense to apply to computerized handling of any biodiversity information; somewhat broader term bioinformatics is often used synonymously with the computerized handling of data in the specialized area of molecular biology. It is the application of informatics techniques to biodiversity information for improved management, presentation, discovery, exploration and analysis.

Biodiversity informatics is a relatively young discipline and it typically based on a foundation of taxonomic, biogeographic, or ecological information stored in digital form. With the application of modern computer techniques, it can yield new ways to view and analyze existing information and can serve as predictive models for information that does not yet exist

BIODIVERSITY vis-a-vis EVOLUTION

Biodiversity as we observe today is an outcome of grand evolutionary march. It is a fact, universally acknowledged that life has been well established a few 100 million years after the formation of earth. Until approximately 600 million years ago, all life consisted of bacteria and smaller single celled organisms. The rapid growth of different life forms started during **Cambrian** explosion. Over the next 400 million years or so global diversity showed little overall trend but was marked by periodic massive losses of biodiversity as described under mass extinction events. It is suggested that the period since the emergence of humans is a part of new mass extinction caused by the impact of human activities.

THREATS TO BIODIVERSITY

According to IUCN Red list, there are 63,837 species with about 19,817 threatened species as of July 19, 2012. 3,947 species are described as critically 'endangered' and 5,766 as endangered. More than 10,000 species are listed as 'vulnerable'. It is speculated that between 0.01 and 0.1 % of all species may get extinct each year. It is relevant to mention that the biodiversity and genetic diversity are dependent upon each other. The diversity within species is necessary to maintain diversity among species and vice-versa. Thus if one type of animal or plant is removed from the system, the cycle can break down and the community may be dominated by a single species. With the ongoing anthropogenic developments, the day is not far off when we will not be able to see most of the available flora and fauna. According to Wilson (1992) HIPPO (an acronym stands for habitat destruction, invasive species, pollution, human over population and over harvesting, are responsible for the loss of biodiversity which is a global crisis and biological extinction has been a natural phenomenon in geological history.

LAW FOR CONSERVATION OF NATURE AND NATURAL RESOURCES

Although there have been different measures since ages to conserve flora and fauna, but of late much public concern has been raiseds to conserve biodiversity by enforcing different provisions of law. The relationship between law and ecosystem is very ancient and has a direct consequence for the biodiversity. It is related to property rights, both private and public. The earliest codified law for the protection and preservation of wild life and environment can be traced back to 3rd century BC at the time of Ashoka the great (King of Magadha). Some of the important rules framed from time to time in India regarding protection of wild life, nature and natural resources are detailed herein as under:

- Wild Life Protection Act, 1887 (10 of 1887) enacted by the British Government in India to prohibit the possession or sale of any kind of specified wild birds, which have been killed or taken into possession during breeding season.

- It has been enshrined in Constitution of India that, "it would be the duty of every citizen to protect land, air and water". Different rules have been framed in India from time to time regarding protection of flora and fauna as detailed herein as under:
  - Wild Birds and Animals Protection Act, 1912 (8 of 1912).
  - Wild Birds and Animals Protection (Amendment) Act, 1935 (27 of 1935).
  - The Prevention of Cruelty to Animals Act, 1960 (59 of 1982).
  - The Wild Life Protection Act, 1972 (53 of 1972).
  - The Wild Life Protection (Amendment) Act, 2002.
  - The Biological Diversity Act 2002 (18 of 2003).
  - The Biological Diversity Rules, 2004.

The United Nations Convention on Biological Diversity (CBD) 1992 recognizes the sovereign rights of states to use their own Biological Resources. The biodiversity Rules were notified in 2004 by the Govt. of India to conserve, sustainable use, fairly and equitably share the benefits arising out of the use of India's biological resources and associated knowledge. In addition, the 1972 UNESCO convention established that biological resources, such as plants, were the common heritage of mankind. Further, convention on Biological Diversity gives sovereign national rights over the biological resources. The new agreement commit signatory countries to conserve biodiversity, develop resources for sustainability and share the benefits resulting from their use. In a Convention on International Trade in Endangered Species of wild Fauna and Flora (CITES) held at Bangkok during March 2013, representatives of 178 countries agreed to add 343 species of plants and animals to CITES appendices I and II. Listing by CITES ensures that trade in them is either banned or strictly monitored.

The Biological Diversity Act -2002 attempts to realize the objectives enshrined in the aforesaid convention. It aims at the conservation of biological resources and associated knowledge as well as facilitating access to them in a sustainable manner and through a just process for the purpose of implementing the objects of the Act; it establishes the National Biodiversity Authority in Chennai. The Act covers conservation, use of biological resources and associated knowledge in India for commercial or research purposes or for bio-survey and bio-utilization. It provides a framework for access to biological resources and sharing the benefits arising out of such access and use. The Act includes in its ambit the transfer of research results and application for intellectual property rights (IPRs) relating to Indian biological resources. The Act also entails foreigners, non-resident Indians, body corporate, association or organization that is either not incorporated in India or incorporated in India with non-Indian participation in its share capital or management. These individuals or entities require the approval of the National Biodiversity Authority when they use biological resources and associated knowledge occurring in India for commercial or research purposes or for the purposes of bio-survey or bio-utilization.

The conservation of biodiversity involves the interaction and collaboration of numerous stakeholders with varied interests and roles. It is important to integrate the multiple stakeholders to determine ways and to begin resolving issues related to biodiversity conservation. Differing perceptions and consequent interests among masses is a challenge which needs to be addressed for effective biodiversity conservation. Thus, biodiversity is a unique and wonderful feature of the living world and time has come to conserve and preserve it for the posterity.

Chapter 18

TAXONOMIC KEYS

The science of describing, naming and classifying **taxon** based on common characteristics features is known as taxonomy. New species are discovered and described in taxonomy, while systematics uses evolutionary relationship to understand biogeography, co-evolution, adaptation and strategies for biological conservation. A taxon (plural, taxa) is a taxonomic unit consisting of population of individuals which are phylogenetically related. The individuals in a taxon show characteristic features that differentiate them from members in other taxa.

A **taxonomic key** is an arrangement of the distinguishing features of a taxonomic group to serve as a guide for establishing relationship and names of the unidentified members of the group. It is an important part of taxonomic studies and is used to identify different types of animals. Without any accurate key, it is not possible to identify particular taxa whereas, in certain other cases, it is only after the examination and description of a particular specimen, the key is constructed based on the description and not from the actual specimens. In such cases, the taxonomic key is treated as mere addition of the descriptions built on characters of the specimens. The key should be modified and expanded so that more and more taxa and their characters are discovered or differentiated and added.

CONSTRUCTION OF TAXONOMIC KEY

Before constructing taxonomic keys, sound knowledge of morphology is essential. One should be well conversant with the technical knowledge of taxonomy. A good style access key requires adequate knowledge regarding characters which are reliable and easily be taken into consideration. Following factors should be taken into account while constructing a taxonomic key:

- Knowledge of phylum, class, order and family of a particular specimen is essential.

- In insects, nature of wings, mouthparts, antennae and legs provide specific features for the construction of taxonomic keys.
- Body shape, position of fins, presence or absence of scales or spines in fin support may help in constructing the keys for identification of fishes.
- Emphasis should also be given to the color of the specimens as the color sometimes vary or may fade in preserved specimens.
- Body size of an individual alone may not be the criteria for its identification.

There are many methods of key construction but the most convenient and universally acceptable one is the **dichotomous key**. The main function of the key is to identify the taxa easily. The key enables the worker to identify particular taxa on the basis of characters. It is relevant to mention that the taxonomic key does not reflect phylogeny. In order to identify a particular specimen, it is very essential to observe and note down the important features of the specimen before proceeding to key construction. While making key only visible characters are taken into consideration. Incomprehensible comparatives should be avoided (e g., legs longer versus legs shorter). To make it clear, comparison can be made as under:

- Legs longer than head width.
- Legs shorter than the head width.
- Legs > 1.5 times head width.
- Legs < 1.5 times head width.   or
- Legs 3.0 mm.
- Legs 2.0 mm or less.

The language of the key should have clear statements. Further, if an insect has to be identified, the following characters may help the taxonomist:

- Body size.
- Shape and structure of wings.
- Length of the antennae, number of antennal segments and type of antenna.
- Position of the head viz., **prognathous, hypognathous** or **opisthognathous.**
- Nature of mouth parts viz., piercing and sucking, biting and chewing, siphoning type etc.
- **Eyes** and **ocelli.**
- **Auditory organs.**
- Number of tarsal segments.
- Structure and type of leg, whether it is adapted for running, swimming, clinging or burrowing.

Following additional tips are of great help, while proceeding through taxonomic key:

- The introductory comments along with abbreviations about the format of the keys should be read carefully.

- Both statements of each taxonomic key should be read.
- If the terms mentioned at any step of the statement are not clear, it should immediately be checked from a good glossary. For example, if it is mentioned that antennae **serrate** or **pectinate**, then the taxonomist should know the meaning of these terms.
- In case of fish scales and fins, and number of antennal or tarsal segments in insects or for that matter any meristic character should be counted from the specimen. instead of making any guess work.
- If the characters are very prominent, examining a few specimens will help to identify them easily. In case of living organisms, it is always better to examine several specimens, if easily available.

Features of Taxonomic Keys

The taxonomic keys are integral part of taxonomic studies and help to identify specimens easily. The taxonomic keys make the identification of specimens easily and are of great help during field survey as well. The taxonomic key should possess following features:
- The key should be dichotomous, alternative characters should be precise.
- The phrases in the key should be separated by colon.
- It must be applicable to all the individuals of the population irrespective of age and sex.
- It must clearly indicate the distinguishing characters.
- Sexual dimorphism present or absent
- Males smaller/ larger than females.

TYPES OF TAXONOMIC KEY

Different types of keys are available depending upon the characters used in identifying a particular taxon.

Dichotomous Key

A taxonomic key consists of a pair of statements that describe the characteristic features of individuals under study. The statements are grouped into 2 or 3 alternative descriptions (such as 1a and 1b) for each trait. It is also relevant to mention that each group of statements refers only to a single trait. Here one of the two statements can be applied to a particular specimen in question. At the end of the applicable statement, there is bold face identification number which takes the taxonomist to next **couplet** until the taxon in question is identified.

To use the key appropriate traits of the unknown organisms are identified while following the steps of the key. The keys that are based on successive choices between two statements are known as dichotomous keys. Such type of key is constructed in a manner using contrasting characters to divide the specimens in the key into smaller groups; each time a choice is made the number of alternative traits is eliminated.

A dichotomous key consists of list of paired statements (each statement is known as lead). The paired leads (the pair of lead together is called a couplet) consist of contrasting descriptions of certain distinguishing features of the organisms. The couplet should not be ambiguous. The leads in the couplet should be parallel. Finally, the specimen under examination is identified. Overall organization of the key is important. Following example will explain the pattern of key:

- Antennae monoliform, short cerci present.
- Antennae not monoliform, cerci absent.

In biology, single access key is also known as **sequential key, analytical key** or **pathway key**, where the author of the key fixes the sequence and the pattern of identification.

Diagnostic Keys

This type of key is based on characters which are most reliable, convenient and available under certain conditions. Multiple diagnostic keys may be preferred for the same group of animals. Diagnostic keys may be designed for field as field guides and may be based on using additional information like geographical distribution or habitat preference of organisms.

Synoptic Keys

These keys are based on the scientific classification. In cases where the classification is already based on phylogenetic studies, the key represents the evolutionary relationship within the group. This type of key often uses more difficult characters which may not be available in the field; as such synoptic keys are not useful for field identification.

Diagnostic versus Synoptic Key

Any single access key manages large set of items into structure that breaks them down into smaller, more accessible subsets with many keys leading to the smallest available classification unit (a species or infraspecific taxon). However, a trade-off exists between keys that concentrate on making identification most convenient and reliable. Such keys are referred as diagnostic keys.

Variation in Single Access Key

The distinction between dichotomous (bifurcating) and polytomous (multifurcating) key is structural one and the identification key software may or may not support **polytomous keys**. In dichotomous keys, one or the other lead statement is true or applicable to a particular species for identification, whereas, in polytomous key, the entire key must be scanned to the end to determine whether more than a second lead may exist or not. If the alternative lead statements are complex (involving more than one character), two alternative statements are significantly easier to understand than couplets with more alternatives. Most traditional single access key use the 'lead-style' where each option consists of a statement only one of which is correct or to be more accurate applicable to the individuals under taxonomic observation.

PRESENTATION OF TAXONOMIC KEYS

The single access taxonomic keys can be presented in different style discussed herein as under:

Nested Style

In this case all the couplets immediately follow their lead at the expense of separating the leads within the couplet. It is relevant to mention that nested style of key gives an excellent overview of the structure of key. A key being moderate, short and is easy to follow. However the nested style with polytomous key is inconvenient because each key has to be scanned to the end to verify that no further leads exist within a couplet.

Linked Style

The lead within the couplet immediately follows each other making polytomous keys easy to record. At the end of each lead some form of pointer (a numbering system, hyperlinks, etc.) creates the connection to the couplet that follows the lead. The taxonomic keys may be of following types:

Bracket Key

The dichotomous key is the best applicable. In this case only two precise alternatives are usually given, which accurately help in the identification of the specimen in hand. This key can be run forward or backward and provides choices to follow. It is quite common among taxonomists and is convenient in identifying the genera and species. For example:

1. Wings opaque 2
   Wings clear..5

2. Antennae serrate 3
   Antennae filiform 4

3. Eyes entire .... complete
   Eyes emarginate .... emarginata

Indented Key

In this type of key, the relationship of different groups is apparent to the eye. It is generally used for short keys, for example,

   A. Wings opaque
   B. Antennae serrate
   C. Eyes entire...complete
   C.C.Eyes emarginate .... *emarginata*
   B.B.Antennae filiform
   C. Large red .... *rufipes*
   C.C.Large black .... *nigripes*

Simple Non-bracket Key

In this type of key, the couplet is often composed of alternatives for immediate comparisons. The specimen to be identified can run forward through this key. The key can also run backward.

Simple Bracket Key

In this key, the number of couplets showing the continuation in the key is shown in parenthesis. While comparing specimens, one can go through this key in both forward and backward directions.

Serial Key

This key is distinguished by having features of both the bracket and the indented keys. Here the species are arranged according to the criteria of key character in common.

Pictorial key

The key consists of graphic representation of common species along with the characters together in a comparative manner. Specimens through this key can be identified even by a lay man. During war time, pictorial keys were used to identify the different types of mosquito larvae.

Branching Key

This key helps in easy and quick separation of small groups of specimens. It is helpful for field study.

Box Type Key

This key is also useful for field workers and it allows quick identification of specimens.

Circular Key

This key contains characters represented in circular manner. It is also helpful for small group of species and is of great help to the field workers. Specimens are often identified with the help of pictures and line diagrams. These pictures are often used in combination with the keys. The pictorial representations are excellent guide in the case of colorful insects, birds with beautiful plumage and mammals with varying coat pattern. Occasionally unidentified specimens are compared with identified specimens.

COMPUTER AIDED KEYS

Taxonomic keys are used to identify animals and plants. Such keys have been printed traditionally on paper in books, scientific journals and other scientific literatures. With the advancement in the field of computer technology, several innovative ranges of computer programs have been developed for publishing biological information. In fact taxonomic keys are now available online. A computer program (TAXOKEY.exe) requiring only a few keystrokes to use, is described as universal taxonomic key in the identification of plants or animal species by IBM compatible personal computers. A DOS text file serves as the database for a key that can be dichotomous or with multiple choices. TAXOKEY can optionally display color or monochrome screen pictures to illustrate the keys. Additional programs are described that are used to make the keys and check for errors as well as convert PCX graphic images for use with TAXOKEY. A second text file can be searched by TAXOKEY for information on particular species descriptions, notes and references. The advantages of online taxonomic keys are detailed herein as under:

- Unlimited number of keys can be taken on laptop directly into the field.
- Several diagrams, illustrations or photographs can be stored in the system, which can help in making comparisons during field trips.
- Without any publication cost, the keys can be modified.
- Frequent correction or modification in the key can be made.

- The different software available for going through the key provides much scope of checking the position of the specimens during field study.

DESCRIPTION OF GENERA

While describing the specimens, following points should be considered.

- Long description should be avoided.
- Characters variable in groups deserve special mention.
- Character which help in describing a particular specimen (taxa) and making it different from others, should be included. The character should also include the type of sex, coloration, and developmental stages along with behavioural, ecological and other biological features.
- The numerical feature of the taxon, like the number of antennal segments, number of spots on the elytra in beetles, spines and scales in fish should also be considered.
- As the type represents the name of the genus, it should carry all the information as much as possible. The description should also carry illustrations which further help in the identification (line diagram, black and white or color photographs) subject to the condition that in most cases the printing charges has to be paid by the author. Color pictures must have sharp focus and the figure should be accompanied by suitable captions.
- The manuscript (MS) submitted for publication in any International Journal of repute should contain original work which has not been published elsewhere and should have essential features of a scientific paper and be written as per guidelines of the journals. Once the MS is submitted for favor of publication, the editorial staff of that particular journal examines the MS to ascertain whether it is strictly as per format of the research journal in which the paper has been submitted. In case, it does not conform to the style of the said journal, it is sent back to the author. In case the paper is found to contain original piece of work which is likely to add to our existing knowledge, it is referred to the referee for their comments. Revisions, if suggested by the referees, have to be done by the authors before the MS is finally recommended for publication. Now most journals invite on-line manuscript for submission and the same is sent to the referees for their comments online.

MERITS AND DEMERITS OF TAXONOMIC KEYS

Taxonomic keys are of great help in identifying the specimens accurately without much difficulty. The taxonomic key consists of a beginning, a series of paired choices and many possible end results. Taxonomic keys are most commonly used by scientists and students of the natural world. A wealth of reliable and efficient identification procedures may be incorporated in good single-access keys. Features that are reliable and convenient to observe most of the time and for most species (or taxa), and also which further provide a well-balanced key (the leads splitting number of species evenly) are always preferred.

It is relevant to mention that multi-access keys largely serve the same purpose as single-access (dichotomous or polytomous) keys, but have many advantages, especially in the form of computer-aided, interactive keys. The user of an interactive key may select or enter information about an unidentified specimen in any order. This allows the computer to interactively rule out possible identifications of the specimens and provides the user with additional information and guidance on what information to enter next. The full-featured interactive keys may readily be equipped with images, audio, video, supplemental text, much-simplified language in conjunction with technical language and hyperlinks to assist the user with understanding of both entities and features.

However, in practice it is difficult to achieve this goal for all taxa in all conditions. Although software exists that helps in skipping questions in a single-access key, the more general solution to this problem is the construction and use of multi-access keys, which allows a free choice of identification steps and are easily adaptable to different taxa (e g., very small or very large) as well as different circumstances of identification (e g., either in the field or laboratory).

Chapter 19

TAXONOMIC PUBLICATIONS

Any research work is not supposed to be complete unless its results are published and brought to the knowledge of other scientists working across the globe. It is important to publish taxonomic research work in some reputed journal, which has world wide circulation. One has to go through the references before attempting to start a new taxonomic work. The taxonomic publications may be in the form of books, pamphlets, journals, articles, chapters, catalogues etc.

TAXONOMIC LITERATURE

Publications in taxonomy range from short description of a new taxon that forms only the part of a page to lengthy **monograph** and **handbook** running into several volumes. Taxonomic publications include identification manuals as well as revision work. The taxonomic publication may stress on the nomenclature, distribution, illustration or life history. There are different kinds of identification tools and taxonomic publications like **atlases**, catalog, **checklists**, **faunas**, **field guides**, **field book**, handbooks, **manuals**, monographs, **synopsis**, **reviews** etc.

Atlas

It provides complete illustrations of the species of a taxonomic group. The atlases represent the taxonomic data; hence the taxa are represented in semi-diagrammatic drawings, full halftones or coloured plates.

Catalog

A catalog is mainly an index to published taxa, arranged in a manner so as to provide a whole series of references for both zoological and nomenclatural purposes. The taxa are generally arranged alphabetically. Its compilation is quite tedious and requires great expertise and critical observation on the part of taxonomist. According to Blackwelder (1967) a catalog has following features:

- The original description references.
- Later references.
- Synonyms with references.
- Range.
- Type locality.
- Types of genus.
- Date about biology, zoogeography, hosts etc.

Following is an example of important catalog:

- Stone et al., 1965. A catalog of the Diptera of America, north of Mexico (Agricultural Research Service, US Deptt. of Agricultural, Washington, D.C.), 1696.

Check List

A checklist is a reference regarding correct nomenclature of the specimens and the arrangement of collections. A checklist varies greatly in its richness. A number of important checklists are available for some better known groups of individuals like butterflies, birds and mammals etc. The list provides citation, type locality, distribution status, habitat, synonyms and comments. Example of an important checklist is as under:

- J.L. Peters et al., (eds, 1931-1987, Checklist of Birds of the World, Vol. 1-16 (Cambridge, Mass Museum of Comparative Zoology).

Fauna

Faunal work represents detailed study of animals of particular areas. The fauna under study may be confined to a particular group or may cover all the animals of an area. Some of the examples of faunal work are being given herein as under:

- Fauna of British India, Taylor and Francis, London- This is a magnum opus and runs into many volumes describing several important groups of animals.
- Fauna of Sikkim, Part-4 : Insects. Kolkata, Zoological Survey of India, 2003, 512 p., tables, $72. ISBN 81-8171-006-1. [State Fauna Series 9]. Details No. 33068
- Fauna of West Bengal: Part 8: Insecta: Trichoptera, Thysanoptera, Neuroptera, Hymenoptera and Anoplura. ed. The Director of Zoological Survey of India, Calcutta, Zoological Survey of India, 1999, 442 p., $55. ISBN 81-85874-21-2.
- The Fauna of India and the Adjacent Countries: Isoptera (Termites) Volume II. Family Termitidae, O.B. Chhotani. 1997, xx, 800 p.

- Faunal Resources of Similipal Biosphere Reserve, Mayurbhanj, Orissa/Ramakrishna, S.Z. Siddiqui and P. Sahu. Kolkata, Zoological Survey of India, 2006, iv, 96 p., tables, figs., plates (pbk). ISBN 81-8171

Field Guide.

Field guides are prepared by the taxonomists to enable persons interested in the study of animal life in nature and to identify the common animals in the field. Some field guides are prepared in the form of pamphlets given to the field workers for periodical check of possible entry of new pests in the given locality or identification of fishes at the site of collection. Following are some of the important filed guides available:

- Peterson Field Guide.
- A Field Guide to the Fishes of Acanthuridae (Sturgeon fishes) and Siganidae (Rabbit fishes) of, 42 p., figs., $13 (pbk). ISBN Andaman & Nicobar Islands/Kamla Devi and D.V. Rao. Kolkata, Zoological Survey of India, 2003, vi 81-8171-017-7.
- A Field Guide to Grouper and Snapper Fishes of Andaman and Nicobar Islands (Family: Serranidae, Subfamily: Epinephelinae and Family: Lutjanidae) P.T. Rajan. Kolkata, Zoological Survey of India, 2001, 103 p., tables, $28. ISBN 81-85874-40-9.
- A Field Guide to Marine Food Fishes of Andaman and Nicobar Islands. P.T. Rajan. Kolkata, Zoological Survey of India, 2003, xiv, 260 p., plates, $83. ISBN 81-85874-96-4. Details No. 30792

Field Book

It provides a complete description about the collected specimens, their habit and habitat, weather conditions and also about the developing stages.

Hand Book

It refers to the field guides, manuals or sometimes comprehensive volume on a group of animals of relatively complete taxonomic study. Examples of some of the important hand books are detailed herein as under:

- Handbook: Indian Amphibians, S.K.Chanda, Kolkata, Zoological Survey of India, 2002, 335 p., figs., maps, $55. ISBN 81-85874-58-1.
- Handbook: Indian freshwater Mollusks, N.V. Subba Rao, 1989, 289 p., $31.00.
- Handbook: Indian Leeches, Mahesh Chandra, 1991, 116 p., $16.00.
- Handbook: Indian lizards, B.K. Tikader and R.C. Sharma, 1992, 250 p., 48 maps, 42 plates, $32.00.

• Handbook: Indian Snakes, R.C. Sharma. Kolkata, Zoological Survey of India, 2003, xx, 340 p., figs., maps, plates, $55. ISBN 81-8171-16-9.

• Handbook on Common Indian Dragonflies (Insecta : Odonata), Tridib Ranjan Mitra. Kolkata, Zoological Survey of India, 2006, viii, 136 p., figs., maps, $35 (pbk). ISBN 81-8171-088-6.

• Handbook on Hard Corals of India, K. Venkataraman, Ch. Satyanarayana, J.R.B. Alfred and J. Wolstenholme. Kolkata, Zoological Survey of India, 2003, xviii, 266 p., $165. ISBN 81-8171-20-7.

Manual

It is published in simple language including characters for species of the specimens. These are meant for the students and layman who can easily identify the animals with the help of such works. Some of the examples of published manuals are as under:

• Pratt, H.S. 1951. A manual of the Common Invertebrate Animals (exclusive of insects). McGraw-Hill Company, New York, 1-854 pp.

• Manual: Identification, collection and preservation of insects and mites of economic importance. Compiled by J.K. Jonathan & P.P. Kulkarni, 1986, 307 p., $25.00.

• Manual on Identification of Schedule Molluscs from India. Ramakrishna and A. Dey. Kolkata, Zoological Survey of India, 2003, 40 p., $17 (pbk). ISBN 81-85874-97-2.

Monographs

Monograph is complete taxonomic publication containing detailed systematic information of all species, subspecies and other taxonomic collections. Following is the example of monographs:

• Frogs of the Genus *Rana*: A Monograph of the South Asian, Papuan, Melanesian and Australian, G.A. Boulenger. Dehra Dun, Bishen Singh Mahendra Pal Singh, 2005, 226 p., $45. ISBN 81-211-0432-7. [Records of the Indian Museum Vol. XX].

• Memoirs of the Zoological Survey of India: Volume15: No. 1: Taxonomy, zoogeography and phylogeny of the genus *Cryptotermes* (Isoptera : Kalotermitidae) from the Oriental region. O.B. Chhotani, $10.00.

• Memoirs of the Zoological Survey of India: Volume: 16 Part 1: Taxonomic studies on some of the Indian non-mulberry silk-moths (Lepidoptera: Saturniidae : Saturniinae). G.S. Arora and I.J. Gupta, 1979, 63 p., plates, $10.50.

• Memoirs of the Zoological Survey of India: Volume 19: Number 1: Geographical Distribution of Odonata (Insecta) of Eastern India/Tridib Ranjan Mitra. Kolkata, Zoological Survey of India, 2002, 208 p., $33 (pbk). ISBN 81-85874-88-3.

Synopsis and Reviews

These include brief summary of current taxonomic knowledge of the group and does not include new taxa. Scattered information is gathered in synopsis and reviews from different sources at one place and thus is useful for future references.

Revision

In taxonomy classification and nomenclature of animals and other organisms is under constant revision. Scientists often form different opinion about the genus to which a particular species belongs to, or whether two or more species with very special features should be classified as species or subspecies. Sometimes taxonomists feel that they have come across a 'new species' and give it a 'new name' but later on find that this species had already been discovered and named by someone else earlier. Under these circumstances the name given earlier takes the precedence and the new name is regarded as junior synonym. When any new taxon is revised, it is published under revision.

• Memoirs of Zoological Society of India: Volume 15: No. No. 3: Revision of Indian crab spiders (Araneae : Thomisidae). B.K. Tikader, 1971, 90 p., $10.67.

• Courtice, Gillian P. and Grigg, Gordon C. (1975-08-01) A Taxonomic Revision of the *Litoria aurea* Complex (Anura:Hyalidae), In South-Eastern Australia. Australian Zoologist, 183: 149-163.

• Bourne, R.A (1973) A taxonomic study of the ant genus *Lasius* Fabricius in the British Isles (Hymenoptera: Formicidae) J. Ent. (B) 42 (i), pp.17-27.

Treatise

Aristotle (384-322 BC) provided the first systematic treatise on animal reproduction, embryology, taxonomy and evolution in his remarkable work *Generation of the Animals* . Sometimes handbooks are referred as treatise for e g.,

• Moore, R.C. (ed.) 1953 Treatise on Invertebrate Paleontology. Geological Society of America and University of Kansas Press.

FEATURES OF TAXONOMIC PUBLICATIONS

Taxonomic work involves study of morphological features and phylogenetic relationships of organisms. In fact, taxonomy is keeping organisms into groups based on set of defined characteristics that is common to all the members of the groups. Taxonomic publications require a lot of detailed study of a particular species. It entails general and specific description which helps to identify the taxon from the other known or similarly closely related groups. It gives a broad idea of the described taxon whereas; the specific description lays emphasis on the characters, which are its major distinguishing features. Following points must be considered carefully while examining a particular specimen.

Diagnosis

During diagnosis most important characters are listed for a particular taxon on the basis of which the taxon under study can be differentiated from other similar or closely related specimen. A given taxon can be differentiated from the other similar or related groups on the basis of most important characters.

Original Description

When a new species, genus or other taxon is discovered, the description published at that time is known as original description. It helps in the subsequent recognition and identification. This serves following two purposes:

- It provides foundation for the new nomenclature under the provisions of Article 10 to 20, of the Code.
- It helps in immediate recognition and identification of the specimens.

Report

It provides more or less complete description of the characters of a taxon without special emphasis on those characters which distinguish it from the other coordinating units. Some of the essential features of taxonomic publications are discussed as under:

- A complete knowledge of the group of the organisms concerned is prerequisite.
- Knowledge of structure and terminology while describing taxa is essential.
- An ability to evaluate differences and similarities.
- The important groups must be selected.
- The words and grammar of the language used should be accurate.
- The description must be clear so that it can be reused and re-checked.

Style of Description

The style of the taxonomic publications should be brief and telegraphic. Articles and verbs of English grammar should be avoided. Series of noun phrases are generally preferred. It implies that the statement should be as brief as possible.

Sequence of Characters

While describing characters, they should be presented in standardized natural order. For example, if the body parts of insects are to be described, then initially the parts of the head region must be discussed first, followed by thorax and abdomen as given in Table 19.1.

Table 19.1 Showing description of characters

Normal Description	Text as per taxonomic norms
The head is one-third longer than it is wide, the antennae are shorter than the body and the outer antennal segments are serrate.	Head one third longer than wide, Antennae shorter than the body, outer segment serrate.

Contents of Description

The specimen is examined by the taxonomist and a general description is prepared along with the characters for its identification. The description should include all possible characters which are useful in distinguishing other taxa at some categorical level. On the basis of description one taxon can easily be distinguished from the other similar known specimen or closely related ones. The description provides a permanent record of the author's ability to observe the characters of specimens under study accurately, record precisely, select and interpret intelligently and express the facts clearly and concisely. The description must include sex and stages of development for e g., nymph, larva and pupa.

While describing species two factors are considered, viz., **diagnosis** and **delimitation**. Diagnosis entails the art and practice of differentiating between two specimens while delimitation involves setting limits to things. It is relevant to add that when a specimen is diagnosed, it is also of imperative necessity to delimit the taxon. Another school of thought suggests that the description should be composite and drawn from different sources, the entire specimen should be described and then the character by which it differs from the rest of the specimen should be discussed. The colour of the specimen should also be described. In many taxonomic groups, brilliant colouration plays a very important role, for e g., colorful birds and metallic colours of insects etc. Following account should accompany the description of the species:

• Place and date of collection and name of the author of the original description (in case of new species, it should be mentioned in the description for e g., *Dysdercus koenigii* new species).

• Habitat.

• Geographical range.

• List of specimens examined during the study.

• Measurements and other numerical data.

• Characters on the basis of which the present taxon differs from its nearest taxon.

• Scientific name of the taxon and its author

• Type.

• Synonymy.

• Discussion.

Illustrations

Taxonomic publications should be adequately illustrated. An out line diagrammatical presentation is best for scientific study. The **Camera Lucida** drawings are best. The illustration should be clear so that it can be reproduced.

Reference and Bibliography

In taxonomic papers, references are published as footnotes when they are sufficiently large in number, they are written as bibliography. The bibliography should contain authors name, date, title, publication or name of the publisher, volume and pages.

Keys

Taxonomic keys help in the identification of specimens by providing diagnostic characters in a series of alternative two choices. After deciding which choice is correct for the given specimen, it is assigned to a particular order. For example, following is the key to order of an adult insect (For details, please see Chapter 18).

1. Wings present ---12
    Wings absent--- 2
2. With long cerci--- Thysanura
    Without long cerci--- 3
2. Large insects, usually 1 inch or more in length (2.54 cm) Orthoptera
3. Small insects, usually ½" in length (1.27cm)---------- 4

FORM AND STYLE OF TAXONOMIC PUBLICATIONS

The taxonomic publication is an important paper, since it is cited for many generations. Before submitting the paper to a journal, the style for that particular journal should be strictly followed. Normally the paper contains following parts:

Title

It is the first part of the paper and contains short words without punctuation marks.

Author's Name

It is written below the title of the paper followed by the address of the institute. Names of authors are usually written in alphabetic order with the senior author's name is given initially.

Abstract and Keywords

A brief summary of the work is given in the beginning. The key words are given after the abstract. It helps in indexing and later search.

Introduction

It deals with the scope of the paper and outlines the reason for the study.

Description

The specimen is described as per taxonomic norms.

Acknowledgements

The author should acknowledge the help received from the place of work, institutes and from the different authorities during the course of study.

ZOOLOGICAL RECORDS

Zoological record (ZR) online, currently published by Thomsen Scientific, formerly BIOSIS and the Zoological Society of London until 2004, is the World's most comprehensive index to Zoological and animal science literature. The ZR covers every aspect of zoology, including biochemistry, biodiversity, behavioural science, biology, biochemistry, cellular and environmental studies, **fossils**, recent and whole animal species, **paleobiology** and systematic taxonomic information. More than 5000 international serials plus approximately 1500 non-serial publications are currently monitored. It also includes professional journals, magazines, newsletter, monographs, books, reviews and conference proceedings. The data base corresponds to the printed index- Zoological Record and includes detail subject indexing in both controlled and language format, complimented by online **thesaurus**. About 72,000 indexed records are added every year. It contains 1.5 millions record archives in electronic formats. It is relevant to mention here that it was first published in 1864. The print copy or electronic format of database can be obtained through on line services. Each one of the millions of records in ZR has been indexed with ZR's specialized system ensuring that searches are both comprehensive and easy to conduct. The description of each and every animal is available in Zoological Record. It has following features:

- It contains nomenclature and original description of animals.
- It can be searched by subject and animal name.
- It contains bibliographic information for journals.
- It contains original description of animal specimens including name change that was available in the print volume.

Polaszek et al., (2005b) have proposed to the International Commission on Zoological Nomenclature, London, UK, to establish a **ZooBank**, an open access mandatory registration system containing descriptions of all new taxa and nomenclatural procedures in animal taxonomy.

It has been proposed that International Commission of Zoological Nomenclature (ICZN) in collaboration with compliers of Zoological Record will act for the development and implementation of ZooBank. It is suggested that Zoo Bank will function as an archived index of zoological names and nomenclatural acts. It shall provide the information about the availability or unavailability of names. The Zoo Bank will act as an open-access, central web-based registry of animal names and taxonomic acts in zoology. The Zoological record database serves to locate the animal with the help of website available for biosis - www.biosis.org.uk. Following is an example of the search result:

Request Details

- Searching for : *Dysdercus koenigii*

Search Results

All names represented in the database with author and or date is currently found in the Zoological Record Animal Named Database. When the name is recorded as new in ZR, the appropriate ZR volume number is tagged. The taxonomic hierarchy provided is taken from the **ZR Thesaurus.**

Classifications currently in use are indicated.

Name   : *Dysdercus koenigii*
Author Date : (Fabricius)
Classification: Pyrrhocoridae
Group   : Hemiptera
        View Taxonomic Hierarchy
Search again *Dysdercus*

- Complete name –Partial name (single epithet), sort by Name

Index to Organism Names

It contains the entire animal, plant and virus name data found within the Thomsen **Biosis** literature databases, Zoological record, **Biological Abstracts** and Biosis Previews which is a complete database on life science. It abstracts and indexes information from more than 5,500 sources all around the world. Over 560,000 new citations are added every year to its data base. It is updated weekly. It contains more than 13 million total records dating back to 1969 (For details, please see Chapter 21).

JOURNALS ON SYSTEMATICS

There are leading scientific journals which publish research papers exclusively on taxonomy, systematics and classification. Some of them are being mentioned herein as under:

- Bulletin of Zoological Nomenclature published by the International Commission on Zoological Nomenclature.
- Journal of Zoological Systematics and Evolutionary Research-Blackwell Science Publications.
- Species Diversity-An International Journal for Taxonomy, Systematics, Speciation, Biogeography and Life History Research of Animals, published by the Japanese Society of Systematic Zoology.
- Systematic Biology Network Newsletter -European Science Foundation.
- Systematic Entomology - Blackwell Science Publications.
- ZooTaxa - International Journal for Animal Taxonomists.
- ZooSystema-It publishes papers on comparative, functional and evolutionary morphology, phylogeny, biogeography, taxonomy and nomenclature.
- Taxa -A biannual publication of the Japanese Society of Systematic Zoology.
- Zoologica Scripta-An International Journal of Systematic Zoology published by the Norwegian Academy of Science.
- Biochemical Systematics and Ecology – Elsevier.
- Biosystematics-An International Journal on animal taxonomy, diversity, ecology and zoogeography, ISSN: 0973-7871(Published by Prof. T.C. Narendran Trust for Animal Taxonomy and Research Council, Calicut, India).
- AGRICOLA Subject Category Codes L700 Animal Taxonomy and Geography-Annual Review of Ecology and Systematics published by Annual Reviews ASC Newsletter Association of Systematics Collections.

REGISTRATION OF NEW NAME IN ZOOLOGY

Biosis, established in 1926, is a non profit organization based in Philadelphia, USA provides a variety of services for those seeking access in life sciences. BIOSIS, UK established in 1980 is a subsidiary of BIOSIS, based in York, England. It compiles Zoological Record. It has developed the Index to Organism Name (ION), a free name search tool that enables any user to check whether a name has been used earlier. BIOSIS is offering; through Zoological record (ZR), to provide a database register of new names in zoology. All names indexed in ZR since 1978 are included in this database. The species described is included in the data base register of new names in Zoology. The authors of new species should provide a copy of their work containing new name or names to Zoological Record (Zoological Society of London and Biosis) under Recommendations 8-A of the International Code of Zoological Nomenclature. However, inclusion of name in this register would not confer or imply the validity of nomenclatural status. Each year ZR selects some, 70,000 items from the life science literature, and extracts some 20,000 new animal names. These names, along with existing names indexed from the literature, are made available to the users through ION in the free resources part of the BIOSIS website. The database serves the resource material needed by those seeking to establish new zoological species, genera or families. The experts in order to ensure that it has not been added previously, twice check the name added in this register. The database register is available on line, free of charge and provides a sound basis for a full nomenclatural repository.

It is essential that the Zoological community agrees on a mechanism to bring all the names together in a central resource. With the help of zoological community and by using existing ZR procedures, a fully comprehensive new animal names database could readily be established.

ELECTRONIC PUBLICATION OF NEW ANIMAL NAMES

Previously the International Commission on Zoological Nomenclature (ICZN) did not recognize on-line-only journals. This means that any name described in an online only journal was not available under the Code and would not be recognized as legally published. A long waited amendments to the Code issued in September 2012 by the International Commission on Zoological Nomenclature now allows publications of new names in online works, provided that the latter are registered with ZooBank, the official register of animal names. According to Krell (2012) the new amendment will speed the process of publishing biodiversity information, improve access to this information through ZooBank and can only help in reducing taxonomic impediment that hinders our discovery and cataloguing of Zoological taxa.

SIGNIFICANCE OF TAXONOMIC PUBLICATIONS

Systematics is communicated through highly standardized publications that are descriptions of taxa (species and their relationships). These descriptions follow Codes (e g., the International Code of Zoological Nomenclature) and have resulted in more than one million printed documents, which unlike most branches of science remain part of the currently accessed body of knowledge.

Taxonomy is perhaps unique among scientific disciplines in requiring access to all of its published literature (unlike many sectors whose literature becomes outdated in few years and is then of primarily historical and not scientific interest). Because taxonomists need to refer to the original descriptions of taxa, information about taxa is cumulative and many species have been described or listed in the literature only once, as such the complete literature is required to ascertain it, not just the most recent papers. It is also necessary to publish the findings in good taxonomic journals. Valdecasas (2011) has proposed a tentative taxonomic index as a standard metric to assess the taxonomic quality of published taxonomic work. The T-index is based on following criteria:

• The description and diagnosis are in strict accordance with the relevant code of nomenclature.

• The standard of description is adequate to contemporary practice in the corresponding clade.

• The journal has unbiased editor and referee systems.

• Any molecular or morphometric evidence that supports a new classification is accompanied by a written description and a diagnosis that is in accordance to the relevant code of nomenclature.

• New taxa hypotheses and full revision work are accorded similar merit.

The discovery of a new taxon is made in the context of another related taxon. For example, new species are diagnosed within a species complex or a genus, following the various codes of nomenclature.

In the age of the internet, it seems obvious to make this entire knowledge accessible to every one interested in natural history and study of taxonomy. It is important the manner in which the published taxonomic information is being disseminated and archived by large international organizations, particularly Encyclopedia of Life and Plazi. The system allows the information to be gathered, indexed and archived on the day of publication.

Chapter 20

COLLECTION AND PRESERVATION OF NATURAL HISTORY SPECIMENS

Taxonomic collections represent the rich biodiversity of different animals of a particular locality. Natural history collection has a major role to play in different aspects of life and contribute significantly towards defeating diseases, addressing environmental pollution, understanding and other scientific studies vital to human society and life on planet earth (Pettitt, 1991, 1994). Museums serve as an excellent source of such repository and are permanent record of fauna and flora including the rare and endangered species. Thus, natural history specimens are an irreplaceable database of species diversity and habitat change.

COLLECTION PROCEDURES

The preparation of material for scientific study involves collection and preservation of animals. This requires a sound knowledge of taxonomy of local flora and fauna. Further, the person going for collections should have the basic training regarding equipments/gadgets used during collection which requires lot of patience, time and energy. The collection procedures vary in different groups of animals and also differ from nature of habitat of the specimens to be collected, which may be terrestrial, aerial or aquatic. Terrestrial collection includes insects, amphibians, annelids, some molluscs, reptiles, birds and mammals. The aquatic collection requires locating animals in fresh water and marine habitat. Fresh water collection includes protozoan, sponges, coelenterates, worms, crustaceans, fish and amphibians etc. whereas others may be obtained from marine habitat.

Time of Collection

The time of collection for different groups of animals varies. It is important to know the habit, habitat and life cycle stages of specimens to be collected and for insects, knowledge about their damaging stages and the host plants is also very essential. Early monsoon provides variety of insects, although some insects are active throughout the year. Some species are only active for a few weeks or day out of the year. Some animals are active in the morning where as, others are at dusk. Some animal species are active during a particular season. For example, during breeding season large number of insect species gathers in groups; while during migration or period of inactivity, different groups of insect may flock.

During period of extreme flowering of many crops and ornamental plants, different insect species may be seen gathering pollens. During growth of foliage, caterpillars of different insects may be observed. It is always better to collect specimens at the time when they are most active. One should be careful not to collect large number of specimens of one kind unnecessarily.

Place of Collection

One can find different groups of animals as per their habitat. So far as insects are concerned they are virtually present in all the habitats. They may be found flying in the air or buried in the soil. They may inhabit fresh water ponds and streams; e g., **flies, caddies flies** and **may flies** or may be found on vegetation, in and around compost, in rotting wood or logs, in leaf mould, in bodies of rotten animals, in cow paddocks, under rocks and pebbles, under bark of the trees. Some insects are found on foliage, flowers and stems or inside the fruits of plants.

Terrestrial Collection

It requires the use of collecting nets, hand gloves, collecting vials, cardboard boxes and containers. For collecting desired animals on land, equipments like insect net, forceps, hand gloves, killing bottles and repellents are also needed. The insect net is used to sweep over the vegetation as well as for catching **butterflies** and other flying insects. The net should be capable of being folded, yet should be strong enough and durable. It can be made with the help of good quality mesh cloth folded conically, stitched and attached over wire-ring and a handle to hold it during sweep operations. Inverted light colored umbrellas or white bed sheet can also be used to spread below the shrubs and trees while they are shaken with a club. The insect that fall on the sheet or umbrella can be collected and narcotized for further preservation.

Aquatic Collection

Aquatic environment fall under two categories e g., fresh water (ponds, pools, lakes and rivers etc.) and marine (sea water).

Fresh Water Collection

Representatives of many phyla including worms, arthropods, molluscs and fish live in fresh water. Reptiles, Mammals and some bird species also inhabit aquatic biota.

Fresh water animals can be collected with the help of **hand nets, plankton nets, dip nets** and small **seine** etc.

Marine Collection

Animals inhabiting marine environment require specialized gadgets for collection. The marine forms may include protozoans, sponges, coelenterates, worms, arthropods, annelids, molluscs, echinoderms, protochordates, fish and mammals. The marine environment may be rough at times with changing tides. Further, many of the marine form inhabit lower strata while others may be found at the surface or just below it.

RECORDING OF DATA

The collected specimen will not serve any purpose, unless it is accompanied by sufficient field notes. As such a **field book** is must for every collection trip. Before collection, it is essential to record data regarding specimens collected. In a large collection, it is easy to assign some number, on the basis of group, season or locality, to the collected specimens, so as to find them easily. Following observation should always be recorded in the field book made up of good quality paper with a medium grade pencil:

- Date of collection.
- Time of collection.
- Stages of collected specimens.
- Number of specimens collected during field visit.
- The host from which the specimens has been collected.
- Altitude or depth from where the specimen has been collected.
- Weather like temperature, humidity, rainfall, sunshine and wind conditions may be recorded.
- Locality should be clearly mentioned. The distance of locality from the nearest landmark is noted, so that in future, if somebody needs to visit the spot one can locate it easily.
- Color of the specimens should be noted down as sometimes the original color of the specimens gets changed due to the vapors of the killing agent. If possible, color photograph of the collected specimen should be taken. This will enable the comparison of collected specimens with the standard ones in the laboratory or museum.

COLLECTING EQUIPMENTS

There are different equipments and methods for collection of variety of individuals. Most animals can be collected carefully with hands. The collected specimens may then be kept in killing jar or containers as per requirement. The specimens may be collected from aquatic or terrestrial habitat.

Aquatic Collection

Such collection involves the use of different types of nets which may either be swiped over the water surface or dipping the net slightly below for collecting the aquatic insects. Following types of collecting gadgets are described herein as under:

Dip Net
A dip net is used as a small seine (Fig. 20.1).

Seine
It is used while holding it inside water to catch insects dislodged from the substrate upstream from the seine (Fig. 20.2).

Dredge Net
It is used to collect samples in streams, ponds, lakes and marine habitat. A dredge consists of a strong net attached to a heavy frame which is pulled along the substratum to collect plants and animals. The net is constructed by making a triangular loop of ¼" metal rod. This is welded to a 3" piece of iron bar which serves as metal rake. The side braces are made from ¼" metal rod attached to the dredge as shown in the figure Fig. 20.3.

Fig. 20.1 Dip net   **Fig. 20.2** Seine   **Fig. 20.3** Dredge net

Terrestrial Collection

Terrestrial specimens can be collected by hand picking while dangerous forms like snakes and scorpions have to be collected with great caution. Insects form the largest class of living organisms on our planet making up to 75% of all the species. More than one million insect species have been described and still many more yet to be described. As such taking their importance into consideration, an account of collection and preservation of insects seems desirable. Instruments like tweezers or forceps, dissecting tools, pocket knife, small camel's hair brush, eye dropper, hand lens, bag or carrying case for the equipments are required. For collection of insects, certain special gadgets are needed which are being described herein as under:

Aerial Net

It is used for collecting flying insects made by stitching the muslin cloth in a conical bag like shape around a wire frame about 30cm in diameter. The wire frame is attached to a stout stick which serves as handle (Fig. 20. 4).

Sweep Net

A sweep net is very effective means of collecting flying insects. It has a short handle and a net bag made up of canvas or other heavy material. It can be swept through dense vegetation and small branches to dislodge small insects. The sweep operation may be carried over several times depending upon the activity of insects (Fig. 20. 5).

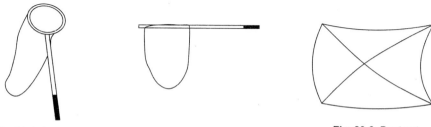

Fig. 20.4 Aerial net   **Fig. 20.5** Sweep net   **Fig. 20.6** Beat net

Beat Sheet

It is a square piece of canvas or cloth sheet used for collection of insects found on shrubs, trees and hanging vegetation. A large bed sheet or paper is placed under the tree; the branch is shaken with the result insects fall over the sheet and then may be picked up and transferred to killing bottle (Fig. 20. 6).

Aspirator

It consists of a tube or bottle open at one or both ends. Rubber cork is placed at the end of the tube. One cork is provided with glass tubing inserted in the tube/bottle and the other end of the tube is provided with rubber tubing. The cork at the other end of the bottle is provided with glass tubing attached to a rubber teat with the help of rubber tubing. While collecting small specimens the end of the glass tube is placed near the specimen and suction is applied through rubber bulb. This creates partial vacuum in the glass tube, with the result, the specimen gets inside the glass tube. Direct suction through mouth is avoided. When the aspirator is not in use the end of the glass tubing should be plucked with little cotton to prevent escape of tiny insects (Fig. 20. 7 and 20. 8).

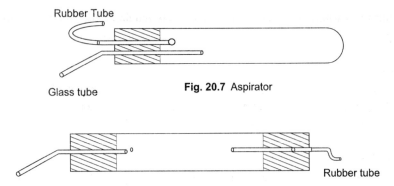

Rubber Tube

Glass tube   **Fig. 20.7** Aspirator

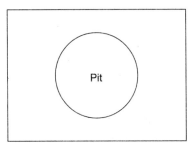

Rubber tube

Fig. 20.8 Aspirator open at both ends

Traps There are different methods for collecting insects with the help of traps. The traps may of following types:

Pitfall Traps There are different types of traps used for collecting insects. It is made by digging small pit and a small container is kept inside. Some preservative like **propylene glycol** is kept in the pit. This trap collects specimens like beetles, spiders, ants and millipedes walking over the ground surface (Fig. 20.9).

Pit

Fig. 20.9 Pitfall trap

Light Traps A light source is hung over a container having a large funnel placed over it, to attract a wide variety of **nocturnal** insects. Once fallen inside the container through funnel, the insects will not be able to come out and may then be collected easily (Fig. 20.10).

Bait Traps These are used for collecting beetles and flies. Fresh fruits like banana, mangoes and melons may serve as good bait.

Berlese Funnel A Berlese-Tulgren funnel consists of a large plastic funnel, a glass jar to accommodate the funnel and a light source fitted above the funnel. About 25 ml ethanol (75%) is kept in the glass jar. This prevents insects to come out of the jar and does not allow predaceous species from feeding over other individuals. It further prevents decomposition of collected specimens (Fig. 20.11). One should be careful during rains while using light source, as it may attract certain venomous insects as well.

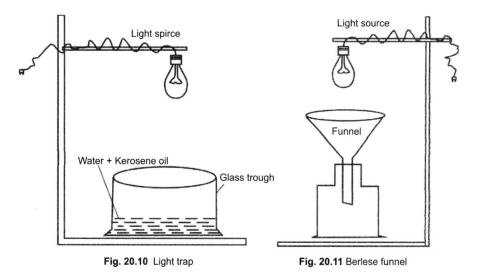

Fig. 20.10 Light trap   **Fig. 20.11** Berlese funnel

Paper Bags It can be made by folding a sheet of paper as shown in the Fig. 20.12. The collected insects can be kept in these paper bags marked with pencil and taken to the laboratory.

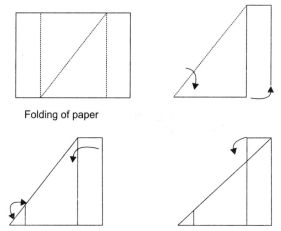

Folding of paper

Fig. 20. 12 Making of paper envelop for keeping insects

NARCOTIZING AGENTS

There are different chemicals used as narcotizing and killing agents like ethyl acetate, amyl acetate, ammonia, para-dichlorobenzene, benzene, chloroform, ethyl chloride, ether, carbon tetrachloride and tetra-chloroethane. Initially potassium cyanide was used as an effective killing agent, however, being deadly poisonous it not generally used now and mild chemicals are preferred. It is relevant to mention, that chloroform makes many insects rigid and brittle. Ethyl acetate is a fair killing agent, as most insects remain relaxed in it for a long time. In a killing jar charged with ethyl acetate, the vapors kill the collected insects quickly without destroying it and also keep the insect soft enough to allow proper mounting suitable for further mounting and collection. There should be proper killing bottles or containers (for large animals) for keeping the narcotized specimens.

KILLING BOTTLE

Once insects are collected, it is better to transfer them immediately in the killing bottle which can be made with half-pint and pint size wide-mouth glass bottle, made up of tough glass with tight fitting screw cap lid. The bottle should be 4" to 5" tall and about 2" in diameter. About 3/8" saw dust should be placed in the bottle and somewhat 3/4" thick layer of freshly prepared plaster of Paris is poured over the saw dust. After it has dried a bit, some holes are made with the help of needles in the layer. The killing agent is poured in the killing bottle, so that it is absorbed by the saw dust and wets it completely. Two or three layers of filter papers cut to size are kept in the killing bottle over the layer of the plaster of Paris. Some crumpled paper ribbons should be placed in the killing bottle, so that jumping insects may not get damaged. A good quality adhesive tape is put around the base of the killing bottle, to prevent the flow of chemicals in case the bottle gets cracked (Fig. 20.13).

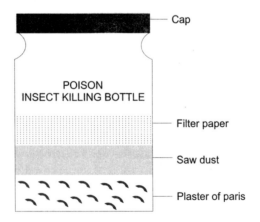

Fig. 20.13 Insect killing bottle

The killing bottle should not be exposed to sunlight as it causes condensation of killing agents on the sides of the bottle which may cause possible damage to the wings, as the insects may stick to it. Large butterflies and moths may be paralyzed by squeezing the thorax, this prevents rubbing off scales by movement against sides of the killing bottle. Plastic containers are not used for making insect killing bottle because the chemicals used in narcotizing the insects will soften the plastic.

Use of Killing bottle

The killing bottle should bear caution label as given in Fig. 20.14. Term 'POISON' should be written over the label in red ink. Sometimes, moisture gets inside the bottle which should be wiped with paper napkins. Large and robust insects like beetles and hoppers should not be placed together so as to avoid possible damage. The killing bottle should be kept out of reach of children and persons who are not aware of the hazardous effect of chemicals.

Fig. 20.14 Caption label

PRESERVATION OF COLLECTED SPECIMENS

After the specimens are collected, the next step is to preserve them. The method of preservation of collected specimens also varies in different groups of animals. Insects are preserved either as dried specimens or kept in liquid.

Preservation in Liquid

Juvenile specimens such as larvae and pupae are collected and kept in small tubes in liquid preservatives. Straight side glass vials are thin walled and often break up easily, hence not recommended. Good quality vials having thick walled glass and neck (homeopathic vials) are preferred. Larvae are killed in boiling water and kept for 1-5 minutes before placing them in alcohol. It prevents them from turning dark. Adult insects are kept in 70-80% ethanol or isopropyl alcohol. Insects with scaly wings or with certain kinds of specific colors if kept in liquid preservatives may get damaged. Heavily chitinized specimens may be studied externally and after being narcotized may be preserved in alcohol. It must be noted that great care has to be taken while preserving soft bodied specimens.

Insects can be preserved in preservatives like Kahle's solution, Carl's solution, chloral hydrate or 80% ethanol. Certain specimens turn dark when preserved in alcohol. In such cases, the alcohol may be replaced 2-3 days after the initial preservation. The identification labels written with water proof black ink should be placed within the vial along with the specimens, additional label may be taped outside the vial.

Emptying of Killing Bottle

The killing bottle should be opened carefully and emptied over a sheet of tough paper and the collected insects are sorted, pinned and grouped order wise.

Preservation of Insects

A brief account of preservation of insects as dry mounted specimen is being discussed herein as under:

Once the insects are collected they should be narcotized immediately, pinned, preserved in temporary storage boxes and allowed to dry completely before transferring them to conventional insect pinning case. Temporary storage box can be made from hosiery boxes. Foam sheet cut to the size of the inside of box should be placed in the box and securely fastened to the base either by some good quality adhesive tape or by pushing pins from all the sides from outside. Some fumigants like naphthalene balls crushed to powder are placed on all the corners of the box within cotton. Different insects like cockroaches, grasshoppers, earwigs, bugs, flies, beetles, butterflies and moths can safely be kept in the temporary storage boxes for short period then finally shifted to insect pinning case.

Relaxing the Collected Specimens

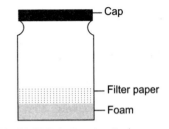

Fig. 20.15 Relaxing chamber

The insects collected if left in the killing jar for quite some time, become brittle. The chamber can be made by keeping little quantity of fine sand in a wide mouth bottle and placing a sheet of thick foam about 1" in diameter cut to the size and placed inside the jar over the sand. Two layers of filter paper cut to the size of the jar are also kept over the foam sheet. Few drops of formalin or carbolic acid crystals are placed in the jar to avoid growth of fungus or moulds. The insects may easily be kept within this relaxing chamber for two to three days depending upon its size (Fig. 20.15).

Mounting of Insects

The spreading boards needed for mounting insects are available with the suppliers of biological goods; however, they can be made as well.

Spreading Board

This is a device used to spread the wings and legs of insects and hold them in desired position until they become dry. The insects are spread over spreading boards. Two stripes of fine grained block boards around 30 cm in length and mounted over a piece of ply board 30 cm x 10 cm around 6 mm in thickness, at a distance of 1.5 cm. The distance between the stripes is optional. It serves as a groove which may be ¼"or ½". The body of the insect can be set within the groove. Two stripes of tough foam are pasted over the block boards' stripes with the help of good quality adhesive as shown in the Fig. 20.16. Medium and large size insects can be directly mounted with the help of pins on the spreading board. Another type of spreading board can be made having adjustable stripes.

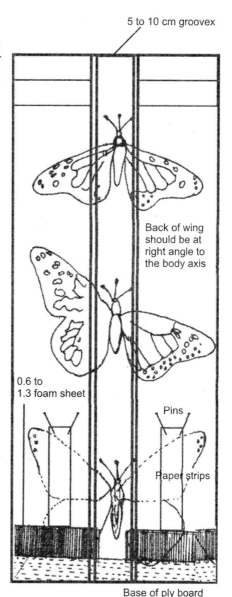

Fig. 20.16 Insect spreading board

Labels on figure: 5 to 10 cm groovex; Back of wing should be at right angle to the body axis; 0.6 to 1.3 foam sheet; Pins; Paper strips; Base of ply board

Pinning of Insects

Special entomological pins (anti corrosive having small diameter with round heads) should be used for pinning the insects. There are about ten sizes of insect pins based on diameter ranging from number 000, 00, 0, 1, 2, 3, 4, 5, 6 and 7, first three sizes are used for direct pinning of small insects. Size 2 or 3 pins are used for most specimens.

Size	Length
000 to 5 .	1.5"
6	1 ¾"
7	2"

Large pin size is required for insects like grasshoppers, giant water bugs and dung beetles. Entomological pins, if not easily available and also being costly, standard sewing needles (No. 9) can be used in lieu of it. Pinning of insects can be done with the help of pinning blocks (Fig. 20.17). Drills are made in the blocks to provide holes of equal depth thus giving the required depth for each step. The holes can be numbered as 1, 2 and 3 and so on to indicate the depth. Large insects are pinned using number 3 pins. Only a quarter of the pins length should remain above the insects.

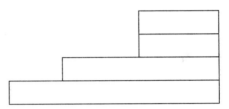

Fig. 20.17 Pinning blocks

The pinning of insects varies in different groups of insects; most insects are pinned eccentrically so that important taxonomic features present on the body of the insect may not be damaged.

Pruthi (1969) has given a good description regarding pinning of insects in different groups of insects (Fig. 20.18).

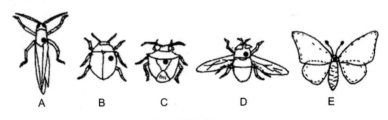

Position of Pin (•)

Fig. 20.18 Position of pinning in different groups of insects
(A-Orthopera, B-Coleopptera C-Hemiptera, D-Diptera, E-Lepidoptera)

• Coleopterans are pinned through right **elytra** or wing cover, close enough to the front, so that the pin passes between the second and third walking legs (through the right wing cover near the base).

• Hemipterans (bugs) should be pinned through **scutellum.**

• Orthopterans (hoppers) should be pinned through the back of the thorax to the right middle line either with one wing spread or both wing back in normal position. The legs should be propped up and antennae brought back along the sides of the body.

• Lepidopterans (butterflies and moths) and odonates (dragon flies) are pinned through the thorax at the thickest point slightly behind the base of the forewings.

• Dipterans and Hymenopterans (flies, bees, and wasps) are pinned through **prothorax** slightly behind the base of the forewings and to the right of the middle line.

Pinning Dowel

A pinning dowel is used to push the specimens down on the pin. It can be made by cutting around a piece of used gel pen refill about 3/8" in length, to have a uniform height of the specimens during display fixed inside a tough foam piece (Fig. 20.19).

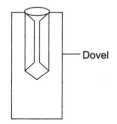

— Dovel

| Data : 08 -VIII- 2008 |
| Locality: Allahabad |
| Order : Coleoptera |
| Family : Coccinellidae |
| Collected by: Ashok Verma |

Fig. 20.19 Pinning   **Fig. 20.20** Insect Label

Labeling

It is very essential to label the collected specimens, as the label provides complete information regarding the specimens. The label should be small (3/8" x ½") so that much space is not occupied by it in the display box. Information on the label can be written with waterproof ink and drawing pens. Date should be written in order dd/mm/yy. In writing date, forms like 08/07/2008 or 7/08/2008 should be avoided. For example, the date 8th July 2008 should be written as 08-Jul-2008 or 08-VIII-2008. After writing the date, locality should be mentioned; followed by the name of order and family and finally the name of the person who has collected the specimen. Now computerized labels are preferred, the correct label may be written as given in Fig. 20.20. The size of the label should not be greater than 18mm x 5mm. While keeping the label with the specimens, consistency should be maintained in the position of the labels. Sometimes two labels the first label being collectors label followed by the taxonomic label. It should be pinned in such a manner that the description can be read easily from the rear of the insect or from the left side. The labels should be kept at uniform height on the pins (Fig. 20.21).

Stuffing of Insects

Sometimes in large insects like cricket and long horned grasshopper, soft abdomen gets partially discolored before getting completely dry. To avoid this, a slit is made along with mid-ventral side of the abdomen then the viscera are removed with the help of a forceps. A small piece of cotton is inserted in the abdomen to give it shape. Insects like dragon fly loose color after being narcotized due to rapid changes occurring in the body.

Display of Small Insects

Some insects being too small may get damaged if pinned, as such double mounting is done through carding.

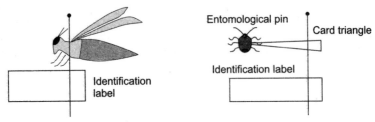

Fig. 20.21 Insect displaying taxonomic label

Fig. 20.22 Displaying carding of small insects

Carding

Small insects can be displayed with the help of triangular piece of card which may be either 20x9 mm or 14x5 mm and by applying water soluble glue on the tip of the card glued to the side of the insect body. Identification labels are then pinned below the triangular card (Fig. 20.22).

Mounting Minute Insects on Slides

Very minute insects belonging to order like **Anoplura** (sucking lice), **Thysanoptera** (thrips), **Collembola** (springtails), **Mallophaga** (chewing lice), **Siphonaptera** (thrips) and **Homoptera** (aphids) are usually mounted on slides. For this, insects should be kept in 5% KOH or lactic acid for de-pigmentation. Then they can directly be mounted in **Arabic-chloral hydrate** medium. After drying of mounting medium, the cover slips should be sealed to prevent absorption of moisture which may discolour the mount. It is an excellent way to preserve and study soft bodied insects.

Storage of Insects

Proper storage of collected specimens is important. Collected insect specimens must be stored in well designed insect pinning cases with tight fitting covers to protect them from insect pests, moulds and dust. Biological homes supply nicely designed insect pinning cases. Wooden boxes are painted inside with solutions of camphor, naphthalene or para-dichlorobenzene in one part each of chloroform and kreosote with six part benzene. There are special cavities in the boxes in which naphthalene, camphor or thymol can be kept. This prevents entry of small insects and prevents growth of moulds. If insects get infected with mould, they should be isolated and cleaned with dry brush and then with a brush dipped in solution of kreosote, alcohol, thymol, camphor and naphthalene. In the storage box, the insects may be grouped order wise. With in the insect pinning case all insects should be mounted at the same distance from the head of the pin. A pinning block may be used as a spacer to help keep specimens and labels at a uniform height.

Protection of Insect Collections

Insect collections are the priceless possession of any museum or institution as such the same must be protected from any damage. The collection must be fumigated off and on. Care must be taken to check the condition of the collection of the specimens and replenish the fumigants kept in the storage box.

Identification of Insects

Identification of collected specimens is very important. The collection would not serve any purpose unless the insects are identified. Even if the specimens are identified, it is always suggested to have an authenticated identification report from an authority.

Dispatching Insects for Identification

Dried insects can be sent by post to the respective identifying agencies for identification. For this small cardboard boxes are prepared. Piece of foam cut to the inner size of the box is placed at the bottom. Pinned insects are arranged over the foam pack. Then a piece of tough card followed by a piece of foam cut to the size of the box is placed over the insect pins to avoid any damage to the specimens. The specimen box is then closed and good quality cellotape is used all around the box. This small box is kept in somewhat another bigger box and space between the two boxes is filled with foam sheets/cotton. A pinch of naphthalene powder in cotton is also placed in the box. A detail description about the field data along with a covering letter should also be placed within the box. Senders name should also be written over the box for easy processing. Smaller insects are kept in plastic vials. Following labels should be fixed over the box before dispatching it.

Preserved Insects for	**Natural History Specimens**
Scientific study	**Handle with Care**

Of No Commercial Value

The best known agencies for correct identification of insects are:
- International Institute of Entomology, 56 Queen The pinning of insects varies in different groups of insects; most insects are pinned eccentrically so that important taxonomic features present on the body of the insect may not get damaged. s Gate, Cromwell Road, London SW7 5 JR, United Kingdom

- Zoological Survey of India, 34, Chittranjan Avenue, Kolkatta-12

In addition, expert taxonomists working in different laboratory from across the globe can also be of great help. Initially agencies used to identify the insects *on gratis*, but now charges have to be paid for the identification.

COLLECTION AND PRESERVATION OF FISH

These can be collected from aquatic biota with the help of seines, nets, minnow traps or other similar traps, hooks and line or electric shockers. After collection, the fishes are brought to the laboratory and sorted according to their shape and size into different groups. Each lot is sorted and assigned the habitat, method of collection and the name of the person who has collected the specimens. The local name of the fish should also be mentioned. Collected fishes are initially fixed in 10% formalin i.e., (one part 40% commercial formalin and nine part water). In larger fish, a cut (5 to 6 inches) should be made in the right side of the abdominal cavity and should be kept in 10% formalin. The formalin should be changed after 2-3 days. Some glycerine is also added. Addition of borax to formalin retards the shrinkage of specimens and softening of bony tissues.

USEFUL TIPS FOR COLLECTION AND CURATION

Natural history specimen collection is quite tedious and requires considerable patience and expertise. Following tips will help in collection excursion:

- Person interested in collecting natural history specimens must be capable of undertaking excursion under extreme conditions and should possess greater mechanical skill.
- Permission from the competent authority must be taken before carrying out collection in restricted area. It is relevant to mention here that persons and sometimes even scientists have been held for not having taken permission to enter in a protected area. It is not only ethical but also a pre-requisite for a scientist to approach the authorities concerned for permission before proceeding for any collection. Section-27 of Indian Wildlife Protection Act-1972 provides penal action for entry without permission to protected area and Section-29 of the aforesaid Act deals with the punishment for removal of wildlife without permission from Chief Wildlife Warden.
- Before going to the field for collection, one should have sufficient boxes, paper bags, collecting equipments, thermo-hygrometer to record temperature and humidity, field book and pencil for taking down observation.
- One should go to the field in full sleeves with shoes and some insect repellent along with a tough stick and an antiseptic cream.

• It is suggested to have a digital camera also to record the insects occurring in nature.

• One should have information about the environmental conditions, security, health and safety at the place of collection.

• Complete information about the habit and habitat of the collected specimens and other related data should be recorded as it is necessary for the proper documentation of the collection.

• Proper curation is also needed by the individuals having expertise in identification of the specimens.

• The collected specimens must be properly housed in museum at a safe place under security. The management of collection and preservation of collected specimens should be handled carefully under the supervision of expert curators.

• There should be networking among important museums so as to enable the curators to have a dialogue among them regarding all aspects of collection including exchange of specimens from one museum to another.

• The collection staff should keep themselves aware of latest development in different aspects of collection.

SIGNIFICANCE OF NATURAL HISTORY SPECIMENS

Natural history specimens have very significant role to play in different aspects of life. The insects can contribute considerably towards defeating disease, combating environmental pollution understanding green house effect. It has been estimated that by the end of the 20th century, between half and one million species would become extinct. The value of natural history collection is enormous but is poorly understood. Natural history collections are an irreplaceable database of information on diversity of species and habitat change. The museum of natural history inspires curiosity, discovery and learning about nature and culture through outstanding research, collections, exhibitions and education.

Blackwelder (1967) is of the opinion that, "when a group of organism interests a student so much so that he decides to do taxonomic work upon it; his first endeavour is to collect, to amass examples of the available kinds, to preserve them in the accustomed way and to use this collection in his studies. Next he will try to identify these species to establish their identity with previously known species......".

According to Hounsome (1984), the basis of all science is that observation by one worker can be verified by others. With taxonomic, distributional and ecological observations, verification is usually possible, if the relevant specimens have been deposited in an accessible and suitable museum collection (Pettitt, 1994).

Ingrouille (1989) has observed, "natural history collections are the foundation upon which the science of systematics is built, and systematics is the alpha and omega of biological sciences from the first naming of newly discovered variation to the incorporation of all knowledge into a system ".

Importance and significance of natural history specimens is being discussed herein as under:

Environmental Studies

Natural history collections have long been indispensable resources for studies of earth's biodiversity and the need to maintain them has recently taken on greater urgency (Davis, 1996, Ponder et al., 2001). Museums offer a unique perspective, providing data over a vast time span ranging from millions of years ago paleontological collections to the present. Three broad areas of study related to species decline and the loss of biodiversity have become crisis disciplines and depend heavily on the baseline information that museum collections offer. These are species' response to habitat loss and fragmentation, biological invasions and the consequences of global climate change.

• Habitat loss (including fragmentation and degradation) is widely considered to be the greatest threat to biodiversity and museum collections allow researchers to document the pace of these changes and their ecological consequences (Shaffer et al., 1998).

• Biological invasions are also considered as an increasingly serious form of global change (Lovei, 1997).

• It has also been acknowledged that global climate change has also threatened the survival of ecological communities and individual species (McCarty, 2001). The effects of global warming have altered the biology of some species of butterflies (Dunn and Winkler, 1999).

Natural history specimens can provide useful data regarding different biological aspects and ecological conditions of the locality from where the specimens have been collected. Further feeding and reproductive behaviour, fecundity, mortality, host-parasite relationships and seasonal migration of different species provide significant information. Biological collections have been particularly useful as sources of information regarding variation in attributes of individuals (e g., morphology, chemical composition) in relation to environmental variables, and provided important information in relation to species' distributions.

Conservation of Nature

According to Pettitt (1997), the collection of research specimens may appear to conflict with the conservation ethics, but it is not the case. Museum curators and conservation agencies are able to advise regulatory measures and actions required to preserve species and ecosystems. According to Oucllet (1985) often the information to define conservation programmes can only be obtained from collections. Distribution and occurrence of a particular specimen may serve as an indicator of climatic conditions of that area and thus help in ecological management. Field biology increasingly reflects growing demand for information and prediction in support of ecological management decisions (Brinkhurst, 1985); but such work needs the objective backing of **voucher specimens**. Voucher specimens are especially important for introduced species or those found in limited habitats and for ecological surveys (Perring, 1977). Mapping the distribution patterns of animals and plants is essential to protect the environment and for the adequate assessment of planning applications. Maps of rare and critical species can be prepared only from museums having voucher specimens. It is relevant to mention that the Royal Society for the Protection of Birds (RSPB) does not favor the use of bird mounts in public displays.

Applications of Biochemical Knowledge

Natural history specimens can provide wealth of unknown or potential information. Level of nuclear fall out, radioactive **Strontium** (Sr) and **Cesium** (Cs) throughout the world can be monitored by analyzing bone tissues collected from the museums since the first nuclear explosion (Challinor, 1983).

Klass et al., (1979) have shown the deleterious effects of pesticides such as DDT on the thickness of the egg shell of birds of prey. The biochemical estimation of feathers has shown the levels of environmental mercury (Thomson et al., 1991a, b). Even a bird 'mount' without head or leg can provide feathers for biochemical analysis or electron microscopy. This shows that all specimens have potential use, whether intact or damaged.

Archaeology and Ethnology

Archaeological excavations provide bones, shells and insects which help to trace the phylogeny of a particular group. Ethnologist also requires bits and pieces of feathers, fur, skin, bone, shells and other botanicals. Thus, identifications would be impossible without extensive reference collections.

Evolution and Behaviour

Natural history specimens help to study the different aspect of evolution like the study of vertebrate phylogeny through comparative anatomy. Classified museum specimens are essential for studying the relationship between different groups of animals which help to analyze the variation within the single species and between the sexes, variation with climate, latitude and with isolation islands, niche variation hypotheses and prey-predator relationships.

Historical Studies

Study of museum specimens can provide sufficient information regarding historical events associated with grand evolutionary march of different animal groups. The history of anatomical preservation and of taxidermy can only be studied with the help of museum specimens.

Service to the Society

The natural history specimens play a significant role in public life. Some of the important contributions of these studies are being discussed herein as under:

Education

Nowadays most museums fulfill the education role with reasonable success; ecological displays help explain the diversity of the life forms that sustain us, show the major patterns of geographic dispersal, and demonstrate the interrelationships between organisms. Interest in natural history specimens can be created by museum exhibits, lectures and publications. Brief notes about these specimens can create public awareness and diversity of the life forms. Involving children and local people in conservation program can help preserve biodiversity in the long run. The exhibits in these museums present natural phenomena, technological innovations and scientific ideas in ways that prompt visitors, interacting with them, to ask themselves questions and reinforce their own learning. Experiences in museums motivate children and adults to become more inquisitive about the natural history specimens.

Medicine

Proper identification of some species of poisonous animals help to work out correct medicine for treatment, this is more important in case of insect and snake bite. Diseases like bilharziasis, bubonic plague, malaria and river blindness (an eye and skin disease caused by a tiny worm called *Onchocerca volvulus,* which is spread by the bite of an infected blackfly) can be controlled, once the agents transmitting the diseases are identified. For example, tick *Ixodes reduvius* and *I. hexagonus* are superficially similar but *I. reduvius* is the causal organism of bovine piroplasmosis (red water fever) while *I. hexagonus* is harmless. Similarly, mite *Trombiculs akamushi* transmits ' rural typhus' while the allied species *T. autumnalis* is harmless.

Health

The strongest link between museum collections and national security is probably in the realm of public health and safety. Collections are often used to track the history of infectious diseases and identify their sources or reservoirs. The most obvious examples are collections of known viruses and bacteria that are stored for comparison with emerging infections. For example, influenza virus from preserved bird specimens in US was compared with that in tissue samples from humans infected in 1918. The studies showed that the virus responsible for 1918 pandemic was more similar to strains infecting swine and humans than to avian influenza, suggesting that the pandemic was not caused by the virus transmitting from birds to humans as previously suspected. Studies have been done regarding about West Nile Virus and Hanta virus. In hotels sometimes mixed and improperly cooked meat is served, which causes diseases like measly pork. It can be identified using reference collection.

Planning

Natural history specimens help in environmental impact assessment (EIA), as certain species may not be available in a particular habitat due to massive anthropogenic development.

Crime Detection

Museum collection can tell the age and race of unearthed human skeletal remains and biological materials like hair, larvae and egg shell from the scene of crime. For this, forensic science comes into action.

Trade and Commerce

Natural history specimens have been a source of inspiration for making patterns on cloths, pottery, calendar, cards and other items. Airplanes hit by birds, can provide pieces of feathers obtained from aero engines which help to identify the bird species soaring at such high altitude and hitting the planes. This may help to devise strategies to prevent accidents during flight.

Recreation

Different authors use museum collections in making coloured plates in their literary contributions. Nature photographers also match the pictures of birds, insects and other animals with reference collections housed in the museums.

Agriculture

By far the greatest role of natural history specimens comes from insects, rodents, fishes, birds and worms etc. Different species of animals attracted to crop can be identified with reference collections. When biological agents threaten an agricultural resource, it is difficult to determine whether they arose naturally, or deliberately? The natural history collections can provide the clues necessary for identify the source of agriculturally harmful organisms. The museums are a forensic resource for determining when and from where a pest, pathogen, or vector was introduced. Identification of household insect pests can also be compared with museum collections. Different varieties of fish can also be studied with the help of natural history specimens to boost aquaculture production.

NATURAL HISTORY COLLECTIONS AS NATIONAL RESOURCE

Natural history collections have attracted the attention of everyone since the dawn of civilization. Many countries of the world have issued commemorative postage stamps on different animals like insects, molluscs, birds, fish, reptiles and mammals highlighting their importance and to draw the attention of masses for their conservation. In India, tiger and peacock have been declared as the national animal and national bird respectively.

According to Pettitt (1986) in British Institutions, the natural history collections represent a great national resource through the activity of FENSCORE (Federation for Natural Science Collection Research). Natural history museums must play an important role in supporting taxonomy and educating the public (Brooke, 2000).

According to Suarez and Tsutsui (2004) to make sure the survival of these assets and the untapped knowledge they contain, the collections must be well curated and maintained. Further, the benefit of these collections to the society must be maximized by stepping up the rate at which this information is entered in databases and made accessible. Ultimately, maintaining and developing the infrastructure of museums will produce long-term benefits.

Natural history collections still have a major role to play in many aspects of life today. It is important to bring taxonomy to the public, as society must know the importance of different groups of animals and their role in nature.

Information from natural history collections about the diversity, taxonomy and historical distributions of species worldwide is becoming increasingly available over the internet. Computerization of collections and development of electronic catalogues are providing new capabilities for curetting collections (Graham et al., 2004). Funding programmes should be also addressed to support this type of collection management and to provide more educational activities between scientists and children, since some of them may belong to the next generation of taxonomists. Science knows no boundaries; collections are not national possessions but assets of the entire scientific world.

Chapter 21

INFORMATION RETRIEVAL

The total number of the species on earth has been estimated to be in the magnitude of 2-50 million. Due to ongoing anthropogenic developments, habitat fragmentation and different ecological factors, there is ongoing loss of **biodiversity** which runs at 10^3 times of its natural pace, eliminating 10^4 species per year. It is estimated that over the next 30 years, 10^6 species are bound to disappear even before they become known.

Once a particular species is discovered, it is ought to be described and published to make it available to all the taxonomists working across the globe. The first published paper about a particular organism finds a place in the record to be used in future. As in the library, books are classified and arranged subject wise, so that they can easily be located. It was Carl Linnaeus (1707-1778), who for the first time recorded information about animals on sheets /separate card so that more information can be added as and where needed and retrieved. Linnaeus compiled the information in alphabetic order for the purpose of later retrieval.

We gather information in order to recollect it, retrieve it, evaluate it, select it, understand it, rearrange it and apply it as and when required. However, with nearly 8.7 millions species it is almost impossible to bear in mind. In recent year vast amount of biological information like genomic and proteomic data, information on biodiversity and natural history and scientific publication has been made available digitally. All these resources require, at least in part, the ability to retrieve information about organisms, typically by the name of the animals. User of this data vary in their knowledge regarding the animals, so that effective interaction with these data sources often require trial and error exploration until the correct taxonomic spellings and appropriate key words are discovered that will allow a successful **query**.

According to Canfora and Cerulo (2004) in recent years information retrieval has become an important subject due to wealth of information available in digital format which has grown exponentially and the need for retrieving such information has assumed a crucial significance. In this context the World Wide Web (www) and the Digital Libraries have also served exceedingly well as they are providing the desired information just at the click of the mouse. These have provided an effective mechanisms and tools to retrieve documents from a very large collection based on the user's requirement. Traditionally, IR has concentrated on finding whole documents consisting of written text, there are still other fields like question-answering IR system, image retrieval etc.

According to Rijsbergen (1979) **information retrieval** (IR) is the scientific discipline that entails the analysis, design and implementation of computerized systems that deal with the representation, organization and access to large amounts of heterogeneous information encoded in digital form. Knowledge organization and classification in information retrieval process examines current efforts to deal with increasing globalization of information and knowledge.

WORKING OF INFORMATION RETRIEVAL SYSTEM

According to Tuominen et al., (2011) exploitation of natural resources, urbanization, pollution and climate changes accelerate the extinction of organisms on earth, which has raised a common concern for maintaining biodiversity. In order to document the existing information about plants and animals an efficient cataloguing of dynamic biological data from different sources is required. The information is also supplemented from literature and natural history collections.

Names and 'ontological taxonomies' of organisms constitute central resource in biodiversity management system. The term **ontology** in terms of computer science refers to individuals, classes, sets, entities, ideas, collections, concepts, functional terms, restrictions, rules and events along with their properties and relations, according to a system of categories. Developing an ontology help to share, reuse and analyze domain knowledge with the help of different web sites. Ontology together with a set of individual instances of classes constitutes knowledge bases which consist of following:

- Cataloguing and metadata.
- Templates for the internal organization of documents.
- Entry relationship approach as the basis for all information organization.
- Database organization, relational databases, object-oriented database and frames.
- Knowledge organizations system classification schemes, taxonomies and ontologies.

An information retrieval process begins when a user enters a query into the system. Queries are formal statements of information needs; for example search strings in web search engines. In information retrieval a query does not uniquely identify a single object in the collection. Instead, several objects may match the query, perhaps with different degrees of relevancy. An object is an entity that is represented by information in a database. User queries are matched against the database information. Depending on the application the data objects may be for example, text documents, images, audio or videos. Most IR systems compute a numeric score on how well objects in the database match the query and rank the objects according to this value. The top ranking objects are then shown to the user.

The biological texts are written, content is indexed in databases and information is searched for using different names and terms from different times and authorities. In biological research, scientific names are used instead of vernacular names. A database is a collection of information that is organized so that it can easily be accessed, managed and updated. The databases can be classified according to the types of content; bibliographic, full-text, numeric and images. Following account explains the different concept using IR system:

Biological Names and Taxonomies

The scientific names are based on binomial nomenclature as proposed by Linnaeus A scientific name often has a reference to original publication where it was first published. Species list catalogues organisms occurring in a certain geographical area which may vary from a small region to the whole globe. The list often follows different hierarchies and the species may be associated with different genera according to the worker who published the list.

TaxMeOn-Meta-Ontology

Being a meta-ontology, **TaxMeOn** defines classes and properties that can be used to build ontologies. The ontology can be used for creating semantic metadata for describing observational data or museum collections. The core classes of TaxMeOn express a taxonomic concept, a scientific name, a taxonomic rank, a publication, an author, a vernacular name and the status of a name. Taxonomic ranks are modeled as individual taxa. Publications include documented source of information. A scientific name and taxonomic concepts associated to it are separated, which allows detailed management of both of them. Different status can be associated to names, such as validity (accepted/synonym), a stage of a naming process (proposed/accepted) and spelling errors. In TaxMeOn, a reference (an author name and a publication year) to the original publication can be attached to a name. The ontology model consists of three parts that serve different purposes:

Name collections

Scientific names and their taxonomic concepts are treated as one unit in the name collection, because there is scope of **vernacular names**. The model supports the usage of multiple languages and dialects of common names. A taxon may have several common names, but only one of them may be recommended or has an official status.

Species list

These have a single hierarchy and seldom include vernacular names. It has more relevance in science than name collections but lack information about name changes. Synonyms of taxa are included in species list and the taxonomic concept is included in a name like in name collection. Taxa occurring in different species lists can be mapped to each other or to research results using the relations.

Biological Research Result

Taxonomic concept is a key element in biological research which can have multiple scientific names (and *vice-versa*) and are used for defining the relations between taxa. The latest research results often redefine concept. Thus the status of a scientific name may change in time as an accepted name may become a synonym.

INFORMATION RETRIEVAL MODELS AND GADGETS

Canfora and Cerulo (2004) have given a detail account of information retrieval consisting of following two essential components:

Information Retrieval Objects

It is generally an artifact that exists in the form of tool or service and responds to answer 'what' the question.

Information Retrieval Models

It consists of set of theories on which the information object is based and responds to 'how' question. It is relevant to mention that the above referred two aspects are related as one object can be based on more than one model and it can serve as a basis for more than one object. On this basis Canfora and Cerulo (2004) have given an account of vertical taxonomy (considering IR models) and horizontal taxonomy (referring to objects).

Vertical Taxonomy

It classifies IR models with a respect to a set of features. It is relevant to mention that modeling the process of IR is complex as many parts of it are vague and difficult to formalize. It is built by exploring two basic features of any IR models. The IR models are characterized by certain aspects.

The Representation The model is adopted to represent both documents and queries. It is an essential component of an IR system and is the representation of information itself. The information can be processed if it is represented in some meaningful way. The information retrieval models can be very complex and the first step is the representation of documents and needs of information. From this representation a reasoning strategy is defined which helps to solve a representation similarity problem to compute the relevance of documents with respect to queries. Representation and reasoning can be used to characterize an information retrieval model

The Reasoning It refers to the framework adopted to resolve representation similarity problems to compute the relevance of the documents with respect to queries. One of the keyword based query representation is discussed as under:

Keyword-base It consists of keywords and the documents containing key words are searched for. These queries are popular as they are easy to express and search. Usually it is a single word but in general it can be more complex combination of operations.

Horizontal Taxonomy

It is derived from an analysis of the application areas of IR. With the help of horizontal taxonomy; information retrieval objects can be classified. An information retrieval object is identified by three components:

Tasks These are concerned with a particular aspect of information retrieval derived from a user point of view. An information retrieval object can support one or more tasks and a task can be stand-alone or it can be integrated in a process to perform a larger task. The task can be classified as under:

Ad Hoc Retrieval It is characterized by an arbitrary subject of the search for short duration. It is typically performed by a researcher doing literature search in a library. A retrieval system's response to casual search is generally a list of documents ranked by decreasing similarity to the query. The internet search engines are examples of information retrieval objects from which one can perform ad hoc search.

Known Item Search It is similar to an *adhoc* search, but the target of the search is a particular document (or a small set of documents) that the searcher knows to exist in the collection and wants to find it. An information retrieval object that performs this task usually implements a precise query language with which a searcher can reach parts of a document with known structure and semantics. For example, in library different articles can be retrieved by a researcher on the basis of the author.

Interactive Retrieval During the interactive task the worker attempts to see how to interact with it and, as a consequence, can modify the current search strategy.

Filtering It combines aspects of text retrieval and text categorization. It processes documents in real time and assigns them to zero or more classes.

Browsing When users are not interested in posing a specific query to the system, but they invest some time in exploring the document space, looking for interesting reference, then they are browsing the space, instead of searching.

Mining It is a process of automatically extracting key information from text documents. It can be language identification, feature extraction, terminology extraction, predominant themes extraction and relation extraction. The IBM text miner is a mining tool integrated with the homonymous text search engine.

Gathering This is an activity pro-active acquisition of information from the possibly heterogeneous sources.

Crawling It is concerned with the activity of selecting new or updating the existing sources of information that will be processed by successive activities. It is also known as indexing process.

Form It refers to the way in which the object is made available to the final user in the form of tool or service. When the object is implemented in the form of software product, then it is referred as tool, whereas, when the object exists only in one or few instances used to deliver some information retrieval services, then it is a service. The examples are the search engines available on the web.

Context The context of an information retrieval object refers to its domain of application. It can be general or specific. A general purpose information retrieval object operates on heterogeneous domains and contents, whereas, context specific system operates on document collections belonging to a specific domain, such as legal and business document and technical papers. Examples are web search engines where the high heterogeneity of information calls for a very general purpose approach, e g., Google, Altavista and Infoseek. Further, the vertical dimension classifies information retrieval models based on representations and reasoning. The horizontal dimension classifies information retrieval objects with respect to the application areas. In recent years information retrieval has assumed an increasing importance due to dramatic growth of the extent of information available in the digital form.

MANAGING ANIMAL DIVERSITY

The taxonomic description of animals has always been quite complex, however, of late extensive use of different software with the help of computers has solved this complexity to a great extent. There are several computer generated description programs available to facilitate checking of the data and **Intkey** used for further data verification and for finding differences, similarities and correlations among taxa. The interactive identification and information retrieval program, intkey offers better and more comprehensive features than any other similar program. These features include:

- To express variability or uncertainty in attributes.
- Defining taxa in terms of nominated sets of characters.
- To specified degree of redundancy.
- Optional display of notes on characters.
- Freedom to carry out operations in any order.
- Generating diagnostic descriptions for specimens or taxa.
- Simultaneous viewing of several illustrations.
- Calculation of ' best' characters for the use in identification of specimens.
- Entry and deletion of attributes in any order during identification.
- Defining key words to represent characters, subsets of characters, taxa and character states.
- Descriptions are brought together; screen display of illustrations of characters and taxa, scaling and scrolling of illustrations.
- The ability to alter treatments of unknown, retrieving free text information (i.e., information not coded in terms of character list).
- Direct handling of numerical values, including ranges of values and non-contiguous sets of values.
- Complete online help, normal and advanced mode of operation and acceptable response time with large sets of taxa.

To produce descriptions, **keys** and Intkey program in different languages need to be translated, as all of the program text (menu, commands, diagnostic messages and help) is in simple text files separate from the program files. English, French, German, Italian and Spanish versions are currently available. The interactive key allows free choice of characters being easy to use and can lead to correct identifications in spite of occasional errors. It can display all the illustrations, full and partial descriptions, diagnostic descriptions, differences and similarities between taxa, lists of taxa exhibiting or lacking specified attributes and distributions of character states within any set of taxa.

Object Oriented Taxonomic Biodiversity Databases

It is relevant to mention that object-oriented (O-O) modeling which includes **O-O system analysis**, designing of software and programming has come to rescue in cases, where complex data prevails. According to Saarenmaa et al., (1995) the object oriented modeling is based on certain concept that makes it possible to build real world models into software that can mimic even the most biological structures. The object oriented models have following features:

- Objects can be classified into hierarchies.
- Objects communicate by sending messages to each other.
- Objects can be associated with each other; they even contain and consist of other objects.
- Objects are encapsulated entities that combine data and behavior (program code) under the same structure.
- Objects can either be classes (templates for examples) or actual instances of these classes or preferably even both at the same time.
- The object oriented modeling provides the tools for modeling complex biological domains such as taxonomy and biological field data logically.

Computer Programs and Internet Resources

There are several systematic and taxonomic websites available which can help in information retrieval. By surfing different sites, one can have access to different animal groups of various continents. Some of the major taxonomic information available on the web is detailed herein as under:

Universal Register

Polaszek (2005 a) has proposed a **universal register** for animal names, which is vital to move taxonomy into 21st century. The registration of animal names is a mandatory requirement for availability according to the Code. Mandatory registration has added advantage of ensuring that all new names and taxonomic methods are checked for compliance with the Code before they are made available. The changes to the Code can be implemented either by making amendments to the existing (4th) edition or in the context of a new (5th) edition.

Zoological Record

The Zoological record (ZR) provides comprehensive coverage with detailed indexing based on a highly developed thesaurus (For detail, please see Chapter 19).

Animal Diversity Web

This site is an online database of animal natural history, distribution of animals, classification and conservation biology.

The Electronic Zoo

It categorizes and lists lot of information on animals and veterinary medicine.

Description Language for Taxonomy (DELTA)

It is a flexible and powerful method for encoding taxonomic descriptions for computer processing that can be used to produce natural language descriptions, conventional or interactive keys, cladistic or phenetic classification and information retrieval systems. The interactive key allows free choice of characters, easy to use and can lead to correct identification in spite of occasional errors. It can display all the illustrations, full and partial descriptions, diagnostic descriptions, differences and similarities between taxa, lists of taxa exhibiting or lacking specified features and distributions of character states within any set of taxa. It has been adopted as a standard for data exchange by the International Taxonomic Database Working Group. The DELTA Mailing List and the DELTA Newsletter are media for discussion of descriptive databases. Topics include, computer programs for taxonomy, data interchange standards, data capture, data analysis, data base design, description printing, expert systems, information retrieval, interactive identification, key making, mapping and taxonomic characters.

Global Index Names (GNI)

It a huge collection of scientific names which are given to both the genus and all of its subsequent species and sub species. Higher levels are also included. Subspecies are not being included. Names are obtained from a number of different but credible sources.

Global Names Architecture (GNA)

It is a system of database, program and web service-a cyber infrastructure used to discover, index, organize and interconnect online information about organisms and their names. Names are included in almost every statement and database about organisms. In the electronic word, names are metadata which can be used to discover and organize information about organisms. The GNA is a communal open environment that manages names so that information about organisms can be managed and serve the need of taxonomists. It is an effort by GBIF and others to develop shared infrastructure and standards to improve the sharing and integration of taxonomic information resources. The GNA aims to identify and redirect the taxonomic information to the taxonomists. It is developing methods that allow more rapid, direct access to the data and a path towards creating a complete list of scientific names that are integrated into a consistent, comprehensive and stable taxonomic framework. Such a framework serves as a global informatics tool for access to information about species

Index to Organism Names (ION)

It is a tool for biologist to check whether a name has been used earlier. It contains the organism names related data gathered from the scientific literature for Thomsen Reuter's Zoological record database. It is the most comprehensive name database, sponsored and hosted by Biosis (Biology Browser providing internet resource guide for Zoology) incorporating name data from the resources of BIOSIS and several other collaborating organizations. Biosis and the Zoological Society of London publish the index to World Zoological literature. The staff of the Zoological Society maintains the Zoological Record. The index currently covers animals (all names reported in ZR since 1978 and including Protista which are generally considered animals), algae, bacteria, fungi and mosses. Biosis is offering through Zoological Record a database register of new names in zoology. The register would provide the basic information required by those seeking to establish which new zoological species, genera or families have recently been described and named.

It is relevant to mention that inclusion of a name in this register would not indicate or imply its validity or any other nomenclatural status. It is an aid for the general bioscience community giving basic nomenclatural and hierarchical information. The ZR volume is used to identify the taxonomic group to which a named individual belongs and the link to further information from ZR or other collaborating organizations. Authors of newly published animal names may use it to check whether their names are correctly reported in the Zoological Record, as recommended by the International Code of Zoological Nomenclature. There is a subset of TRITON the 'Taxonomy Resource and Index to Organism Names' which provides full index and bibliographic data for newly published and changed animal names reported in Zoological Record since 1978.

The Global Biodiversity Information System (GBIF)

It is an international non-profit organization which provides free and universal access to data regarding the world's biodiversity. A number of countries and different organizations share this information system. The GBIF On line Resource Centre (ORC) is a global on-line service that provides easy access to documents, best practices, tools and links relevant for GBIF Participants and their Biodiversity Information facilities. This system allows advanced functional facility such as support for diverse resource types, wide thematic scope, different ways of accessing resources, enabling community contributions, different levels of resource access and multilingual support. GBIF in partnership with the Encylopedia of Life, the Catalogue of Life Partnership and many others organizations which either provide or require access to information about names including following:

• Developing the Global Resources Discovery System (GBRDS) to serve as a common name - resource registry system capable of serving the requirements of multiple data mobilisation initiatives.

- Working with the Biodiversity Information Standards group to define a series of exchange standards for taxonomic/nomenclatural data suited to different user communities.

- Offering the Integrated Publishing Toolkit as a taxonomic and nomenclatural data publishing system that, in combination with the GBRDS, provides a distributable publishing and exchange framework for the GNA

- Developing best practice guides on the mobilization, use and application of taxonomic data in biodiversity informatics.

- Hosting workshops and providing support for GNA activities.

- Developing taxonomic indexing services, tools and applications that draw upon the GNA.

TaxonTree

Cynthia et al., (2004) have given a very detail account of visualization for taxonomic and phylogenetic trees. The TaxonTree provides several ways to search, choosing Latin name or common name or **synapomorphies** queries data within TaxonTree help to visualize it more easily. While Animal Diversity Web (ADW) searches all text available at the ADW, search results are presented as a node-link diagram, where user can see the path back to the root to clarify the name results and can browse nearby nodes to provide further biological context.

Expert Center for Taxonomic Identification (ETI)

It develops and produces scientific and educational computer databases and the world biodiversity databases.

SYS Tax Database System

This database is an integrated concept based system for systematics and taxonomy storing biodiversity data. It stores a number of 'concepts' of a taxon regarding its systematic position and its synonyms, literature and information about botanical gardens and multimedia objects.

Bayesian Analysis in Molecular Biology and Evolution (BAMBE)

It is free software package for the Bayesian analysis of phylogenies and includes programs for analyzing aligned DNA or RNA sequence data and allows data sets with gaps or indeterminate sites.

Mantis

It is a biological database manager featuring the ability to store taxonomic and specimen data, images and sounds, management of citations, specimen loans and addresses.

TreeBASE

According to Roderic (2007) the phylogenies stored in TreeBASE provide a wealth of information on phylogeny. It also serves as a resource for studies on the relative merits of different source of data, the shape of evolutionary trees and methods for querying trees. It is currently the only available large scale phylogenetic database. Its utility is hampered by lack of taxonomic consistency; both within the database and with names of organisms in external genomic, specimen and taxonomic databases. It differs from other phylogenetic databases, such as PANDIT and TreeFam, in being primarily a collection of evolutionary trees for organisms rather for gene families. Although it contains only a small fraction of the evolutionary trees published to date, the database is continually growing in part, because a number of journals either require or encourage authors to submit their data sets and trees to TreeBASE. In addition to supporting simple text searches to retrieve data, TreeBASE has tools for searching based on tree similarity and for constructing super tree.

Pandora

It is database system for taxonomy and biodiversity of Washington Seattle.

BIONet International

It is global network for taxonomy.

Diana

It is a software management system for DELTA.

Platypus

It is a database package for taxonomists to manage taxonomic, geographical, ecological and bibliographic information.

Taxis

It is a taxonomic database management system designed for biologists (professionals as well as amateurs).

U Bio Portal

Information about organisms is often linked to name. This can create problems in information retrieval because one taxon can have many names and the same name can refer to many taxa. U Bio is working on tools which would provide biological informations that address these problems. The U Bio Taxonomic Name Server acts as a name thesaurus. Names have many different classes of relationships that can be used to organize and retrieve information that is explained along with names. These classes are divided into two inter-connected services.

OCEAN

The collection of the Zoological Institute at Russian Academy of Science is one of the largest in the world and contains over 100,00 samples of 26,000 species of marine invertebrates and over 1,60,000 specimens of 8,700 species of marine and freshwater fishes and fish like vertebrates of the world. The 'OCEAN' consists of four main tables: the taxonomic table containing the name and nomenclature of the taxa, the geographical table including the data of field books and catalogues of museum collections, the ecological table and a bibliographic table.

Systema Naturae 2000

It consists of an up-to-date historical cross referenced classification of life anywhere, anytime based on original reliable scientific literature. The main classification is a compilation of the scientific literature available to the author and reflects the most recent insights. Here each taxon is presented at its original position in the main as well as in the original classification. It also deals with both extant and extinct taxa.

Taxonomicon

It is biodiversity information system that contains information from a multitude of sources. Information from each original and authentic source remains intact and also provides a historical account. The taxonomicon combines following four major components:

The Index of Life
It attempts to enumerate the names of the world's past and present biota. The list is completed with nomenclatural issues like authorship, synonyms, homonyms and the common names. It allows accurate synonymy at the circumscription level by combining the scientific name of the taxon with the sources in which it appeared.

The Tree of Life
It shows multiple alternative and historical classifications. Separate classification based on taxa and clades have been given. Synonyms and alternative entries can easily be included in or excluded from the tree.

The Web of Life
It shows many types of interrelationships between species and allows more complex constructions.

The Facts of Life

It provides information on the taxa. The information consists of mainly structured metadata, i.e., no observation of specimen data. Currently, the Taxonomicon focuses on geographic distribution, habitat and geologic distribution, but it allows much wider array of properties of different types, e g., conservation status, body size, pH and salinity ranges, population size etc., that may be used as input by many other systems.

Tree of Life Web Project (ToL)

It is a collaborative effort of biologists and nature enthusiasts around the world. The project provides information about biodiversity, the characteristics of different groups of organisms and their evolutionary history. ToL pages are linked one to another hierarchically, in the form of the evolutionary tree of life. It is a collection of information regarding the biodiversity about each species and for each group of organisms, living and extinct. Starting with the root of all Life on earth and moving out along the diverging branches to individual species, the structure of ToL project, thus illustrates the genetic connections between all living things.

Species 2000

It is a federation of database organizations across the world that compiles the *Catalogue of Life*, a comprehensive checklist of the world's species, in partnership with the Integrated Taxonomic Information System (ITIS). It is an autonomous non profitable federation of taxonomic database organizations involving taxonomists through out the world being a participant in GBIF and recognized by UNEP. Its goal is to collate a uniform and validated index to the world's known species. The aim of Species 2000 project is to create a validated checklist of the entire world's species (plants, animals, fungi and microbes) by bringing together an array of global species databases covering each of the major groups of organisms. The creation of Species 2000 was initiated by Frank Bisby and colleagues at the University of Reading in UK in 1997. The *Catalogue of Life* was first published in 2001. While administrators and member organizations of Species 2000 are located across the world, the secretariat is located at the University of Reading.

Each database covers all known species in the group using a consistent taxonomic system. The participating databases are widely distributed throughout the world and current number is 52. The existing global species database presently account for some 60% of the total known species. As such substantial investment in new databases will be needed for full coverage of all taxa to be achieved. The progamme in partnership with the Integrated Taxonomic Information System (ITIS) of North America currently produces the Catalogue of Life. This is used by the Global

Biodiversity Information Facility (GBIF) and Encylopedia of Life (EoL) as the taxonomic backbone to their web portals. There are two regional programmes:

• Species 2000 Europa   It is working with global and regional databases based in Europe.

• Species 2000 Asia-Oceania It is working to promote taxonomic studies in Asian countries.

ITIS (Integrated Taxonomic Information System)

ITIS provides an automated reference database of scientific and common names for species. It refers to the partnership of federal agencies and other organizations from the United States, Canada and Mexico with data experts from around the world. The ITIS classification is a true Linnaean nested hierarchy. The ITIS database is an automated reference of scientific and common names of biota of interest. ITIS is a part of the US National Biological Information Infra Structure (NBII) and an associate member of GBIF. ITIS and its international partner, Species 2000, cooperate annually to produce the Catalogue of Life, a checklist and index of the world's species. ITIS and the Catalogue of Life is core to the Encyclopedia of Life initiative announced in May 2007.

ITIS provides basic scientific information on the nomenclature, taxonomy and common names of large number of individuals in English, French and Spanish. It also acts as gateway to additional information by providing users with an 'Innovative Search Portal'. This portal automatically uses ITIS information to boost the relevance of specific queries in several major Internet Search Engines. It also facilitates access to other specialized databases in the area of biotechnology, genomics, botany, entomology, bibliographies or to numerous collections of biological specimens and observation data. As of April 2012, it contains over 719,000 scientific names, synonyms, and common names for terrestrial, marine, and freshwater taxa from all biological kingdoms (animals, plants, fungi, and microbes).

ITIS couples each scientific name with a stable and unique taxonomic serial number (TSN) as the 'common denominator' for accessing information on such issues as invasive species, declining amphibians, migratory birds, fishery stocks, pollinators, agricultural pests, and emerging diseases. It presents the name in a standard classification that contains author, date, distribution and bibliographic information related to the names. It allows comparing a list of taxon names to the scientific name, rank and author. This is a completely new version of the old tool. The ITIS news is now available through an RSS (Rich Site Summary) feed.

ZooBank

It is part of the Global Names Architecture and is supported by the U.S. National Science Foundation ZooBank provides a means to register new nomenclatural acts, published works, and authors. It is the official ICZN online registry for scientific names of animals, an authority for zoological nomenclature, and an information hub providing a dynamic web nomenclature for all other web-bioinformatics. According to Wilson (1992) a leading authority on biodiversity, ZooBank is a major step towards completing the Linnaean enterprise, which is essential for mapping earth's still poorly known fauna. Time has come to ensure access and integration of taxonomic information, for this the ZooBank and other systems of taxonomic data integration on the web. With the firm foundation of an authority on scientific names, the rest of biology will be immensely strengthened, and humanity correspondingly benefited. The progress report of ZooBank for the first quarter of 2013 is now available. It contains the following informations:

- 148,625 total registrations
- 92,543 registered Nomenclatural Acts
- 18,555 registered Authors
- 37,527 registered Published Works

The Catalogue of Life

ITIS (an organization of North America) and its international partner, Species 2000 (global network based in UK and Japan), initially worked independently started collaborating in June 2001cooperate to annually produce the Catalogue of Life, a checklist and index of the world's species. ITIS and the Catalogue of Life is core to the Encyclopedia of Life initiative announced May 2007. EOL will be built largely on various Creative Commons licenses. The Catalogue holds essential information on names, relationships & distributions of 1.3 million species of plants, animals, fungi and micro-organisms.

The latest edition of the catalogue of Life (August, 2013) lists 1,426,888 species for all kingdoms, from 135 databases with information on 109,882 infraspecific taxa and also includes 1,036,570 synonyms and 411,816 common names with coverage of 74% of the estimated 1.9 million species known to science. The animal check list is published every year and can be cited and used as a common catalogue for comparative purposes by many organizations. It plans to prepare a comprehensive catalogue of all known species of organisms on earth. This catalogue is complied with sectors and is provided by 95 taxonomic databases from across the world. This combined annual checklist has become well established as a cited reference used for data compilation and comparison.

The information retrieval system and the geographic information not only make the work of the zoologists easier in gathering the data on species and manually mapping these data to quickly visualize the information on the occurrence of the animals from the collection of the museums. This kind of software and databases will be helpful in the analysis of the long-term changes of fauna composition among different regions. This will also help to consider the various theories regarding distribution of fauna over different continents, using maps with geological reconstruction. To ensure research of excellence in taxonomy, the taxonomic information (e g., Zoological Records, primary literature, with a specific mention of Zoological Records) should be freely available to the wide taxonomic community. Ongoing anthropogenic development resulting in loss of biodiversity requires improvement in taxonomic information management. The need of the hour is to prepare up-to-date database of taxonomic study clearly utilizing the latest available techniques so that the same can be retrieved as and when required.

Chapter 22

RECENT TRENDS IN TAXONOMY

Taxonomy is the science of identifying, naming, describing and classifying animals on the basis of similarities and differences and placing them in the system developed by Linnaeus (1758). This has been the way of managing vast diversity of organisms. This approach to classification was satisfactory as long as species were regarded as static and immutable. Therefore acceptance of the evolutionary relationship among species by incorporating genealogical information in the system of classification has been followed. The advent of molecular systematics, large-scale relational data basing, innovative imaging system and other advances in bioinformatics has revolutionized the concept of animal taxonomy. The traditional morphological criteria derived from skins, bones, teeth, feathers, wings etc. can be amplified by **DNA** analysis and other molecular **biological markers**, chromosome analysis and by an increasing knowledge of behavior.

Now many of the species are known by single or a few specimens and also much emphasis is being laid on the subspecies and populations. With the advent of molecular and biological techniques mentioned above, newer vistas are being explored and are being applied for classifying the organisms. The integration of all such new methods deserve following description:

MORPHOLOGICAL CHARACTERS

Morphological characters like structure of genitals and wing **chaetotaxy** are some of the characters which have now been studied with the help of scanning electron microscope (SEM). The SEM study provides finer detail of the minute characters of insects. The three dimensional picture reveals either to unknown details of the specimen. The structure can be magnified up to 10,000x times.

EMBRYOLOGICAL STUDY

Developmental stages of different groups of animals also help in the classification. For example, the egg shell of fruit fly, *Dacus oleae* and *Ceratitis capitata* (Diptera : Tephritidae) show superficial similarity in shape and size, however, SEM study help to differentiate the two species. The eggs of the species are distinguished on the basis of anterior pole. The classification of white flies is based on the basic of structure of pupae. In trematodes, the structure of **cercaria** larva varies in different groups.

ECOLOGICAL APPROACH

Initially Aristotle classified animals on the basis of habitat; e g., they were divided into three-category viz., aquatic, terrestrial and aerial. However, it is well established that every species has its own ecological niche and it differs from its nearest species in different factors like food preference, breeding season, tolerance to various physical factors etc. On the basis of ecological differences, the species can be differentiated from each other. For example, some of the *Anopheles* species can be segregated from each other on the basis of habitat (Table 22.1).

Table 22.1 Distribution of different species of *Anopheles*

S. No.	Species	Habitat	Water
1.	*A. melanoon*	Rice fields	Cool water
2.	*A. maculipennis*	Cool running water	Cool water
3.	*A. labranchiae*	Mostly warm water	Brackish Water

BEHAVIOURAL PATTERN

Behaviour also helps in the taxonomic study. The comparative behavioural pattern is useful in the classification of crickets, bees, wasps, beetles, frogs and fishes etc. Closely related species of birds can be differentiated on the basis of their song recorded with the help of sonograph. Different species of termites can be identified on the basis of nest they build. Patterns of pupal coverings also help in the identification of various insect groups.

CYTOTAXONOMY

In cytotaxonomy, cytological studies including the chromosomal studies are conducted which help to understand the individuality of the chromosomes the

karyotype, chromosome number, size and morphology. The chromosomal taxonomy is useful in determining the phylogenetic relationship of the taxa, as well as in the differentiation of **sibling** or **cryptic** species (group of species that appear very similar). It may be added that the species in a cryptic complex are typically very close relatives or sibling species and in many cases cannot be easily distinguished by molecular phylogenetic studies. Therefore it requires very meticulous study. On the basis of number and morphology of the chromosomes, two synonym species of earwig of the genus *Labidura* (Dermaptera) and fruit fly (Diptera) have been separated. On the basis of number of chromosomes, phylogenetic relationship among the various families of order Trichoptera has also been worked out. It is relevant to mention that in some insect orders viz., Coleoptera, Odonata and Diptera, the chromosome number is fairly constant, whereas, in other groups like Lepidoptera, Trichoptera, scorpions and fish, the chromosome number shows marked variations.

The nature of chromosomal change may not truly reflect the extent of genetic change. The closely related species may show considerable chromosomal rearrangement whereas; reproductively isolated species may show similarities in chromosomal structure and may differ only in their gene content. Species and higher taxa of animals may differ in the structural organization of the chromosomes, which may show **paracentric** and **pericentric** inversions, presence of supernumerary chromosomes, diffuse or concentrated chromatin and different banding patterns. The phylogeny of horse, ass and zebra (*Equus*) has been traced with the help of chromosomal studies. Similar events during the process of spermatogenesis in insect order **Mallophaga** and **Anoplura** provide support for affinity between these two groups. Salivary gland chromosomal studies of certain Dipteran larvae have also helped in the construction of dendrograms of related species.

Chromosomal studies help in the comparison of closely related species, e g., sibling species are often far more different in chromosomal organization than they do appear in external morphology. In higher taxa, chromosomal analysis has helped in the establishment of phyletic relationships. Increase or decrease in the number of the chromosome may also provide some basis for phylogeny. There are different cytological methods which may help in the identification of different species discussed herein as under:

MOLECULAR PHYLOGENY

It is very difficult to analyse the evolutionary history of animals because representative of nearly 24 phyla of animals appeared within short span of time before and during Cambrian period and have since evolved along separate lines. Thus it appears that all the clades on the phylogenetic tree are long and branched so closely at their base; as such it is difficult to determine their relationships.

Systematics facilitates classifying the individuals into categories based on overall similarities. Further, phylogenetic analysis resulted in unravelling the branching pattern of evolution and helped in the construction of **family tree**. It was observed that classification based strictly on cladistics is somewhat complex and inconvenient too and sometimes also contravenes the basic principles.

To overcome this drawback, molecular techniques are now employed which use the degree of similarity between individuals and thus help to deduce the relationship on the basis of molecular sequences between the different groups of animals. Certain relationship between organisms which otherwise can not be inferred from morphology, can easily be deduced by comparison of macromolecular data viz., **genes** (DNA) and **proteins** (For details, please see Chapter 8).

Molecular phylogeny uses such data to build relationship tree that shows the probable course of evolution of various organisms. It has been suggested that the conserved sequences, such as mitochondrial DNA are expected to accumulate mutations over time and assuming a constant rate of mutation provide a molecular clock for dating divergence. Thus nucleotide sequences drawn from several organisms are used to work out the phylogenetic tree. The steps are the individuals' mutations like substitutions, insertions etc. necessary to transform the sequence of one species into another.

The molecular phylogeny utilizes structure of molecule to gather information on evolutionary relationship. The result so obtained is expressed with the help of **phylogenetic tree**. Trees can be constructed from nucleotide sequences viz., **transversions** (a subset of nucleotide sequences, amino acid sequences, **indels**, **insertions** and **deletions**) etc. and mitochondrial nucleotide sequences. Trees can also be created from subsets of nucleotide sequences, if the trees are identical, the process is reliable. Molecular phylogeny provides new, powerful and independent tests of theory of evolution. Evolution requires molecular phylogeny to be consistent with classical phylogeny. It is also predicted that all parts of genome should evolve in parallel and exhibit the same taxonomic pattern. In all aspects molecular phylogeny amply confirms the theory of evolution.

CHEMOTAXONOMY

The physiological function of an individual is controlled by different biochemical reactions going inside the body. The hormones and enzymatic machinery represent the sequence of evolution and the relationship among the different groups of organism. Proteins are more closely controlled by genes and less subject to natural selection and thus are reliable indicators of genetic relationship. Thus the species can be differentiated on the basis of amino acid sequences in the proteins. Advances in analytical instrumentation especially chromatography in all its forms, followed by electronic detection methods have also helped a lot. The biochemical characters have been found to be of great help in taxonomic studies. Some cases are discussed herein as under:

- On the basis of quantitative analysis of amino acids (ascorbic acid), phylogenetic relationship between different orders of birds has been demonstrated. In some birds, it is produced in the liver and in other cases both liver and kidney showed its presence, while in third group it is absent from both the glands. It is presumed that, the ancestral enzyme system was first introduced in the liver and somehow transferred to the kidney and finally in passerine birds it is completely lost.

- Biochemical studies have shown that similar **terpenoids** are present in bacterium *Plocanium cartilageneum* (Phyllophoracea : Gigratinales) and *Microcladia* (Ceramiacea : Ceramiales). This shows that the algae possess the capacity to extract terpenoides from sea water. However, according to the concepts of chemotaxonomy, the terpenoids are produced internally and are specific to different orders of algae.

- Halogenated **monoterpenes** were found in species of *Schottera* and *P. cartilageneum*. Morphology of *Sacchromonospora* reveals eight strains isolated from the soils of Yunan province in China.

- In order to determine the taxonomy of the isolated cultures, the morphology and other associated features like cell wall composition, **phospholipids**, **menaquinones**, **mycolic acids**, physiological characteristics and DNA-DNA hybridization in comparison with the standard species of *Saccharomonospora viridis* 3306 and two published species *Sacc. yunanensis* 4650, *Sacc. viridis var. fujinensis* 350 were studied.

- All the strains of *Saccharomonospora* viz., 3306, 4650, 10, 13, 23, N, 4022, 4029, and 4153 produce predominantly single ovoid spore with warty surface at the tips of simple unbranched **sporophore** of variable length. The different strains possess following features:

   ▪ Strains 3306, 4650, 10, 13, 23, N, 4029 and 350 have cell wall composition IV; phospholipids type II, menaquinones MK-9(H4) so that it is inferred that they should belong to the *Saccharomonospora*.

   ▪ Strain 4153 has no aerial mycelium, cell wall composition type II, containing **xylose**, thus it should belong to *Micromonospora*.

   ▪ Strain 4022 is very similar to the type strain *Saccharomonospora* in morphology and having phospholipids, menaquinones and **mycolic acids**. However an important difference exists in the cell wall composition; it contains xylose and **madurose** but not arabinose. In this case the pattern of absorption of infrared light by biological products depends on their chemical composition and can therefore yield many useful taxonomic features. This technique has been mainly applied to micro- organisms.

HISTOCHEMICAL METHODS

Histochemical analysis of protein, free amino acids, enzymes, carbohydrates and nucleic acids help in the determination of similarities and differences between different groups of animals.

COMPARATIVE SEROLOGY

This method is used to compare the proteins of different individuals and is based on the principle of **antigen-antibody** reaction where protein of one individual will show stronger antibody reaction to the proteins of closely related organisms than to the proteins of the distant related ones. Thus when an antigen (usually a protein) is injected into an animal, it stimulates the recipient to produce antibodies, which agglutinate the donor's proteins. This is the basic concept of antigen-antibody reaction. Serological resemblances can best be expressed as overall similarity of the structure of the relevant proteins in which large number of differences and resemblance are reflected in the reaction of anti sera. However it must be made clear that study of only one protein may not give the correct picture of the relationship between different groups of individuals.

NUMERICAL TAXONOMY

It deals with the classification of taxonomic units based on their characters states into different groups by numerical methods. It is based on the principles of Adanson (1757) in which the maximum characters are given equal weight and different taxa can be constructed due to diverse character correlation in various groups. In this case, numerical affinity or similarity between taxonomic units is assessed. It is also sometime unreasonable to give equal importance to all the characters including special adaptations, parallel evolution, genetic and developmental phenomenon. Further as the biological taxonomist face lot of problems while accepting complex mathematical and statistical methods, they do not prefer to use it (For details, please see Chapter 9).

MOLECULAR TAXONOMY

Classification schemes for groups where phylogeny has long been debatable can now be worked out with the help of molecular techniques. It is based on similarities in the possession of macromolecules between different organisms. Organisms can be classified on the basis of distribution of small molecular weight compounds such as **amino acids, terpenes, flavonoids** etc. It also includes the study of molecules like DNA, RNA, **polysaccharide** and proteins. The DNA is extracted from the organism and made to hybridise *in-vitro* with the cell lines of other organisms and then studied for base sequencing. Evidences support the notion that combined data (morphology + molecular studies) analysis provides a more robust estimate of phylogenetic relationship (Sivaramakrishnan et al., 2011).

In *Saccharomonospora,* the colour of the aerial **mycelium** and the substrate mycelium is not stable. It is variable even on the same medium and physiological characteristics reveal no obvious difference among all the strains. Result of DNA-DNA hybridization has shown that, there is high DNA **homology** between strain 3306 and strain 13, 23 and 10. On the basis of these similarities, these strains belong to *Saccharomonospora viridis.* There is difference between strain 4022, 4153 and 3306 in cell wall composition and DNA homology.

Different molecular techniques are useful at various taxonomic levels. The study of mitochondrial DNA and that of electrophoretically discovered different enzymes, help in comparison of populations and closely related species, the distantly related species and higher taxa are compared with the study of proteins and nucleic acid sequencing, immunological methods and DNA hybridization techniques. The study of structure of proteins, nuclear DNA, **molecular DNA,** ribosomal nucleic acids and **r-RNA** has been of great help in unraveling the relationships of lower **eukaryotes** and various groups of **prokaryotes**. The sequencing of proteins and nucleic acids provide enough information regarding molecular pattern of genome of a particular species. By comparing the sequence of base pairs of two homologous DNAs and RNAs, the number of mutational differences can also be worked out.

BARCODING LIFE

The classification and identification of organisms is the oldest and most universal phenomenon of biology. Over the past 250 years, more than 1.8 million species of animals, plants and other organisms have been described. This represents only small number of estimated 1-100 species of eukaryotes alive today. With the usual taxonomic procedures, it will take another 1,500 to 15,000 years to complete the global inventory of life. In fact, the usual taxonomic process is fraught with the following shortcomings:

• It is rather difficult to recall the names and diagnostic morphological features of species, once they have been formally described.

• It is quite tedious to differentiate closely allied species and often the experts find it quite confusing to complete the task.

- It is very difficult to obtain accurate identification of some insects.

To overcome the above referred problems, rapid, accurate and globally accessible procedures for species delimitation and identification is required. With the advent of efficient DNA amplification and sequencing techniques combined with advances in computing and information technology, a DNA based system of species identification is available. The DNA barcoding involves the use of specified DNA sequence to provide taxonomic identification for a specimen. The tool of DNA barcoding shows great potential for studying the systematics of different species groups.

Basics of Barcoding

Joseph Woodland and Bernard Silver in late 1940s invented a system of four white lines on a dark background, for classifying instruments and other items and this technique was issued on October 7th, 1952 as US Patent #2,612,994. It was later developed into the Universal Product Code (UPC) in the early 1970s. Modern UPC symbol use an 11 digit series of lines, each representing 10 numerals for a total of 1011 unique combinations. In contrast to UPC system, in which a completely unique commercial barcode (which only differ by one digit from the other barcodes) is consciously assigned to each product, distinctively generated barcodes are developed through the accumulation of random mutations between reproductively isolated groups of organisms. A barcode is a machine readable digital tag, usually a series of stripes, which encodes information about the item to which it is attached. The barcode helps to identify different items. The barcode can also include some systematic or 'taxonomic information', yielding data not only about type but also on attributes such as origin, major classification and date of description (Blaxter, 2004).

In practical terms a DNA barcoding program involves determination and comparing the nucleotide sequences of several hundred base pairs from a particular gene region to provide an immediate diagnosis of species barcode arranged in a library after cataloguing. For this, a standard region of the mitochondrial genome has been employed to provide species-specific DNA barcodes. The **mitochondrial DNA** (mt DNA) is preferred because it is present in hundreds of eukaryotic cell. The greater number of mitochondria makes it far simpler to recover mitochondrial genes than their nuclear counterpart from small quantity of tissues or when DNA preservation is poor. A 648-base pair region near the 5′ end of the ubiquitous **cytochrome** *c* **oxidase I** (COI) gene has been proposed as potential barcode.

DNA barcoding is a taxonomic method which uses a short genetic marker in an organism for example, mitochondrial DNA to identify a particular species quickly. It is based on relatively simple concept on the fact that mitochondria of most eukaryotic cells contain the mt DNA, which has been observed to have relatively fast mutation rate. This results in significant variance in mt DNA sequences between species and a comparatively small variance within species. This new technology makes use of short but specific DNA tags or 'barcodes' to distinguish one species from another. It uses a small part of the mitochondrial genome, 650 to 750 bases of the cytochrome *c* oxidase I gene (COI) to provide a unique fingerprint for each species.

Herbert et al., (2003) have proposed that with a modest 2% per million year rate of sequence evolution, a 600 bp (base pair) segment of DNA will theoretically provide 12 diagnostic nucleotide differences between any two species that have been separated by only one million years.

It is relevant to mention that a combination of high inter specific and low intra specific variation in COI divergence is crucial for the implementation of a DNA barcoding system. Herbert et al., (2003) carried out a follow up study in which they performed more than 13,000 pair wise comparisons based on COI sequences from 2238 animal species (447 genera, 11 phyla) in Gene Bank. They reported that about 8% of such pairs showed more than 8% sequence divergence and that more than 98% of pairs exhibited more than 2% divergence. On the other hand, individuals from the same species exhibited a level of sequence variation averaging less than 0.3%.

Herbert et al., (2003) created cytochrome oxidase profile at following three different taxonomic levels:

• For each of seven dominant animal phyla using 100 COI, amino acid sequences obtained from the Gene Bank database, each from different family and representing all available databases.

• For eight of the largest insect orders using COI amino acid sequences each from a single representative of different family.

• Using COI DNA sequences from single individuals representing 200 closely related species of moths covering five families, collecting around Guelph (Ontario, Canada).

It has been found that mitochondrial gene viz., (COI) ranges in length from 510 - 530 amino acids among different animal species. The cytochrome *c* oxidase (COI) is divided into two sections (COI-5′; COI-3). The COI-5′ section contains 215 amino acids in *Drosophila melanogaster* and extends from the amino acid terminal to the beginning of M 6; the sixth membrane segment. The COI-3′ section which extends from the point to the carboxyl terminal of the protein has 233 amino acid long chains in *D. melanogaster*. The COI protein has a critical metabolic function that is conserved across all life, where oxidative metabolism is going on; there is substantial variation in its amino acid composition.

Applications of DNA Barcode

DNA barcode has been successfully applied to different groups of animals. Some notable examples are being discussed herein as under:

Identification of Birds

Herbert et al., (2004 a) sequenced DNA barcodes of 260 of the 667 bird species of North America and carried out COI sequence and found that the sequences were either identical or most similar to sequence of the same species. Variation in COI sequence between species (interspecific) averaged 7.93%, whereas, variation within species (intraspecific) averaged 0.43%. In few cases there were deep intraspecific divergence indicating possible new species.

Delimiting Cryptic Species

DNA barcode has also been used to delimit **cryptic species**. A neotropical skipper butterfly, *Astraptes fulgerator* is a cryptic species complex, due to subtle morphological differences as well as an unusually large variety of host plants. It is very difficult for taxonomists to completely delimit the species.

Herbert et al., (2004 b) sequenced the COI gene of 484 specimens from north-western Costa Rica. The sample included at least 20 individuals reared from different species of food plant, extremes and intermediates of adult and caterpillar colour variation. Herbert et al., (2004 b) concluded that *Astraptes fulgerator* consists of 10 different species in north western Costa Rica. These results highlight the potential of DNA barcoding in the discovery of new species when used in conjunction with traditionally collected data.

However, it is relevant to mention that new molecular techniques, DNA barcoding and so on, has simplified the identification of species but the basic information that can be provided by this method is only a small part of what taxonomy provides. Barcoding is a powerful tool but it can not replace the taxonomist who must integrate information from multiple sources.

MOLECULAR OPERATIONAL TAXONOMIC UNITS (MOTU)

With the advancement made in the field of molecular sequence techniques, it is possible to define and diagnose molecular operational taxonomic units (MOTU). Floyd et al., (2002) introduced the concept of Molecular Operational Taxonomic Unit (MOTU) to define those entities identified in a molecular context. To be more precise MOTU is a subset of an OTU that represents the more comprehensive assemblage. Of late, one of the most widely used molecular approaches in species identification is DNA barcoding (Herbert et al., 2003). Following a strict operational workflow, this technique reveals sequence variation among taxa at short specific genomic regions (Borsienko et al., 2008). In DNA barcoding, the designation MOTU has been widely used to describe 'clusters of sequences (that act as representatives of the genomes from which they are derived) generated by an explicit algorithm'.

It is quite difficult to identify **meiofauna** taxa viz., animals with a body size ~1 mm (or less), and variety of nematodes constituting meiofaunal specimens. Blaxter (2003, 2004) and Blaxter et al., (2004) have generated DNA barcode data sets for meiofaunal specimens using barcode sequences (referred as **molecular operational units** or MOTU). They have also used the **nuclear small subunits** (nssu) as a marker and have also tested nssu along with cytochrome oxidase subunits (COXI), with equivalent resolution.

The molecular barcoding system has also greatly helped in working the phylogeny of most of the animal groups. Thus the molecular phylogenetics attempts to determine the rates and patterns of changes occurring in DNA and proteins and to reconstruct the evolutionary history of genes and organisms. In DNA barcoding literature MOTU can designate different situations, Floyd et al., (2002) have interpreted it as belonging to three distinct groupings:

M1—a group of unidentified organisms sharing similar sequences.

M2—a group of organisms within a species that is distinct at the molecular level from other members of the species (Kerr et al., 2009).

M3—a group of organisms from different species those are similar at the molecular level (Ferri et al., 2009).

A careful analysis on the basis of MOTU system would help in rapid and effective identification of most taxa including those not encountered earlier and it also helps in the investigations of the evolutionary pattern of diversity. Taxa defined by MOTU methods can be used for standard taxonomic and ecological surveys. By comparing the barcode sequence with a data base of sequences from specimens identified to Linnaean taxa before sequencing, the anonymous survey specimens can be placed within the known taxonomic framework and the biology of the organism from which they are derived (Floyd et al., 2002; Blaxter and Floyd 2003 and Blaxter, 2004).

INTEGRATED OPERATIONAL TAXONOMIC UNITS (IOTU)

Nowadays, molecular techniques are widespread tools for the identification of biological entities. However, until very few years ago, their application to taxonomy provoked intense debates between traditional and molecular taxonomists. To prevent every kind of disagreement, it is essential to standardize taxonomic definitions. In order to regulate an integrated approach for taxonomy, a new concept of Integrated Operational Taxonomic Unit (IOTU) has been introduced by Galimberti et al., (2012). This concept links different data sources in taxonomy, allowing morphological, ecological, geographical and other characteristics of living beings to be better combined with molecular data. IOTUs are defined by molecular lineages that have further support from at least one more part of the 'taxonomic circle' (DeSalle et al., 2005). It paralleled the Molecular Operational Taxonomic Unit (MOTU). The latter is largely used as a standard in many molecular-based works (even if not always explicitly formalized). However, while MOTUs are assigned solely on molecular variation criteria, IOTUs are identified from patterns of molecular variation that are supported by at least one more taxonomic characteristic. From a systematic point of view, IOTUs are more informative than the general concept of OTUs and the more recent MOTUs. According to information content, IOTUs are closer to species, although it is important to underline that IOTUs are not species. Overall, the use of a more precise panel of taxonomic entities increases the clarity in the systematic field and has the potential to fill the gaps between modern and traditional taxonomy. It is hoped that the use of IOTUs should help to add more to the species definitions.

CATALOGUING ANCIENT LIFE

Lambert et al., (2005) examined the possibility of using DNA barcoding to assess the past diversity of the earth's biota. The COI gene of a group of extinct Ratitae birds and moa (endemic to New Zealand) were sequenced using 26 sub fossil moa bones. Each species sequenced had a unique barcode and intraspecific COI sequence variance ranging from 0 to 1.24%. To determine new species, standard sequence thresh hold of 2.7% COI sequence difference was set. This value is 10 times the average intraspecific differences of North American birds.

CONSTRAINTS OF DNA TAXONOMY

Some workers stress the fact that DNA barcoding does not provide reliable information about the species level. On the other hand, another school of thought feels that it is gross over simplification of the science of taxonomy. Recently diverged species might not be distinguishable on the basis of their COI sequences; however, it has been suggested that about 96% of eukaryotic species can be detected with barcoding, though most of these would also be resolved with traditional means; the remaining 4% do however, pose problems, which can lead to errors that are unacceptability high when relying on DNA bar coding. It is recommended that DNA barcoding needs to be used alongside traditional taxonomic tools and alternative forms of molecular systematics, so that problems can be identified and errors detected, with the help of traditional taxonomy as well as molecular taxonomy. All mt DNA genes are maternally inherited and hybridization leads to misleading results. Thus for correct interpretation of result traditional taxonomy along with molecular taxonomy should be studied together.

Systematic zoology has undergone revolution during the past 25 years. A major step has been the formulation of an explicit and objective methodology for determining phylogenetic relationships that result in proven hypotheses. Innovative biochemical, cytogenetic and developmental approaches to phylogenetic relationships have broadened the scope of systematic Zoology. The diversity increases the potential for generating and testing hypothesis of phylogenetic relationship and temporal evolution.

The recent trend of taxonomy is based on the studies on phylogeny, ontogeny, adaptations including divergence and convergence, population biology, genetic and biochemical similarities and behavioural patterns. Advances in molecular systematics have provided new insights about the relationship among organisms. Recent molecular techniques such as **polymerase chain reaction** (PCR) allow easy analysis and comparison of DNA structures which has enabled grouping of all life forms into three domains: bacteria, archaea and eukarya.

CONVENTION ON BIODIVERSITY

During Earth Summit held in Rio de Janeiro in 1992, the Convention on Biological Diversity (CBD) came into existence. The three goals of this convention - conservation of biological diversity, sustainable use of its components, and fair and equitable sharing of the benefits arising from the use of genetic resources have become prime points so far as biodiversity is concerned. However, already at the Second Meeting of the Conference of the Parties (COP) to the CBD, it was realized that taxonomic information, taxonomic and curatorial expertise and infrastructure are insufficient in many parts of the world, especially in developing countries. Hence, such inadequate information was anticipated to be one of the key obstacles in the implementation of the Convention, in particular of Article 7 on identification and monitoring. In order to overcome this taxonomic impediment, subsequent COP's endorsed consecutive recommendations and established the Global Taxonomy Initiative (GTI).

THE GLOBAL TAXONOMIC INITIATIVE (GTI)

This has been created to remove or reduce the 'taxonomic impediment'. The GTI has been established by the Conference of the Parties (COP) to address the lack of taxonomic information and expertise available in many parts of the world and thereby to improve decision-making in conservation, sustainable use and equitable sharing of the benefits derived from genetic resources. The GTI has been developed by governments, under the Convention on Biological Diversity and is implemented by several stakeholders including governments, non-government and international organizations, as well as taxonomists and the institutions where they work. Taxonomy is important for all types of ecosystems and therefore the initiative is an important issue applicable to every effort under the Convention. The GTI aims to support implementation of the work programmes of the Convention.

PARATAXONOMY

Often, an alternative to time consuming full identification process, sorting specimens to recognisable taxonomic units (RTUs) is proposed. In this case, the non-specialists group the specimens based on just external features into RTUs. It is now widely accepted method in the conservation biology and species based ecology, (Oliver and Beattie, 1993; Krell, 2004).

INTEGRATIVE TAXONOMY

It is the biological discipline that identifies, describes, classifies and names extant and extinct species and other taxa. Nowadays, species taxonomy is confronted with the challenge to fully incorporate new theory, methods and data from disciplines that study the origin, limits and evolution of species. Padial et al., (2010) has proposed integrative taxonomy as a framework to bring together the concept and methodological developments that directly bear on what species are, how they can be discovered and how much diversity is on the earth. Accordingly taxonomy needs to improve species discovery and description and to develop novel protocols to produce the much-needed inventory of life in a reasonable time. Thus, if a new species is discovered, it would be formally described under the Linnaean system and then the activities of taxonomists and molecular biologists would be integrated.

The arrangement of taxa and the establishment of the similarities between them (taxonomy) is an essential aspect for all other studies in biological diversity. In present times taxonomic (and phylogenetic) research faces numerous challenges and there is enormous innovation. There are new methods of observing, comparing and analysing characters, some of them non-invasive both concerning molecular and morphological techniques. New kinds of characters are being considered in taxonomic work; most importantly, DNA sequences are increasingly explored for purposes of species distinction. Finger printing techniques such as DNA sequencing, single nucleotide polymorphism, nuclear and plastid microsatellites and amplified fragment length polymorphism are used by biologist to study the evolutionary relationships, species delimitation, hybridization and population dynamics. Investigators of the demographic history of closely related populations or species can use several nuclear DNA sequences to test specific hypotheses of how past geological events influence observed patterns (Knowles et al., 2007).

The failure in the progress and advancement of taxonomy so far is due to the fact that many taxonomists tend to work alone. Instead, taxonomists should be in the centre of all activities concerning species identification, be it by the use of molecular methods for comparisons with other similar taxa and their evolutionary relationships. Now time has come to bring taxonomists, molecular phylogenetists and computer programmers on a common platform for integrated study.

Chapter 23

SIGNIFICANCE OF TAXONOMY AND BIOSYSTEMATICS

Taxonomy deals with the identification, nomenclature and classification of animals in definite order following certain rules. It lays emphasis on their diversity, habit and habitat, life cycle, spatial and geographical distribution. **Biosystematics** is the study of biological diversity and its origin and the interrelationships among organisms, species, higher taxa or other biological entities. It is the scientific study of the kinds and diversity of animals, the relationship between them, the evolutionary pattern and the processes involving comparative study of living and fossil species. In biosystematics, cytogenetic or biochemical data is used to assess taxonomic relations especially within evolutionary framework. It also takes into account genes and the evolution of taxa including intrinsic traits and ecological interactions. The word systematics stem from the Latinised Greek word 'Systema' and can be defined as the classification of living organisms into hierarchical series of groups emphasizing their similarities and differences. In other words biological **systematics** is the study of the diversification of living forms, both past and present, and the relationships among living things through time. Relationships are visualized as evolutionary trees (synonyms: cladograms, phylogenetic trees, phylogenies). Systematics includes the following:

- *Taxonomy* - naming, describing and classifying all living and fossil forms.
- *Phylogenetic Analysis*-the diversity of life, discovery of relationships of organisms, reconstruction of evolutionary lineages, and pattern of evolutionary history including the physical and biological environmental settings in which changes have taken place.
- *Biogeography*-the mapping of the distributions of species.
- *Ecology*- an understanding of the habitats and environmental factors that control the distribution of species.

According to Vecchione et al., (2000) systematics can be divided into two fundamental levels:

- Species
- Higher- level relationships

Species level questions address the basic natural groups of organisms whereas, higher-level systematics attempts to assemble species into hierarchies of related groups. Initially, relationships have been based simply on morphological similarity. Modern systematics attempts to infer evolutionary relatedness based on shared characters that are derived rather primitive over the evolutionary scale.

To be more precise, taxonomy involves discovering, identifying and describing new species, while systematics utilizes evolutionary relationships to understand biogeography, co-evolution and adaptations. It is relevant to mention here that identification is not classification. Identification involves placing an individual into an already existing classification scheme. It is the taxonomy and systematics which acts as fulcrum of almost every aspect of biological and environmental science. In brief taxonomy refers to the theory and practice of classifying organisms or putting in order while in systematics, information about the population, species and higher taxa are gathered.

SYSTEMATICS AND THEORETICAL BIOLOGY

Taxonomists have laid the foundation of evolutionary biology as it has helped in the establishment of role of **natural selection** in the process of **evolution** and variation. Systematics is fundamental to biology because it is the foundation for all studies of organisms by showing how any organism relates to other living individual. Taxonomy has been instrumental in the establishment of species concept and the process of **speciation** which has further led to the identification of different groups of animals along with their ecological study. Initially it was believed that **mutation** alone was responsible for evolution. However with the rise of taxonomy it was held that the natural selection is the guiding force behind evolution. It is relevant to mention that **mimicry** and associated evolutionary phenomenon are the outcome of natural selection. Taxonomy has also helped in the development of behavioural science.

SYSTEMATICS AND APPLIED BIOLOGY

Systematics is essential to identify species that are important for human health and food production. It has played an important role in the field of applied biology viz., agriculture, medicine, public health, **biodiversity** conservation and management of natural resources. In addition to its role in the identification and classification, taxonomy has helped in the study of all aspect of organisms, the role of lower and higher taxa in the economy of nature and in the evolutionary history. It is also important for protection, conservation and functioning of ecosystem. Systematics thus plays an important role in the synthesis of different kinds of knowledge and also deals with the diverse forms of life. Some of the important contribution of systematics is being detailed herein as under:

Green Revolution

To boost agriculture production, there has been an extensive use of pesticides for the control of insect pests of agricultural importance. However before any control operation against insects is initiated, it is necessary to identify the insect **pests** and study their **bionomics.** It is well known that every species has its own ecological niche and differs from other species so far as the host selection, breeding potential and resistance to different insecticide is concerned. This information is of great help to make insect control operations successful. Further proper identification of several soil inhabiting bacteria helps in the enrichment of the soil fertility. It is generally believed that only vegetable crops are prone to insect attack, but much of our timber is also lost due to invasion of insects on forest belt and fiber. Further, fruit and cash crop are also liable for insect damage as such sound knowledge of ecology and life cycle of insect pests is quite essential.

Biological Control

Continuous and indiscriminate use of pesticides has resulted in the toxicity to non-target organisms, development of resistance in different insect pests, environmental pollution and health hazards associated with the residual effects of pesticide. Therefore development of alternate plant protection technologies has become an imperative necessity; with the result emphasis is being laid on the use of **integrated pest management** (IPM) strategies including the use of biological agents which is much more economical and safe. Unless the different biological agents are identified properly, they can not be successfully used against insect pests. Hence systematics is very helpful for the proper identification of **parasites** and the **predators**.

Economic Entomology

Biosystematics has played an important role in the field of economic entomology. Insects and **vectors** of public importance can be identified before any control measure is initiated. A famous case of malaria prevalent in Europe has been investigated by the taxonomists. The mosquito *Anopheles maculipenis* Meign was prevalent throughout Europe; **malaria** was restricted to few local populations and huge funds were allocated for the control of this mosquito. However it was discovered by Hackett and Missiroli (1937) and Bates (1940) that there were several **sibling species** of *A. maculipennis,* which were responsible for the transmission of malaria. Thus the control operations were initiated against the target species and malaria was brought under control.

Quarantine

Often insect pests along with passenger's baggage and different items of trade and commerce are introduced in other country during travel and transportation. There is Department of Quarantine and Plant Protection under the Ministry of Agriculture, Government of India at the entry point like aerodrome and ports etc., where the consignment is checked and if found to contain any such pathogen/ pest is then kept under isolation for stipulated period and only then it is allowed to check out.

Management of Wild life

There is much stress on the conservation and management of wild life resources in different countries. In India proper identification of different animals helps in their categorization under **Wild Life Protection Act**, 1972, of India (amended till date).

Trade and Commerce

Several useful insects provide products like honey, silk, lac and dyes. Correct identification and biology of these beneficial insect species can help to produce disease resistant varieties and thus help in the production of such products on large scale. Introduction of many exotic and **larvivorous** species of fishes has helped to boost aquaculture to a record level and India is now 4th country in the world in terms of fish production.

Ecological Problems

Biosystematics also plays a very important role in minimizing the environmental pollution to a great extent. There are variety of insects, **pathogens** and **micro-organisms** which are responsible for many economic injury and lots of **pesticides** and other chemicals are used to control the same. This has resulted in the development of many environmental issues.

Therefore, knowledge of taxonomy is essential to understand the dynamics of pest populations and its relation with the other components of the ecosystems. Many species of **phytoplanktons** and **zooplanktons** are regarded as **indicators** of pollution and fluctuations in their population density indicate the extent of pollution in the aquatic biota. Proper identification of such indicator species is of great help in detecting and monitoring the level of pollution. The role of biosystematics in biology can be summed up as under:

• There are more than one million species known today. This has been made possible only with the help of taxonomy. Thus, biosystematics helps to provide an overall estimate of biodiversity on this planet.

• With the help of biosystematics, genetic and phylogenetic relationship between different organisms can be traced out with much ease. Without having a sound knowledge of taxonomy, it is very difficult to work out the lineage of animal diversity.

• Several evolutionary events in the history of animal kingdom can be traced out easily with the help of taxonomy.

• Taxonomy is of immense help in analyzing the observations related with different fields of biology like ethology, physiology, genetics, biogeography and geology.

• Biosystematics helps in the development of database of different life forms by proper collection and preservation of natural history specimens.

TAXONOMY AND SYSTEMATICS

Taxonomy and systematics as a branch of biological sciences has a significant role in the study of animal diversity. Taxonomy has contributed a lot in this direction, detailed herein as under:

• With the help of systematics, the earth's biodiversity can be catalogued and conserved to a great extent and the status of threatened species can be ascertained.

• Universal rule of nomenclature of animals allows taxonomists and other interested to know them easily and share their knowledge about a particular group. Sound study of taxonomy helps to identify a particular specimen easily.

• Taxonomy also helps in revealing numerous interesting evolutionary phenomenon of nature.

• The characters shared by certain taxa and the biological causes of the differences between them are assessed.

• Knowledge of systematics helps to understand **bio-prospecting** the discovery of biological indicators, sustainable agriculture and the judicious utilization of biological resources.

• Systematics helps in the survey of flora and fauna and micro-organisms, identification of individuals and naming new species. It also helps to understand biological variation, biogeography, evolutionary biology and host-parasitic relationship. The data used for these studies are derived from: morphology, anatomy, genetics, embryology, physiology, biochemistry, cytology, immunology, embryology, ecology, geography, paleontology and breeding experiments.

* The field of molecular systematics involving the study of DNA and protein sequences is developing fast as an important new discipline.

* Systematics deals with populations, species and higher taxa and also aims to analyze the causes of variations within taxon.

* Study of systematics helps to draw relationship between different groups of organisms which assist in reconstruction of **phylogeny**. This helps in assessing the biological knowledge and resources which further serve the human beings.

NATURAL HISTORY COLLECTIONS

The foundation of systematics is based on permanent reference collection. The preserved specimens serve as authentic record of variation in genetic make up and body form, the past and recent geographical distribution of organisms, their adaptation in response to changed environmental conditions etc. A sound knowledge of natural history specimens enables us to know about the different groups of animals and their incidence in particular habitat. From the collection of specimens, one can know the habit, habitat and biology of different groups of animals. Scientific research in systematics also helps in unraveling the following issues:

* It helps to provide new approach to forensic sciences.

* With the help of systematics guided analysis of vaccine efficiency can be done.

* With the help of identification of organisms, it is easy to locate the origin of emerging diseases and their vectors e g., Hanta virus, HIV, Nipach virus, West Nile Virus).

* It allows the identification of invasive pest species as well as potential biological control agents.

* It is used to reconstruct the evolutionary history of functional changes in gene sequences that are linked to patterns of development (e g., color blindness) or disease (e g., cancer).

It is relevant to mention that systematics has provided basis for theoretical biology as well as applied biology and has played an important role in unraveling the mystery of different evolutionary processes and of course the causes of variations and diversity. According to Wheeler and Valdecasas (2005) taxonomy is not a minor discipline of biological science but rather a fundamental and absolutely necessary discipline. It is high time that taxonomists around the globe come forward to promote and defend meaningful taxonomy. New comers to taxonomy or related fields need to learn the theory and traditions and unique requirements of taxonomy and accord to taxonomy the same considerations, which are due to any other discipline. According to Wilson (2000) a new global systematics initiative is under the way on following front:

• All species program aims to measure the available biodiversity in all the three recognized domains, the Bacteria, Archea, and Eukarya by using new technology.

• The Encyclopedia of Life, expanding the all-species program by providing an indefinitely expansible pages for each species and containing information either directly available or by linkage to other databases.

• The Tree of Life representing the reconstructed phylogeny of life forms in ever final detail with particular emphasis on genomics. A good taxonomy ultimately relies upon:

• The recognition of characters that unite related taxa.

• The detection of unique characters that diagnose a taxon.

• The comparison of characters that differentiate related taxa.

• The description and ordering of those characters in such way that each taxon can be recognized by other workers.

TAXONOMY vis -a- vis BIODIVERSITY CONSERVATION

According to Wilson (1992) the natural history collections serve as comprehensive evidence of species and clade diversity. Moreover, proper identification of organisms is necessary to monitor biodiversity at any level (Vecchione and Collette, 1996 a, b). Considerable attention has been given to the possibility of conducting an all taxa biotic inventory (ATBI) of an ecosystem (Yoon, 1993).

It is essential to understand taxonomy for the implementation of the recommendations passed in Convention on Biological Diversity (CBD) held at Rio de Janeiro during Earth Summit in 1992. About 170 countries are signatory to the United Nations Convention on Biological Diversity. Under the Convention, countries are obliged to monitor biodiversity and to promote sustainable development. Initially it does not appear to have much to do with taxonomy, but one of the Articles of the Convention requires every signatory country to identify and monitor the biodiversity of the organisms within its borders. This can not be achieved without sound knowledge of taxonomy thus, proper conservation of biodiversity is not possible unless due emphasis is given to species identification.

The biodiversity is the totality of genes, species and ecosystem in a region. It encompasses the diversity of life on earth; it includes all forms of life and the ecosystem of which they are part. It forms the basis for sustainable development. Moreover, the concept of **hot spots** allows recognition of some priority areas for immediate conservation. Hot spots in the biodiversity context refer to the areas, which are severely threatened and prone to human activities and at the same time support a unique biodiversity and contain outstanding examples of evolutionary process of speciation and extinction. Unless one has a sound knowledge of taxonomy, it is very difficult to assess the biological diversity of a particular habitat.

Taxonomy thus helps to identify economically important species and endangered animals and livestock used in agriculture and for global food security. It also helps to address alien species. Systematics is also quite important in understanding to explain the earth's biodiversity and could be used to preserve and protect endangered species.

As per Global Taxonomy Initiative, "Inadequate taxonomic information and infrastructure coupled with declining taxonomic expertise hinders our ability to make informed decisions about conservation, sustainable use and sharing of the benefits derived from genetic resources ".

According to modern concept of taxonomy, the taxa can be verified by molecular methods and the differences in the **ethology** and **autecology** of the species are assessed. The planet earth is the home for rich and diverse array of organisms. The first and foremost task before any conservation operation requires the proper recognition and identification of the species. The biodiversity is simple, unique feature of beautiful living world and can not be assessed or conserved properly without the help of taxonomy.

Scientists estimate that there are actually about 13 million species, though estimates range from 3 to 100 million. Due to loss of species and biodiversity at the current rate of extinction, it is apprehended that during the next half-century, one third of all organisms will be lost. The condition of museums of natural history specimens continue to deteriorate year after year for want of trained curators and taxonomists and this is quite alarming. The function of taxonomy is to establish identity and together with systematics it plays an important role in assessing the upper levels of biological organization from organisms to ecosystem, the mapping and analysis of biodiversity and the development of **Tree of Life** all the way from genes to species.

The biodiversity today is under threat due to environmental degradation, loss of habitat, over-exploitation and introduction of exotic species. A proper identification of threatened, endangered or declining species would help to chalk out proper conservation strategies for them. This can only be possible with clear understanding of taxonomy. Thus, taxonomic knowledge is very important in the management of different types of ecosystems.

Global biodiversity is being lost at an unprecedented rate as a result of human activities, and decisions must be taken now to combat this trend. It is very important to identify correctly the native and invasive species. Taxonomy provides basic understanding about the components of biodiversity which is necessary for effective decision-making about conservation and sustainable use. Taxonomic information is essential for agencies and border authorities to detect, manage and control of **Invasive Alien Species** (IAS). Effective control and management measures can only be implemented when exotic species are correctly and promptly identified.

Networking and sharing of experiences, information and expertise can aid in lowering the costs associated with IAS and reduce the need for eradication programmes with early detection and prevention. When eradication is needed, taxonomists can offer expertise that is central to developing the most effective yet economic and eco-friendly eradication measures.

Increased capacity-building (especially for developing countries) is necessary to identify, record and monitor invasions; provide current and accessible lists of potential and established IAS; identify impending threats to neighbouring countries; and to access information on taxonomy, ecology, genetics and control methods. It is vital that adjacent countries and all countries in a particular manner can recognize invasive species and agree on their nomenclature. Baseline taxonomic information on native biota at the national level is also important to ensure that IAS can be recognized and distinguished from naturally present species.

TAXONOMIC MINIMALISM

According to Beattie and Oliver (1994) biological surveys are in increasing demand while taxonomic resources continue to decline. How much formal taxonomy is required to get the job done? The answer depends on the kind of job but it is possible that taxonomic minimalism especially (i) the use of higher taxonomic ranks (ii) the use of **morphospecies** rather than species (as identified by Latin binomials) and (iii) the involvement of taxonomic specialists only for training and verification, may offer advantages for biodiversity assessment, environmental monitoring and ecological research.

As such, formal taxonomy remains central to the process of biological inventory and survey but resources may be allocated more efficiently. For example, if formal identification is not required resources may be concentrated on replication and increasing sample sizes. Taxonomic minimalism may also facilitate the inclusion in these activities of important but neglected groups, especially among the invertebrates and perhaps even micro-organisms.

TAXONOMY: THE PRESENT SCENARIO

Taxonomy as a branch of biology is very exciting and challenging that requires meticulous study. It is relevant to mention that taxonomy is not a minor branch of biological science but rather a fundamental and absolutely necessary discipline. Modern biological taxonomy strives to determine the evolutionary or phylogenetic relationship between different groups of organisms. Features of organisms that reflect evolutionary relationships are most useful for biological relationship. However, it must be made clear that mere similarity is not always a reliable indicator of evolutionary relationship. Knowledge of living species can help to use earth's natural resources in a sustainable manner. It is high time to make species exploration, discovery and description on priority basis.

By the end of the 18th century, the estimated number of species on earth was about one million. Today's more accurate estimates place the number of species between five and 30 million. Linnaeus provided a solution to the first bioinformatics crisis and now we find ourselves at another turning point where our world's changing biodiversity due to human impacts is increasing the demand for taxonomic information and expertise. It is relevant to mention that use of internet for the progress and popularization of taxonomy, as well as for free and easy access to taxonomic and biodiversity information is helping a lot to the professionals and amateurs.

Discovering and describing how many species inhabit the earth remains a fundamental quest of biology even when we are entering the "phylogenomic age" in the history of taxonomy with so many important issues facing us viz., invasive species, climate change, habitat destruction and loss of biodiversity in particular, the need for authoritative taxonomic information is higher than ever (Zhang, 2011).

The modern systematics entails to differentiate individual organisms and establish the basic unit's viz., species and arranging these units into a logical hierarchy that enables easy and simple recognition on the basis of similarity. Further, it also establishes the relationship between all levels of hierarchy. Systematics uses taxonomy as a primary tool in understanding organisms and its interrelationship with other individuals.

It is very important to know the living organisms around us, their accurate identification and classification is very important (Kapoor, 1998). Without taxonomy, one would not be sure of the identity of organisms. Moreover, there would be no meaningful genome projects and development in medical science would be seriously affected.

As Kapoor (1998) has pointed out, taxonomy is essential in theoretical and applied biology (agriculture and forestry, biological control, public health, wild life management, mineral prospecting through the dating of rocks by their enclosed fauna and flora, national defence, environmental problems, soil fertility and commerce, etc). About 1.8 million species have been described since Linnaeus and it is generally estimated that only around 10% of the world's biota has so far been described (Wilson, 2000; Disney, 2000).

Obviously, taxonomy plays the major role and its importance as basic science for the remaining disciplines should be taken into consideration. However, although society has a growing need for credible taxonomic information in order to allow us to conserve, manage, understand and enjoy the natural world, support for taxonomy and collections is failing to keep pace (Wheeler et al., 2004).

The discipline of taxonomy traditionally covers three areas of stages: alpha (analytically phase), beta (synthetic phase) and gamma (biological phase) taxonomy (Kapoor, 1998; Disney, 2000). Alpha taxonomy is the level at which the species are recognized and described; beta taxonomy refers to the arrangements of the species into a natural system of lower and higher categories, and gamma taxonomy is the analysis of intraspecific variations, ecotypes, polymorphisms, etc.

A shortage of taxonomic information and skills and confusion about the 'species concept' cause problems for conservationists (Mace, 2004; Wheeler et al., 2004). Species conservation needs taxonomic solutions such as a set of practical rules to standardize the species units included on regional and global species lists. Related to this, a new term like **phylogenetic diversity** (total amount of evolutionary history represented by a species or group of species) is being used (Faith et al., 2004; Isaac et al., 2004). The solutions require a new kind of collaboration among conservation biologists, taxonomists and legislators, as well as an increased resource of taxonomists with relevant and high-quality skills (Mace, 2004).

It is claimed that the rate of species description is being impeded by the lack of cyber infrastructure resources (Wheeler, 2008). Other factors, such as lack of funding, lack of recognition for taxonomic work, the small number of researchers and the distractions from research due to additional teaching load, administrative work etc. are also likely a reason of slow rate of species descriptions.

There has been much discussion of the **taxonomic impediment** and of speeding up descriptive taxonomy (Ebach et al., 2011). Brown (2013) has suggested for automating the 'material examined' section of taxonomic papers to speed up description. Computer generated programs like MANTIS, DELTA, LUCID are used for this.

Taxonomy in fact provides knowledge about species - their identity, how they live and how they interact with others and the environment. The different individuals in the ecosystem enable us to understand the functional role of biodiversity and help with the diagnosis of exotic pests and disease organisms.

According to Dov Por (2007) each paper and research report requires for a **taxonomic affidavit**. Therefore, a correct identification of the species studied should be given in the material and methods section, the taxonomist who has identified the specimen should be duly acknowledged and location of the museum in which the studied specimen is deposited should also be mentioned. It is suggested that systematics must embrace comparative biology and evolution, not speed and automation. Further, the systematists have come under a volley of criticism because of the alleged inadequacy of the 'traditional' taxonomic paradigm to curb 'biodiversity crisis' and expeditiously make available the products of systematic research-usually species name-to the professional biological 'user' community (Carvalho et al., 2008).

Taxonomy is constantly reinventing itself through new tools and improved theories and methods to improve its efficiency and time has come to develop and create interest in taxonomic studies. The future of taxonomy is not bleak, linkages between science and society means that the role of taxonomy is more important than ever before. It would not be out of place to quote Alfred Russell Wallace that "......the most perfect collections possible in every branch of natural history should be made and deposited in national museums, where they may be available for study and interpretations". If this is not done, posterity will never forgive us.

BIBLIOGRAPHY

Adanson, M. (1757) *Histoire naturelle du : coquillages : avec la relation abrégée d'un voyage fait en ce pays, pendant les années* 1749, 50, 51, 52 & 53. Par M. Adanson ; ouvrage orné de figures. Publication Info: A Paris: Chez Claude-Jean-Baptiste Bauche Contributed By: Smithsonian Institution Libraries Harvard University, MCZ, Ernst Mayr Library.

Adanson, M. (1763) *Famillies des Plantes*. 1. Paris, Vincent.

Agassiz, L. (1848) *Solduri sumptibus Jent et G*assman,531.www.biodiversitylibrary. org/pdf3/0054796000016933).

Agler, B. et al., (2011) *Eighteen years of thermal marking in Alaska*. Abst. In Otolith Workshop. Juneau, Alaska: Decoding the Otolith. April 19-20.

Allen, J.A. (1897) *The Merton Rules*. Science **6** (131) : 9-19.

Aschlock, P.D. (1979) *An evolutionary systematist's view of classification*. Syst. Zool. **28**: 441-450.

Aschlock, P.D. (1985) *A revision of the Bergidea group: A problem in classification and biogeography*. J. Kansa Ent. Soc. **57**: 675-688.

Avise, J. C. and R. M. Ball Jr. (1990) *Principles of genealogical concordance in species concepts and biological taxonomy*. In: D. Futuyma and J. Atonovics, (eds.); *Surveys in Evolutionary Biology*. 45-67. Oxford University Press, Oxford.

Bates, M. (1940) *The nomenclature and taxonomic status of the mosquitoes of the Anopheles maculipennis complex*. Annals of the Entomological Society of America XXXIII: 343-356.

Baun, D. (2008) *Trait evolution on a phylogenetic tree: Relatedness, similarity, and the myth of evolutionary advancement*. Nature Education. **1** (1).

Beattie, A.J. and Oliver, I (1994) *Taxonomic Minimalism: Trends in Ecology & Evolution*. Vol.**9** (12) 488-490.

Berger, J. (1997) *Population Constraints Associated with the Use of Black Rhinos as an Umbrella Species for Desert Herbivores*. Conservation Biology. **11** (1) 69–78.

Bertrand, Y. et al., (2006) *Systematics and Biodiversity*. **4** (2) 149-159

Bessey, C.E. (1908) *The taxonomic aspect of the species*. Amer. Nat. **42** : 218-224.

Bibby, C. J. et al., (1992) *Putting biodiversity on the map: priority areas for global conservation*. International Council for Bird Preservation, Cambridge, United Kingdom.

Blackith, R. E. and Blackith, R.M. (1968) *A numerical taxonomy of orthopteroid insects*. Aust. J. Zool. **16** : 111-131.

Blackwelder, R. E. and C.L. Boyden.(1952) *The Nature of Systematics.* Syst. Zool. **1**: 26-33.

Blackwelder, R.E. (1962) *Animal taxonomy and the new systematics.* In: B. Glass (ed.); *Survey of Biological Progress.* **4** : 1-57.

Blackwelder, R.E. (1967) *Taxonomy: A Text Book and reference book.* 698. John Wiley and Sons. Inc.

Blaxter, M. (2003) *Molecular Systematics : Counting angels with DNA.* Nature. **421**:122-124.

Blaxter, M. (2004) *The promise of molecular taxonomy.* Phil. Trans. R. Soc. B**359**:669-679.

Blaxter, M. and R. Floyd, (2003) *Molecular taxonomics for biodiversity surveys: Already a reality.* Trans. Ecol. Evol.**18** : 268-269.

Blaxter, M., et al., (2004) Utilizing the new nematode phylogeny for studies of parasitism and diversity. In: R. Cook and D.J. Hunt, (eds.); *Nematology Monographs and Perspectives.* 615-632.

Blunt, W. (2001) *Linnaeus : The Complete Naturalist.* Princeton University Press. Princeton, New Jersey. 264.

Bolton, B. (1985) *The ant genus Triglyphothrix forel a synonym of Tetramorium Mayr.(Hymenoptera : Formicidae).* Journal of Natural History **19** : 243-248.

Borisenko, A.V. et al., (2008) *DNA barcoding in surveys of small mammal communities: a field study in Surinam.* Mol. Ecol. Res. **8** : 471–479.

Brinkhurst, R.O. (1985) *Museum Collections and Aquatic Invertebrate Environmental Research.* In: E.H. Miller, (ed.); *Museum Collections: Their Roles and Future in Biological Research.* **25**: 163-168. Brit. Columbia Prov. Mus. Occ. Paper.

Brooke, M. L. (2000) *Why Museums Matter. Trends in Ecology and evolution.* **15**:136-137.

Brown, W.L Jr. and Wilson, E.O. (1956) *Character displacement.*Syst.Zool.5:49-64.

Brown,B.V.(2013) *Automating the "material examined" section of taxonomic papers to speed up species descriptions.* Zootaxa **3683** (3):297-299.

Cain, A.J. and Harrison, G.A. (1958) *An analysis of the taxonomist's judgment of affinity.* Proc. Zool. Soc. London. **13** : 85-98.

Camp, W. H., and Gilley, C. L. (1943) *Structure and Origin of Species.* Brittonia 4, 324–395.

Canfora, G. and Luigi Cerulo. (2004*) A taxonomy of information retrieval models and tools.* Journal of Computing and Information Technology.**12** (3) 175-194.

Caraway V. et al., (2001) *Assessment of hybridization and introgression in lava-colonizing Hawaiian Dubautia (Asteraceae : Madiinae) using RAPD markers.* AM.J. Bot. **88** : 1688-1694.

Caro, T. et al., (2004) *Preliminary assessment of the flagship species concept at a small scale.* Anim.Conserv.**7**: 63-70.

Carvalho, M. R. de et al., (2008) *Systematics must embrace comparative biology and evolution, not speed and automation.* Evol. Biol. **35** : 150-157.

Cavalier-Smith, T. (1998*) A revised six-kingdom system of life Biological Reviews.* Vol.**37** (3) 203-266.

Challinor, D (1983) *Director's Dilemma.* New Scientist. **97** : 907.

Chapman, R. N. (1928) *The quantitative approach of environmental factors.* Ecology. **IX** (1) : 1-40.

Cuvier, G. (1829). *Le regne animal distribute après son organization,* Paris. Crochard et., Cie., p.16.

Cynthia, S. et al.,(2004) *Visualizations for taxonomic and phylogenetic trees.* Bioinformatics. **20** (17) 2997-3004.

Daly, H. V. (1985) *Insect morphometrics.* Ann.Rev. Entomol. **30** : 415-438.

Darwin, C. (1859*) On the Origin of Species by Means of Natural Selection.* In : Mayr, E. (ed.); *The Preservation of the Favoured Races in the Struggle for Life.* 502. London, John Murray, Cambridge, Harvard University Press.

Darwin, C. and Wallace, A. R. (1858) *On the tendency of species to form varieties, and on the perpetuation of varieties and species by natural means of selection.* J. Proc. Linn. Soc. London. Zool. **3** : 45-62.

Davis, P. (1996) *Museums and the Natural Environment: The Role of Natural History Museums in Biological Conservation.* London: Leicester University Press.

DeSalle R, et al., (2005) *The unholy trinity: taxonomy, species delimitation and DNA barcoding.* Philos Trans R Soc London [Biol] **360** : 1905–1916.

Dietz, J. M., et al., (1994) *The effective use of flagship species for conservation of biodiversity: the example of lion tamarins in Brazil.* In: P. J. S. Olney et al., (eds.); *Creative conservation: Interactive management of wild and captive animals.*32-49.Chapman and Hall, London.

Disney, H. (2000) *Hands on taxonomy.* Nature. **405**: 307.

Dobzhansky, T. (1937) *Genetics and the origin of species.* (New York: Columbia University Press).

Dobzhansky, T. (1970) *Genetics of the Evolutionary Process* (New York: Columbia University Press) 71-93.

Dov Por, Francis (2007) *A "taxonomic affidavit": Why it is needed? Integrative Zoology* **2**:57-59.

Drummond, A., et al., (2006) *Relaxed phylogenetics and dating with confidence.* PLOS Biology. **4** (5) : 3 88.

Du Rietz, G.E. (1930) *The fundamental units of biological taxonomy.* Vensk. Bot. Tidskr. **24** : 333-428.

Dubois, A. (2011) *A zoologist's viewpoint on the Draft BioCode.* Bionomina **3** : 45-62.

Dubois, A. (2005) *Proposed rules for the incorporation of nomina of higher-rankcd zoological taxa in the International Code of Zoological Nomenclature. 1. Some general questions, concepts and terms of biological nomenclature.* Zoosystema, **27**: 365-426

Dunn, P.O. & Winkler, D.W. *(1999) Climate change has affected the breeding date of tree swallows throughout North America.* Proceedings of the Royal Society of London B, 266, 2487-2490.

Ebach, M.J. et al. (2000) *Impediments to taxonomy and users of taxonomy accessibility and impact evaluation.*Cladistics.**27** : 550-557.

Ebach, Malte C. et al., (2011) *Impediments to taxonomy and users of taxonomy: accessibility and impact evaluation.* Cladistics. **27** (5) 550-557.

Ehrlich, P. R and Raven, PH (1969) *Differentiation of Populations.* Science.**165** : 1228-1232.

Eldredge, N. and Gould, S. J. (1972) Punctuated equilibria: An alternative to phyletic gradualism. In: T.J.M. Schopfed. (ed.); *Models in Paleobiology..* 82-115.Freeman Company, San Francisco.

Emery,C. (1915) *Contributtoalla conosoenzadelle formiche delle isole italiana.* Ann.Mus. Civ.Stor.Nat.Genova. **6** : 244-270.

Faith, D. P. et al., (2004) *Integrating phylogenetic diversity, complementary, and endemism for conservation assessment.* Conservation Biology., **18** : 255-261.

Ferri, E, et al., (2009) *Integrated taxonomy: traditional approach and DNA barcoding for the identification of filarioid worms and related parasites (Nematoda).* Front Zool **6** : 1. doi: 10.1186/1742-9994-6-1.

Fleishman, E. R. et al., (2001) *Empirical validation of a method for umbrella species selection.* Ecological Application **11**: 1489-1501.

Floyd, R., et al., (2002) *Molecular barcodes for soil nematode identification.* Mol Ecol **11**: 839–50.

Galal, F. H. (2009) *Comparison of RAPD and PCR-RFLP markers for classification and taxonomic studies of insects.* Egypt. Acad. J. Biolog. Sci. **2** (2) : 187-195.

Galimberti, A., et al., (2012) *Integrated Operational Taxonomic Units (IOTUs) in Echolocating Bats: A Bridge between Molecular and Traditional Taxonomy.* PLoS ONE 7(6) : e40122. doi:10.1371/journal.pone.0040122.

Gardiner, B .G. (1982) *Tetrapod classification.* Zool. J. Linn. Soc. London **74** : 207-232.

Gaston, K. J. (2000) *Global Patterns in biodiversity.* Nature **405** : 220-227.

Gaston, K. J. (2006) *Biodiversity and Extinction: Macroecological patterns and people.* Progress in Physical Geography..**30** (2) 258-269.

Godfray, H. C. J (2002b) *Towards taxonomy's glorious revolution.* Nature. **420** : 461.

Godfray, H. C. J and Knapp, S. (2003) *Taxonomy for the twenty first century.* Phil. Trans. R. Soc. Lond. vol. 359 .1444: 559-569.

Godfray, H. C. J. (2002a) *Challenges for taxonomy.* Nature. **417**: 17-19.

Godfrey, L. and Marks, J. (1991) *The nature and origin of Primate species.* Yearbook of Physical Anthropology 34:39-68.

Graham, C. H. et al., (2004) *New developments in museum based informatics and applications in biodiversity analysis.* Trends in Ecology and Evolution. **19** : 497-503.

Hackett, L. W. and A. Missiroli. (1937) *The varieties of Anopheles maculipennis and their relation to the distribution of malaria in Europe.* Rivista di Malariologia. **14** : 45-109.

Haeckel, E. (1866) *Generelle Morphologie der Organismen. Reimer, Berlin.*

Harrison,S. et al., (2004) *Does the age of exposure of serpentine explain variation in endemic plant diversity in California.* International Geology Review. Vol.(46).xxx-xxx.

Harrrison, S. (1959) *Environmental determination of phenotype.* Systematic Assoc. **3**: 81-86.

Hennig, W. (1957) *Systematik und Phylogenese.* Ber. Hundertzahrfeier, Deutsh. Entomol. Ges.50-70. Berlin.

Hennig, W. (1966) *Phylogenetic Systematics* (trans. D.D. Davis & R. Zanger). University of Illinois Press, Urbana.263.

Herbert, P. D. N., et al., (2003) *Biological identifications through DNA barcodes.* Proc. R. Soc. B 270:313-321.

Herbert, P. D. N. et al., (2004 a) *Identification of birds through DNA Barcodes.* PloS Biol. **2** (10):e312.

Herbert, P. D. N. et al., 2004a) *Ten species in one: DNA barcoding reveals cryptic species in the neotropical skipper butterfly Astraptes fulgerator.* Proc. Natl. Acad. Sci. USA **101**: 14812–14817.

Hermida, M.C.,et al.,(2002) *Heritability and evolvability of meristic characters in a natural populations of Gastrosteus aculeatus.* Can. J. Zool. **80** (3):532-541.

Heywoods, V. H. and Watson, R. T. (1995) *Global biodiversity assessment.* Cambridge Univ. Press.

Heywoods, V. H (1967) *An Introduction to Numerical Taxonomy.* Academic Press, N.Y. London.

Hill, M.O. (1973) *Diversity and evenness: A unifying notation and its consequences.* Ecology. **54** : 427-432.

Hounsome, M.V., (1984) *Research: Natural Science Collections.* 150-155. In: J.M.A. Thompson. (ed.); *The Manual of Curatorship.* Butterworths & Museums Association, London.

Hunter, M. Jnr.(2002) *Fundamentals of Conservation Biology,* Blackwell Science, Massachusetts,U.S.A

Huxley, J.S. (1939) *Clines: An auxillary method in taxonomy.* Bijdr.Dierk. **27**: 491-520.

Huxley, J.S. (1940) *The New Systematics.* 583. Clarendon Press, Oxford.

Inger, R.F. (1961) *Problems in the application of the subspecies concept in vertebrate taxonomy.* In: Blair, W.F. (ed.); *Vertebrate Speciation.* Univ. Texas Press, Austin., 262-285.

Ingrouille, M. (1989) *The decline and fall of British Taxonomy.* New Scientist. 7 Jan. 1989: 64.

International Commission of Zoological Nomenclature (1999) *International Code of Zoological Nomenclature,* adopted by the International Union of Biological Resources International Commission on Zoological Nomenclature. 4th edition. International Trust for Zoological Nomenclature. London.

Isaac, N. J., et al., (2004) *Taxonomic Inflation: Its influence on macroecology and conservation.* Trends. Ecol. Evol., **19** (9) : 464-9.

Izsak, J. and Laszio, Papp. (2000) *A link between ecological diversity indices and measures of biodiversity* (www.sciencedirect.com).

Johannsson, O.E. and Minns, C.K.(1987) *Examination of association indices and formulation of a composite seasonal dissimilarity index.* Hydrobiologia. **150** : 109-121.

Johnsingh, A. J. T., and Joshua, J. (1994) *Conserving Rajaji and Corbett National Parks: the elephant as a flagship species.* Oryx, **28** : 135–140.

Johnson, M.W. (1979) *Insect pest management strategies for control of Liriomyza sativae Blanchard (Diptera : Agromyzidae) on pole tomatoes in southern California.* Dissertation. University of California.

Kapoor, V.C. (1998) *Principles and Practices of Animal Taxonomy.* Science Publishers.

Kerr, K.C.R., et al., (2009) *Filling the gap- COI barcode resolution in eastern Palearctic birds.* Front Zool. **6** : 29. doi: 10.1186/1742-9994-6-29.

Kiauta, B. (1967) *Distribution of the chromosome numbers in Trichoptera in the light of phylogenetic evidence.* Genen en. Phaemen. **12** : 110-113.

Klass, E. E., et al., (1979) Avian egg shell thickness: variability and sampling. In: Banks, R.C, (ed.); *Museum Studies and Wildlife Management.* Selected Papers. 253-261. Smithsonian Institution Press for National Museum of Natural History, Washington.

Kleiman, D. G., and Mallinson, J. J.C. (1998) *Recovery and management committees for lion tamarins: partnerships in conservation planning and implementation.* Conservation Biology. **12** : 27–38.

Knowles, L.L., et al., (2007) *Coupling genetic and ecological niche models to examine how past population distributions contribute to divergence.* Current Biology, **17**: 940-946.

Kopp M. (2010) *Speciation and the neutral theory of biodiversity: Modes of speciation affect patterns of biodiversity in neutral communities.* Bioessays. **32** (7) : 564-70.

Krell, F.T. (2004) *Parataxonmy vs Taxonomy in Biodiversity studies-pitfalls and applicability of 'morphospecies' sorting.* Biodiversity and Conservation.**13** : 795-812.

Krell, F.T. (2012) *Electronic publication of new animal names-An interview with Frank-T Krell, Commissioner of the International Commission on Zoological Nomenclature and Chair of the ICZN ZooBank Committee.* Evol. Biol. 12:184.

Lambert, D.M., et al., (1987) *Are species self defining?* Syst.Zool.**36** : 196-205.

Lambert, D.M., et al., (2005) *Is a large scale DNA-based inventory of ancient life possible?* J. Heredity. **96** (3) : 279-284.

Landres, P. B., et al., (1988) *Ecological uses of vertebrate indicator species: a critique.* Conservation Biology **2** : 316–327.

Lincoln, R. et al., (1998) *A dictionary of Ecology, Evolution and Systematics.* Cambridge University Press, Cambride.

Linnaeus, C.(1758) *Systema Naturae,* 10th ed. Stockholm, Laurentii, Salvii.

Lovei, G. L. (1997) *Global change through invasion.* Nature. **388** : 627-628.

Mace, G.M. (2004) *The role of taxonomy in species conservation.* Philosophical Transactions of the Royal Society of London Series B-Biological Sciences. **359** : 711-719

Magurran, A. E. (1988) *Ecological Diversity and Measurement* (Croom Helm,London).

Mason, H.L. (1950) *Taxonomy, systematic botany and biosystematics.* Madrono.**10** : 193-208.

Mayr, E. (1931) *Birds collected during the Whitney South Sea Expedition.12: Notes on Halycon chloris and some of its subspecies.* Amer. Mus. Novitates. **469** : 1-10.

Mayr, E. (1942) *Systematics and the Origin of Species from the view point of a Zoologist.* 334. Columbia University Press. New York.

Mayr, E. (1953) *Concepts of classification in higher organisms and micro-organisms.* Ann. N.Y. Acad. Sc. **56** : 391-397.

Mayr, E. (1957) *The species problems.* Amer. Assoc. Adv. Sc. Publ. No.50. Washington, DC.395.

Mayr, E. (1961) *Cause and effect in biology: Kinds of causes, predictability and teleology are viewed by a practicing biologist.* Science, **134** : 1501-1506

Mayr, E. (1963) *Animal Species and Evolution.* Cambridge, Mass: Belknap Press. ISBN 0-674-03750-2.

Mayr, E. (1968) *The role of systematics in biology.* Science. **159** : 595-599

Mayr, E. (1969) *Discussion: Footnotes on the philosophy of biology.* Philosophy of Science **36** : 197-202.

Mayr, E. (1969) *Principles of Systematic Zoology,* 475..New York, Tata-McGraw Hill, Publishing Co. Ltd., New York.

Mayr, E.(1981) *Biological classification towards a synthesis of opposing methodologies.* Science. **214** : 510-516.

Mayr, E. (1982a) *The growth of biological thoughts: Diversity, Evolution, and Inheritance,* Cambridge, M.A. and London: The Belknap Press, New York. ISBN 0-674-36446-5.

Mayr, E. (1982c) *Processes of speciation in animals.* In: Barrigozzi, C. (ed.); *Mechanisms of Speciation.* 1-19. New York, Alan R. Liss.

Mayr, E. (1991) *More Natural Classification. Nature* (London). **353** : 122.

Mayr, E. (2001) *What evolution is?* New York: Basic Books.ISBN.0-405-04426-3.

Mayr, E., et al., (1953) *Methods and Principles of Systematic Zoology.* McGraw Hill. NewYork.

Mayr, E., Aschlock, PD (1991) *Principles of Systematic Zoology.* McGraw Hill, New York.

McCarty, J. P. (2001) *Ecological consequences of recent climate change.* Conservation Biology **15** : 320–331.

McKenzie, D. H. et al., (1992) *Ecological indicators. Volumes 1 and 2.* Elsevier Science Publishers, New York.

McKenzie, D. H., et al., (1992) *Dike emplacement on venus and on Earth.* Journal of Geophysical research 97:doi.:029/92 JE01559 issn.148-0227.

McNeill J. (1978) *Purposeful phenetics.* Syst. Zool. **28** : 465-482.

Michner, C. D. and Sokal, R. R. (1957a) *A quantitative approach to a problem in classification.* Evolution.**11** : 30-162.

Michner, C.D. and R. R. Sokal, (1957b) *A statistical method for evaluating systematic relationships.* Microbiol. **17** : 201-226.

Minelli A.(2013) *Zoological nomenclature in the digital era.* Front Zool. 2013 Feb 4;10(1):4. doi: 10.1186/1742-9994-10-4.

Moore, J.A. (1993) *Taxonomy –What's in a name? Science as a way of knowing: The Foundations of Modern Biology:* Harvard University Press. Convention In Biological Diversity-Belgian Clearing House Mechanism.

Mora, C., et al., (2011) *How many species are there on Earth and in the ocean?* PLoS Biol. (8): e1001127. doi: 10.1371/journal.pbio.1001127.

Myers, N. (1988) *Threatened biota: 'Hot spots' in tropical forests.* Environmentalists. **8** : 187-208.

Myers, N.(1990) *The biodiversity challenge: Expanded hotspot analysis.* Environmentalists **10** : 243-256.

Nagaraju, J.; et al., (2001) *Comparison of multilocus RFLPs and PCR-based marker systems for genetic analysis of the silkworm, Bombyx mori.* Heredity. **86** : 588-597.

Naik, V.N. (1984) Taxonomy a synthetic discipline. In: *Taxonomy of Angiosperms.* Tata-McGraw Hill, Publishing Co. Ltd.

Nelson, G. and Platnick, N.L. (1981) *Systematics and biogeography: Cladistics and Vicariance.* Columbia University Press, New York.

Oliver,I. & Beattie, A.J. (1993) *A possible method for the rapid assessment of biodiversity.* Conservation Biology. **7** : 562-568.

Oucellet, H. (1985). *Museum collections : perspectives.* Brit. Columbia Prov. Mus. Occ. Paper **25** : 215-218.

Padial, J.M., et al., (2010) *The integrative future of taxonomy.* Frontiers in Zoology. *7* :16 doi:10.1186/ 1742-9994-7-16.

Padian, K.(1999) *Charles Darwin's Views of Classification in Theory and Practice.* Systematic Biology, **48** : 352-364.

Pandey, S.K.and Nautiyal, P. (1997) *Statistical evaluation of some meristic and morphometric characters of taxonomic significance in Schizothorax richardsonii* (Gray) and *Schizothorax plagiostomus* (Heckel). Indian Journal of Fisheries. **44** (1)75-79.

Patterson, C. (1988) *Homology in classical and molecular biology.* Mol.Biol. Evol. **5** : 603-625.

Pawson, M.G. and J. R. Ellis. (2005) *Stock identity of Elasmobranchs in the Northeast Atlantic in relation to assessment and management.* E.journal of Northwest Atlantic Fisheries Science. **35** :.Art.38:1-22.

Perring, F.H. (1977) *The role of natural history collections in preparing distribution maps.* Mus. J. **77** (3): 133.

Peterson, H.E.H. (1978) *More evidence against speciation by reinforcement.* S.Afr. J. Sci. **74**: 369-371.

Peterson, H.E.H.(1985) The recognition concept of species. In: E.S. Vrba, (ed.);*Species and Speciation.* No. **4**: 21-29. Transvaal Museum. Pretoria :Transvaal Museum Monograph.

Pettitt, C. (1991). *What Price Natural History Collections or 'Why do we need all these bloody mice'?.* Museum Journal. **91** (8) : 25-28.

Pettitt, C. (1994) Using the Collections. In : Stansfield, G. (ed.); *Manual of Natural History Curatorship.* London: HMSO pp.xviii+306 ISBN: 0-11-290531-7.

Pettitt, C. (1997) *The Cultural impact of Natural Science Collections.* In: J.R. Nudds and C.W. Pettitt, (eds.); *The Value and valuation of Natural Science Collections.* Proc. International Conference, Manchester, London. 94-103. xii+276p.

Pettitt, C.W. (1986) *Collections Research in the United Kingdom.* In: R.B. Light et al., (eds.); *Museum Documentation Systems.* Butterworths, London. 221-229.

Pharmavati, M. et al., (2004) *Application of RAPD and ISSR markers to analyse molecular relationships in Grevillea (Proteaceae).* Austr. Syst. Bot. **275** : 59-67.

Platnick, N.I. (1979) *Philosophy and the transformation of cladistics.* Systematic Zoology. **28** : 537-546.

Poczail, P. et al., (2008) *Analysis of phylogenetic relationships in the genus Solanum* (Solanaceae) *as revealed by RAPD markers.* Plant Syst.Evol. **275** : 59-67.

Polaszek, A et al., (2005 b) *Zoo Bank: The open access register for zoological taxonomy:* Technical Discussion Paper. Bulletin of Zoological Nomenclature **62**(4) : 210-220.

Polaszek, A. (2005 a) *A universal register for animal names.* Nature. **437**: 477 doi:10.1038/437477a; Published online

Ponder, W. F., et al., (2001) *Evaluation of museum collection data for use in biodiversity assessment.* Conservation Biology **15** : 648–657.

Prakash, R. (2000) *Molecular evolutionary tree.* Everyman's Science. **XXXV**: (2) 63-68.

Pruthi, H.S. (1979) *Text Book of Agricultural Entomology.* ICAR, New Delhi. 977.

Purvis, A. and Hector, A. (2000) *Getting the measure of biodiversity.* Nature **405** : 212-219.

Ricketts, T. H. et al., (1999b) *Terrestrial eco-regions of North America: a conservation assessment.* Island Press, Washington, D.C., USA.

Ricotta, C. (2002) *Bridging the gap between ecological diversity indices and measures of biodiversity with Shanon's entropy: Comment to Izsak and Papp.* Ecological Modelling. **152**(1) 1-3.

Rijsbergen, C.J. (1979) *Information Retrieval.* Butterworths, London.

Roderic, D.M. (2007) *TBMap: A taxonomic perspective on the phylogenetic database TreeBASE.* BMC Bioinforamtics. **8** : 158-167.

Rogers, D. J. and T.T. Tanimoto. (1960) *A computer program for classifying plants.* Science.**132** : 1115-1118.

Rohlf, F.J. and Sokal, R.R. (1967) *Taxonomic structure from randomly and systematically scanned biological images.* Systematic Zoology. **16** : 246-260.

Saarenmaa, H. et al., (1995) *Object oriented taxonomic biodiversity databases on the World Wide Web.* In: Kempf, A. and Saarenmaa, H., (eds.); *Internet Applications and Electronic Information Resources in Forestry and Environmental Sciences.* Workshop at the European Forest Research Institute, Joensuu., Finland, August 1-5.

Sales, E., et al., (2001) *Population genetic study in the Baleaaric endemic plant species Digitalis minor (Scrophulariaceae) using RAPD markers.* Am. J. Bot. **88** : 1750-159.

Scott-Ram, N.R. (1990) *Transformed cladistics, taxonomy and evolution.*Cambridge University Press.0521-340861.

Shaffer HB, Fisher RN, Davidson C. (1998) *The role of natural history collections in documenting species declines.* Trends Ecol. Evol. **13** : 27–30.

Simberloff, D. (1998) *Flagships, umbrellas and keystones: Is single species management passed in the landscape era?* Biolgical Conservation **63** (3) 247-257.

Simpson, E.H. (1949) *Measurement of diversity.* Nature. **163** : 688

Simpson, G.C. (1945) *The principles of classification and a classification of mammals.* Bull. Amer. Mus. Hist., **85** : 1-vi.1-350.

Simpson, G.G, (1961) *Principles of Animal taxonomy,* 247. Columbia University Press., New York.

Sivaramakrishnan, K.G., et al., (1980) *Emerging trend in molecular systematics and molecular phylogeny of mayflies (Insecta : Ephemeroptera).* Journal of Threatened Taxa.**3** (8) :1975-1980.

Sneath, P.H.A. (1962) *The construction of taxonomic groups.* 289-332. In: G.C. Ainsworth and P. H. A. Sneath, (ed.); *Microbial Classification,* 12th Symposium of the Society for General Microbiology,. Cambridge University Press, Cambridge.483 p.

Sneath, P.H.A. and Sokal, R.R. (1973) *Numerical taxonomy: The Principles and Practice of Numerical Classification.* 573. WH Freeman, San Francisco.

Sneath, P.H.A.(1957) *The application of computers to taxonomy.* J. Gen.Microbiol. **17** : 201 -226.

Sokal, R. R. and Rohlf, F. J. (1966) *Random scanning of taxonomic characters.* Nature **210** : 461-462.

Sokal, R. R., and Michener, C.D. (1958) *A statistical method for evaluating systematic relationships.* Univ. Kansa. Sc. Bull. **38** : 1409-1439.

Sokal, R. R., and Rohlf, F.J. (1962) *The comparison of dendrograms by objectives methods.* Taxon **11**: 33-40.

Sokal, R. R., and Sneath, P. H. A. (1963) *Principles of Numerical Taxonomy.* 359. W.H. Freeman, San Francisco.

Sørenson, T. (1948) *A method of establishing groups of equal amplitude in plant sociology based on similarity of species content and its application to analyses of the vegetation on Danish Commons.* Biol.Skr. **5** (4) : 1-34.

Souframanien, J. and Gopalkrishna. (2004) *A comparative analysis of genetic diversity in blackgram genotypes using RAPD and ISSR markers.* Theor. Appl. Gent. **109** : 1687-1693.

Stolte, K. W. and Mangis, D. R. (1992) *Identification and use of plant species as ecological indicators of air pollution stress in Nat'l Park Units.* In: McKenzie, D. H. et al., (eds.); *Ecological Indicators vols 1 and 2, International Symposium, Fort Lauderdale, Florida USA.* 373–392.Elsevier Science Publishers Ltd., New York.

Strickland, H.E. (1878) *Rules for Zoological Nomenclature.* John Murray, London.

Suarez A V, and Tsutsui ND (2004) *The value of museum collections for research and Society.* BioScience. **54** (1) 66-74.

Sylvester-Bradely, P.C. (1951) *The Subspecies in Paleontology.* Geol. Mag. **88**:88-102.

Sylvester-Bradely, P.C. (1956) *The Species Concept in Paleontology.* Syst. Zool.. Publ., London., No.**2**: p 145.

Tautz, D. et al., (2002) *DNA points the way ahead in taxonomy.* Nature **418** : 479.

Templeton, A.R. (1989) *The meaning of species and speciation: A genetic perspective..* In: D. Ottte and J.A. Nadler (eds.); *Speciation and Its Consequences.* 3-27. Sunderland, Mass, Sinauer Associates.

Thompson, D.R., et al., (1991 a) *Mercury accumulation in great skuas of known age and sex and its effect upon breeding and survival.* J. Appl. Ecol. **28** : 672-684.

Thompson, D.R., et al., (1991b) *Historical changes in mercury concentrations in the marine ecosystem of the north and north-east Atlantic Ocean as indicated by sea bird feathers.* J. App. Ecol. 29:79-84.

Thuiller, W. (2007*) Biodiversity: Climate Change and the Ecologist.* Nature **448** : 550-552. doi:10.1038/448550a; Published online [1st]August 2007.

Tiwari, B.K., et al., (2004) *The biology of triploid fish.* Reviews in Fish Biology and Fisheries.**14** : 391-402.

Touminen,J. et al., (2011) *Biological Names and Taxonomies on the semantic web-Managing the changes in Scientific conception.* 255-269. In: *The Semantic Web Conference*; ESWC2011,Heraklion Crete,Greece.May29-June2'2011.Proc.PartII.

Valdecasas, A.G. (2011) *An index to evaluate the quality of taxonomic publications.*

Vecchione ,M. et al., (2000) *Importance of assessing taxonomic adequacy in determining fishing effects on marine biodiversity.* ICES Journal of Marine Sciences, **57**: 677-681.

Vecchione, M. and B. B. Collette. (1996 a) *Fisheries agencies and marine biodiversity.* Annals of the Missouri Botanical Garden, **83** (1):29-36.

Vecchione, M. and B.B. Collette (1996 b) *The central role of systematics in marine biodiversity problems.* Oceanography. **9** (1):44-49.

Vences, M., S. et al., (2004) *Natural colonization or introduction? Phylogeographical relationships and morphological differentiation of house geckos (Hemidactylus) from Madagacar.* Biological Journal of the Linnaean Society. **83** : 115-130.

Vilatersana, R. et al., (2005) *Taxonomic problems in Carthamus (Asteraceae) RAPD markers and sectional classification.* Bot. J. Linn. Soc. **147** : 375-383.

Wagele, J.W. (2005) *Foundations of Phylogenetic Systematics.* Verlag Dr. Friedrich Pfeil. München, Germany.

Warne, K. (2007) *Tribute: Organization Man.* Smithsonian (Magazine) May 2007 105-11.

Western, D. (1987) *Africa's elephants and rhinos: flagships in crisis.* Trends in Ecology **2**:343–346.

Wheeler, Q. D. and A. G. Valdecasas. (2007) *Taxonomy: Myths and Conceptions.* Anales del Jardine Botanico de Madrid **64** (2):237-241.

Wheeler, Q. D. and A. G. Valdecasas. (2005) *Ten challenges to transform Taxonomy.* Graellsia **61**: 151-160.

Wheeler, Q. D. et al., (2004) *Taxonomy: Impediment or Expedient?* Science, **303**:285.

Wheeler, Q., (2008) The new taxonomy. In : A. Waren (ed), *The Systematics Association Special Volume Series.* 237. CRC Press, Boca Raton, London, New York.

Wheeler, Q.D. (2002) *Taxonomic triage and the poverty of phylogeny.* Phil. Trans. R. Soc. Lond. B *29* April 2004. **359**- 1444: 571-583

Wheeler,Q.D. and Valdecasas, A.G. (2007) *Taxonomy: Myths and Misconceptions.* Annales-Jardine Botanico De Madrid. **64** (2): 237-241.

Whittaker, R. H. (1972) Evolution and measurement of species diversity. Taxon.**21**:213-251.

Wiktor, A. (2001) *The slugs of Greece (Arionidae, Milacidae, Limacidae, Agriolimacidae Gastropoda Stylommatophora).* Vol. **8**. Irakleio: Natural History Museum of Crete, Hellenic Zoological Society. Fauna Graeciae.

Wiley, E. O. (1978) *The evolutionary species concept reconsidered.* Systematic Zoology.**27** : 17-26

Wiley, E.O. (1980) *Phylogenetic Systematics and vicariance Biogeography.* Systematic Botany **5** (2).194 -220.

Williams, J.G.K., et al., (1990) *DNA polymorphisms amplified by arbitrary primers are useful as genetic markers.* Nucleic Acids Res. **18** : 6531-6535.

Williams, K.D. and Moffitt, S.D. (2011) *Quantifying reader accuracy for thermal mark identification of Pacific salmon through the use of single blind pre-season test samples.* Abst. In Otolith Workshop. Juneau, Alaska: Decoding the Otolith. April 19-20, 2011

Wilson EO (2003) *The Encyclopedia of life.* Trends in Ecology and Evolution.18:77-80.

Wilson, E. O. (1987) *The earliest known ants: an analysis of the Cretaceous species and an inference concerning their social organization.* Paleobiology **13** (1) : 44-53.

Wilson, E.O. (1992) *The diversity of life* (Norton, New York).

Wilson, E.O. (2000) *On the future of Species.* Conservation biology. **14** (1) :1-3.

Wilson, E.O. and W.L. Brown, Jr. (1953) *The subspecies concept and its taxonomic application.* Syst. Zool. **2** : 97-111.

Yoon, C.K. (1993) *Counting creatures great and small.* Science **260** : 620-622.

Zhi-Qianz, Zhang (2011) *Animal Biodversity: An Introduction to higher-level classification and taxonomic richness.* Zootaxa. **3148** : 7-12.

Zhi-Qianz Zhang (2012) *A new era in Zoological nomenclature and taxonomy: ICZN accepts e-publication and launches ZooBank.* ZooTaxa. **3450** : 88.

Zuckerkandl, E. and L. B. Pauling. (1962) Molecular disease, evolution and genetic heterogenecity. In : M. Kasha and B., M. Pullman, (eds.); *Horizons in Biochemistry.* 189-225.

Zwickl D.J. ,D. M. Hills. (2002) Increased taxon sampling greatly reduces phylogenetic error. Syst Biol. 2002 (**51**) : 588-598.

Following websites have been consulted

www.fao.org/sd
www.ovid.com
www.sciencedirect.com
www.sciencedaily.com
http://animal.about.com
www.scienificamerican.com
http://evolution.berkely.com
http://understandingevolution.com
www.catlogueoflife.org.annualchecklist/2013
www.answer.com/topic/international-code-zoological-nomeclature

AUTHOR INDEX

GENUS INDEX

SUBJECT INDEX